INTERNATIONAL CENTRE FOR MECHANICAL SCIENCES

COURSES AND LECTURES - No. 308

STRUCTURAL OPTIMIZATION UNDER STABILITY AND VIBRATION CONSTRAINTS

EDITED BY

M. ZYCZKOWSKI

TECHNICAL UNIVERSITY OF CRACOW

SPRINGER - VERLAG WIEN - NEW YORK

Le spese di stampa di questo volume sono in parte coperte da
contributi del Consiglio Nazionale delle Ricerche.

This volume contains 111 illustrations.

Printed in Italy

In order to make this volume available as economically and as rapidly as possible the authors' typescripts have been reproduced in their original forms. This method unfortunately has its typographical limitations but it is hoped that they in no way distract the reader.

ISBN 3-211-82173-2 Springer Verlag Wien-New York
ISBN 0-387-82173-2 Springer Verlag New York-Wien

PREFACE

Optimal design of structures leads, as a rule, to slender and thin-walled shapes of the elements, and such elements are subject to the loss of stability. Hence the constraints of structural optimization usually include stability constraints. Loading parameters corresponding to the loss of stability are, in most cases, expressed by eingenvalues of certain differential equations, and hence the problems under consideration reduce to minimization of a certain functional (volume) under eigenvalues (critical loadings) kept constant. Briefly, wa call such problems "optimization with respect to eigenvalues", though in many cases the eigenvalue problems are not visible explicitely.

Optimal design under vibration constraints is related to that under stability constraints because of at least two reasons. First, the vibration frequencies are also expressed by eigenvalues of some differential equations, and hence the relevant problems belong also to optimization with respect to eigenvalues. Second, in nonconservative cases of structural stability we usually have to apply the kinetic criterion of stability, analyzing the stability of vibrations in the vicinity of the equilibrium state: hence both problems are directly interconnected in such cases.

The course on structural optimization under stability and vibration constraints was given in Udine, 20 - 24 June 1988, by five researchers particularly active in this field, namely prof. M. Zyczkowski (coordinator) and prof. A. Gajewski from the Technical University of Cracow, Poland, prof. N. Olhoff from the University of Aalborg, Denmark, prof. J. Rondal from the University of Liège, Belgium, and prof. A. P. Seyranian from the Institute of Problems of Mechanics in Moscow, U.S.S.R.

Part I, by M. Zyczkowski, deals just with optimal structural design under stability constraints. It gives first a general introduction to structural optimization, discussing typical objectives, design variables, constraints and equations of state. Then a chapter is devoted to optimization of shells in the elastic range, but the remaining chapters deal with optimal design under inelastic stability constraints, mainly creep buckling constraints (columns, trusses, arches, plates, shells).

Part II, by A. Gajewski, is devoted to optimal design both under stability and vibration constraints. Basic optimization methods are discussed in detail, and then particular attention is paid to multimodal optimal design resulting from simultaneous analysis of buckling in two planes (columns and arches). Further problems discussed in this part are: optimization of a plate under nonconservative forces, viscoelastic column compressed by a follower force optimized with respect to its dynamic stability, and a plane bar system in conditions of internal resonance.

Also part III, by N. Olhoff, deals with optimal structural design under both stability and vibration constraints, namely with optimal design of conservative mechanical systems with respect to fundamental and higher order eigenvalues. A unified variational approach for optimal design of one-dimensional continuum systems with respect to simple fundamental eigenvalue is presented first, and then extended to multimodal design and higher order eigenvalues. Examples concern natural frequencies of axial, torsional and transverse vibration of rods and beams, critical whirling speeds of rotating shafts, and structural buckling loads.

Part IV, by J. Rondal, is devoted just to optimal structural design under stability constraints, namely to optimal design of thin-walled bars and beams. In this case not only single buckling modes must be taken into account (local plate buckling, flexural column buckling, torsional buckling, lateral buckling), but also interactive buckling between single modes. Effects of imperfections are studied in detail, and the results are presented in various efficiency charts. This part of the present volume is particularly oriented towards engineering applications.

Part V, by A. P. Seyranian, deals with optimal structural design under both stability and vibration constraints, and mainly with flutter constraints. Sensitivity analysis is used as a tool of optimization: it is presented both for discrete cases and for distributed cases. Gyroscopic systems, nonconservative problems of elastic stability, aeroelastic stability of panels in supersonic gas flow, and bending - torsional flutter of a wing are considered in detail.

A reader of the present volume is expected to be familiar with basic problems and methods of structural optimization, as well as with fundamentals of structural stability and vibrations.

Finally, the efforts of the CISM Rector, Prof. S. Kaliszky, CISM Secretary General, Prof. G. Bianchi, the Editor of the Series, Prof. C. Tasso, and all the CISM Staff in Udine are gratefully acknowledged.

Michal Zyczkowski
Politechnika Krakowska

CONTENTS

PART IV
by J. Rondal

PART V
by A. P. Seyranian

PART I

M. Zyczkowski
Technical University of Cracow, Cracow, Poland

ABSTRACT

Part one consists of seven chapters corresponding to seven lectures delivered. The first chapter gives general introduction to structural optimization, discusses typical objectives, design variables, constraints and equations of state. Chapter 2 applies the concept of local shell buckling to optimization of elastic shells under stability constraints. The remaining chapters are devoted to optimization with respect to plastic or creep buckling: trusses, columns, arches, plates and shells are optimized, in most cases with rheological properties of the material allowed for. The last chapter gives a short survey of recent results, obtained within the years 1984 - 1988.

1. GENERAL INTRODUCTION TO STRUCTURAL OPTIMIZATION

1.1. Formulation of optimization problems

Extensive use of computers resulted in a rapid progress in structural analysis. However, the next and more advanced step consists in replacement of analysis by synthesis, namely by optimal structural design.

Mathematical problems of optimal structural design - as of most optimization problems - consist of four basic elements: design objective, control variables or decisive variables called here design variables, constraints, and equations of state. We look for upper or lower bound (maximum or minimum) of the design objective specified as a function or functional of design variables, functions or vectors (sets of functions or of parameters) under certain constraints. These constraints are usually expressed in terms of some other variables, called state variables or behavioural variables; they may also appear in the objective function and are interrelated and related to design variables by the equations of state.

A proper choice of the design objective and of constraints constitutes the most important point of the design philosophy. As a "design objective" or a "constraint" we understand here a notion or an idea rather than its mathematical expression. Such a notion may be insufficiently defined or its definition may be non - unique. Then a certain criterion is needed to specify the objective function or the constraint function and to formulate uniquely the relevant mathematical optimization problem. Sometimes even substitutive criteria are introduced. Such criteria used in structural optimization will be discussed in detail in Sec. 1.2. and 1.4. The above remark refers to deterministic approach; in probabilistic approach such criteria specifying objectives and constraints are even more necessary.

We discussed here design objectives and constraints in the same manner. Indeed, in some optimization problems the role of the design objective and of a constraint may be interchanged. Such an interchange is called sometimes alternate equivalent, dual or mutual formulation of the problem (though in mathematical programming the term "dual" is used also for other interchanges).

It is convenient to present a structure under consideration as a point in an abstract design space, with the design variables serving as coordinates. Any solution satisfying the constraints and equations of state is called an acceptable or feasible solution. In general, optimum is then not achieved. To find an optimum we have to derive a relevant necessary condition, called the optimality condition. Such conditions in structural optimization were discussed in detail by W.Prager [1,2], W.Prager and J.Taylor [3], L.Berke and V.B.Venkayya [4], M.Save [5], C.Fleury and M.Geradin [6], Z.Mróz and A.Mironov [7]. Sufficient conditions, important from the theoretical point of view, are often much more difficult to be formulated and employed. A distinction between the local extremum and a global extremum inside the whole admissible domain is particularly important here.

The first paper on structural optimization is ascribed to Galileo Galilei [8], 1638, who discussed optimal shape of beams. At present, the number of papers devoted to optimal structural design, exceeds five thousand. They are discussed in detail in monographs, textbooks and survey papers; an extensive list of sources is given in the monograph by A.Gajewski and M.Życzkowski [9].

1.2. Design objectives and their criteria

The cost is the most typical objective in many optimization problems. However, this objective is not uniquely defined and various criteria must be introduced to specify the relevant objective function.

In structural optimization the total cost consists, as a rule, of three elements: material costs, manufacturing costs and exploitational coasts. If the material is preassigned, then the material cost is proportional to the volume of the structure. Similarly, exploitational costs are in many cases proportional to the mass, and hence to the volume of the structure (first of all we quote here aircrafts and rockets, but also all vehicles etc.). Hence, if the manufacturing costs are not particularly essential, then volume of the structure constitutes a reasonable substitutive criterion for the cost as a design objective.

It should be noted that even if the volume of the structure is chosen as a design objective, a further specification is sometimes necessary. For example, if the structure consists of various materials, then an introduction of certain weight coefficients is justified: they are different for material costs (unit prices) and for exploitational costs (specific weight). A reasonable criterion to formulate the objective function should then be introduced.

In the simplest case the volume is a functional determined as follows

for a bar structure

$$V = \sum_{i=1}^{n} \int_{0}^{l_i} A_i(s)ds \tag{1.1}$$

and for a surface structure

$$V = \sum_{i=1}^{n} \iint_{S} h_i(\alpha,\beta)dS , \tag{1.2}$$

where the summation extends over n bars or over n segments of a surface structure, A_i denotes the cross-sectional areas, s – the variable along the axis of a bar or arch, h_i – the thicknesses, and α and β – the parameters of a plate or shell. For a single bar with a constant cross-section minimization of V is reduced to minimization of the cross-sectional area A. However, (1.1) and (1.2) hold only for straight, plane or (approximately) for slightly curved elements; for highly curved elements more exact expressions must be introduced and then minimization of V is not equivalent to minimization of A (the first one is better justified in engineering applications).

Mathematical expressions for the volume (1.1) or (1.2), are very simple indeed. Sometimes it is much more difficult to find a criterion formulating the objective function or functional e.g. if the beauty (aesthetics) of a structure is assumed as the design objective (or one of the design objectives, e.g. for a cupola). However, even here one can imagine such mathematical criteria.

If several objective functions are considered, we speak about multi- objective optimization. In this case the term multi-criterial optimization is equivalent. If however, one design objective is specified by various criteria, then a single-objective, simultaneously multi-criterial optimization takes place (e.g. for various criteria of the cost).

We mentioned in Sec. 1.1. that sometimes a constraint may be interchanged with the design objective. Since the variety of constraints is large, the variety of design objectives increases then accordingly.

1.3. Design variables

A fairly large variety of design variables may appear in structural optimization problems. Most of them refer to the shape. For classification purposes we gather typical design variables for individual structural elements in Table 1, being an extension of that proposed by W.Krzyś and M.Życzkowski [10,11], M.Życzkowski and A.Gajewski [12]. The symbol n stands here for a finite number of parameters to be determined (n degrees of freedom, vectors as design variables), 0 - the relevant design problem does not exist, ∞ - one function of one spatial variable should be evaluated, $n \times \infty$ - n functions of one variable, ∞^2 - one function of two variables, and so on. The symbols containing infinity lead to distributed parameter structural optimization (functions as design variables).Of course, in practical applications only some of the design variables listed are free to be determined, and some are restricted or preassigned, and then the distributed parameter optimization is often reduced to finite dimensional optimal design or parametric optimal design.

Explanations to Table 1 :

A1. As the magnitude of cross-section we understand here the design of characteristic dimension along the axis of the bar.

A2. Shape of a solid cross-section may be determined for example by a function $r = r(\theta)$ in polar coordinates.

A1 and A2. These elements of design may be considered separately only for geometrically similar sections. In general,we have to classify them jointly as ∞^2 (e.g. $r = r(\theta,z)$ in cylindrical coordinates).

A4. Shape of the boundary is here represented by the length of the bar.

B2. Shape of a thin-walled cross-section may be, in general, determined by two functions, e.g. $r = r(\theta)$ and $h = h(\theta)$, where h denotes wall thickness.

C3. Symbol n refers here to optimal lay-out of a truss of frame.

E1. Magnitude of cross-section is represented in surface structures by the thickness, in general function of two variables.

E2 - G2.Shape of the cross-section of plates or shells might be considered in sandwich or multilayer structures.

In some cases the equations of state are partial differential equations to be solved in domains with unknown boundaries - just these boundaries are described by functions subject to optimal evaluation as design variables. These cases are called "optimal shape-design", though, in fact, any variable shown in Table 1 influences the shape. The relevant positions are marked by an additional frame.

Certain design variables of other types should also be

TABLE 1. Quantitative classification of design variables related to shape

Design variables / Type of elements	(1) Magnitude of cross-section	(2) Shape of cross-section	(3) Shape of axis or middle surface	(4) Shape of boundary	(5) Mode of support	(6) Mode of reinforcement or of stiffening
(A) Straight bar with solid cross-section	∞	∞	0	1	n	n
(B) Thin-walled straight bar	∞	$2 \times \infty$	0	1	n	n
(C) System of straight bars	$n \times \infty$	$\boxed{n \times \infty}$	n	n	n	n
(D) Arch, plane curved bar	∞	$\boxed{\infty}$	∞	1	n	n
(E) Plate, panel	∞^2	0	0	$\boxed{\infty}$	∞	n
(F) Cylindrical shell	∞^2	0	∞	$\boxed{\infty}$	∞	n
(G) Arbitrary shell	∞^2	0	∞^2	$\boxed{\infty}$	∞	n
(H) Three-dimensional body	0	0	0	$\boxed{\infty^2}$	∞^2	n

mentioned here: we may look for optimal non-homogeneity (Young's modulus, yield-point stress, specific weight, coefficient of thermal expansion, etc), or optimal anisotropy. The above variables may be termed "material design variables". Moreover, we may look for optimal loading distribution, optimal distribution of temperature or other scalar or vector fields.

1.4. Constraints and their criteria

From the mathematical point of view the most important classifications are: constraints which may be expressed as equations or as inequalities; constrains in the form of algebraic, differential or integral equations (with appropriate boundary conditions) or relevant inequalities. Constraints in the form of equations reduce the number of independent design variables (except isoperimetric, integral constraints) and their number cannot exceed the number of design variables, whereas constraints in the form of inequalities determine an admissible domain of solutions; the number of such constraints may be arbitrarily large.

From the engineering point of view the constraints of structural optimization may - in most cases - be classified in two groups: behavioral (performance, exploitational) constraints and technological constraints. Behavioral constraints refer e.g. to strength, stiffness, stability, vibration of structures under a given system of loadings, or several systems of loadings (multi-purpose optimal design), loadings variable in time or moving loads etc. They will now be discussed in detail.

The constraints imposed on strength of a structure under a given loading system or loading program are usually considered as the most important constraints of structural optimization. Indeed, strength constraints have to keep the structure always on the safe side. However, various approaches are possible here and the relevant criteria are based on diverse hypotheses.

At least, two-step hypotheses are needed. The first step refers to uniaxial state of stress. Criterion of strength may then be formulated using either purely phenomenological approach, or fracture mechanics, or allowing for necking phenomena etc. But even in the simplest phenomenological approach we may combine it either with the assumption of linear elasticity, or plastic hardening specified by a certain schematization of the stress-strain diagram, or with creep phenomena.

The second step needs a hypothesis how to transpose the theoretical or experimental results obtained for uniaxial stress to multiaxial stress. In any case a so-called failure hypothesis must be introduced here. Such a hypothesis determines reduced stress σ_{red}, equivalent – inasmuch as strength or inelastic behaviour is concerned – to the stress in uniaxial state. This reduced stress is used either to formulate directly the strength constraints or to develop a theory of inelastic deformations preceding final rupture.

In the simplest, phenomenological approach combined with linear elasticity, the criterion expressing the strength constraints under multiaxial stress is written as follows

$$\sigma_{red}(x_i) \le k \,, \tag{1.3}$$

where k denotes admissible stress in uniaxial tension. In most cases a decrease in design dimensions of a structural element results in an increase of stresses. Then to obtain minimal volume the stresses should be as high as possible,

$$\sigma_{red}(x_i) = k \,, \tag{1.4}$$

either throughout body as a whole, or, at least, along certain surfaces or certain lines in the body. Structural elements satisfying (1.4) are called the elements (or structures) of uniform strength. Eq. (1.4) may be regarded as a substitutive criterion for the strength constraint. This substitutive criterion may lead to correct solutions of optimization problems, or to false results e.g. if geometry changes of the structure are allowed for. The above problem is discussed by R.Razani [13], T.P.Kicher [14], K. Reinschmidt, C.A.Cornell and J.F.Brotchie [15], V.P.Malkov and R.G.Strongin [16], R.H.Gallagher [17].

A related concept of structures of uniform creep strength was introduced by Yu.V.Nemirovsky and B.S.Reznikov [18] (stationary creep), M.Życzkowski and W.Świsterski [19] (non-stationary creep due to geometry changes).

Criterion (1.3) has a local character. On the other hand, criteria of strength based on the theory of perfect plasticity have a global character, e.g.

$$P \le \frac{\bar{P}}{n} \,, \tag{1.5}$$

where $\bar{P}$ denotes the limit load-carrying capacity of the structure, n – safety factor. But even here a certain counterpart of structures of uniform strength may be introduced. D.C.Drucker and R.T Shield [20,21] proved that constancy of the rate of dissipation per unit volume D throughout the structure is a sufficient condition for optimality. It implies, of course, full plastification of the body at plastic collapse. In many cases D cannot be constant, but full plastification remains as a first step towards optimality, A.Zavelani - Rossi [22]. Shapes of full plastification were investigated in detail by Z.Kordas and M.Życzkowski [23], Z.Kordas [24], and were used for further optimization by B.Bochenek, Z.Kordas and M.Życzkowski [25].

Other criteria for the strength constraints in perfect plasticity are based on the maximal carrying capacity (if geometry changes are allowed for, J.Skrzypek and M.Życzkowski [26]), decohesive carrying capacity (K.Szuwalski and M.Życzkowski [27]), etc.

Behavioural constraints imposed on stability of a structure are of fundamental importance for the present lecture notes and will be described in a separate section, 1.6.

The notion of stiffness of a structure is even worse defined than that of strength. A certain appropriately chosen criterion is necessary here. In a perfectly rigid structure, supported against rigid-body motion, the following quantities vanish: displacements; strains; velocities; strain rates; elastic strain energy; dissipated energy, and so on. Hence, in a deformable structure, any of these quantities may represent compliance and may serve as a criterion of stiffness.

Consider first an elastic structure under the loadings constant in time. Then the velocities, strain rates and dissipated energy vanish even here, but the remaining quantities listed may be used for the formulation of stiffness constraints. Displacement $\underset{\sim}{u} = \underset{\sim}{u}(x_i)$ forms a vector field; to achieve a limitation of displacements inside the body as a whole we have to introduce a certain norm and to impose constraint on the norm

$$\| \underset{\sim}{u}(x_i) \| \leq u_{adm} , \qquad (1.6)$$

where u_{adm} denotes a prescribed admissible displacement (scalar). The norm in (1.6) should be understood as used twice: first, as a norm of the vector (length of the vector seems to be the most appropriate here, but maximal absolute value of the components is also being used); second, as a norm of a function of coordinates. Chebyshev's norm $\| \underset{\sim}{u} \|_{\infty}$ or Gaussian norm $\| \underset{\sim}{u} \|_2$ are most often in use. For example, most Standard Codes introduce limitation on maximal deflections of beams; such maximal deflection should be understood as Chebyshev's norm of deflection functions. Similar criteria of stiffness may be obtained by introducing norms for strains, but they are not in use: strains are not as representative, not as visible as displacements. In other words, strains are "to local" to be used as stiffness criteria; however, generalized strains (e.g. curvature of the beam) are employed.

On the other hand, elastic strain energy has a global character, is a scalar in contradistinction to displacements and strains, and is often being used as a criterion of stiffness (rather as a criterion of compliance, which is equivalent). This idea was introduced to optimal structural design by Z.Wasiutyński in 1939, [28]. The relevant constraint has the form

$$\frac{1}{2} \iiint_V \sigma_{ij} \varepsilon_{ij} \, dV \leq L_{adm} , \tag{1.7}$$

where L_{adm} denotes a certain admissible value of the elastic energy, responsible for compliance. Sometimes only the constraint (1.7) is called the compliance constraint.

Considering inelastic structures we may use velocities, strain rates or dissipated energy to formulate stiffness constraints; dissipated energy was used for the first time by W.Prager [29] who discussed optimal design in creep conditions; M.Życzkowski [11] employed generalized strain rate (rate of curvature) for the same purpose.

It is often required to have the lowest natural frequency of the structure bounded from below, and such a requirement may be used as an optimization constraint; it was first introduced by F.I.Niordson [30]. Sometimes several frequencies or even the whole spectrum of frequencies is employed, N.Olhoff [31], V.A.Troitsky [32].

The constraints listed above belong to behavioural or exploitational constraints and may be interchanged with the design objective. Moreover, some constraints may determine minimal or maximal dimensions, minimal or maximal values of material constants etc. Such constraints are called technological or side constraints. Minimal dimensions are, in fact, often related to manufacturing process,but also other factors may be important here, e.g. allowance for corrosion starting from the surface of the body. Minimal internal cross-sectional area of a pipe-line may be required in view of hydrodynamic reasons.

Moreover, manufacturing reasons may require that the

cross-section of individual members are the same: such a requirement belongs to technological constraints.

1.5. Equations of state

Equations of state relate design variables to state variables or interrelate state variables. Sometimes they are very simple, e.g. if the cross-sectional area of a bar $A(x)$ is the design variable, and the strength constraint is imposed on the stress in simple tension σ due to the normal force $N(x)$, then

$$\sigma(x) = \frac{N(x)}{A(x)} \tag{1.8}$$

may be called the equation of state. In general, equations of state in structural optimization comprise, first of all, basic equations of the mechanics of deformable bodies, namely equilibrium equations, compatibility equations and constitutive equations. Moreover, we classify here equations determining loss of stability, fracture, vibration frequency, etc.

Equations of state may also be regarded as a particular type of optimization constraints. They belong to equation-type constraints, whereas those discussed in Sec. 1.4 belong mostly to inequality-type constraints.

1.6. Stability constraints in structural optimization

In fact, the whole theory of structural stability may be regarded as an introduction to formulation of stability constraints in structural optimization. A proper formulation of those constraints should protect the structure against any possible form of the loss of stability; this feature brings often essential difficulties to correct optimal design. Most papers deal with bifurcation loads for perfect elastic structures (buclking, divergence); other typical constraints refer to flutter, bifurcation loads for elastic-plastic structures, creep buckling in various formulations, snap-through of perfect or imperfect structures etc.

Constraints imposed on the lowest buckling load of perfect structures,corresponding to the first buckling mode, belong to the case of eigenvalues as constraints. Variation of the shape of structure in optimization process affects also all other eigenvalues and may result in lowering of a higher-order eigenvalue beneath the first one. Then the result of unimodal optimal design is false and multimodal optimal design should be employed.

Very often the constraints act in opposite directions. For example, when designing a thin-walled tubular cross-section for pure torsion, the strength constraint tends to increase the mean radius R and to decrease the wall thickness h; conversely, the wall stability constraint results in decreasing R and increasing h. Similar situation may also take place if global and local stability conditions are employed, e.g. when designing ribbed structures. In such case the optimum is, as a rule, reached at the boundary of the admissible domain in the space of design variables, if both constraints are satisfied in the form of equations. The relevant design is called "simultaneous mode design", SMD (F.R.Shanley [33,34], L.Spunt [35]). To a certain degree it resembles multimodal optimal design, but now various constraint types interfere.

Indeed, simultaneous mode design leads often to optimal shapes, but should cautiously be used. Such structures are very sensitive to imperfections which cause here mode interaction: e.g. global stability influences local stability and vice versa (A. van der Neut [36]). General considerations on the effects of mode interaction on optimal design of structures are due to J.M.T.Thompson and W.J.Supple [37,38]. Hence, in such cases we should either perform structural optimization allowing for initial imperfections or optimize perfect structures with sufficiently raised safety factor.

In most cases the loss of stability has an integral character: a structure looses its stability as a whole. However, in certain particular cases the effects of the loss of stability are rather concentrated and, as a reasonable approximation, we may consider that phenomenon as a local one. Such a situation takes place e.g. when considering stability of shells, mainly of cylindrical shells under non-homogeneous stress distribution, A.S.Volmir [39]. Then the stability condition derived for a homogeneous state of stress, σ_{ij} = const., constant radius of curvature ρ, uniform thickness h and homogeneous material characterized for example by Young's modulus E, expressed in terms of stresses in the form

$$f(\sigma_{ij};\ \rho,\ h,\ E) = 1 \tag{1.9}$$

may also be applied to non-homogeneous conditions for the most dangereous point, and the relevant "local" stability constraint takes the form

$$f\,[\sigma_{ij}(\theta,\ z);\ \rho(\theta,\ z),\ h(\theta,\ z),\ E(\theta,\ z)] \le 1\ , \tag{1.10}$$

where θ and z denote cylindrical coordinates. Further, requiring (1.10) to be satisfied in the form of an equation at any point θ, z, we obtain a condition similar to that of uniform strength; the shapes thus obtained are called the "shapes of uniform stability" (M.Życzkowski, J.Krużelecki [40]). This idea leads, in principle, to infinitely many equally possible buckling modes. Of course, mode interaction may occur and a suitable safety factor should be introduced.

Creep buckling constraints show some peculiarities. Almost any theory of creep buckling introduces a certain notion of critical time t_{cr}. Hence, three equivalent interchanged formulations of structural optimization are possible here: besides the basic formulation of minimal volume V as a design objective with a given loading P and a given critical time t_{cr} as constraints, we may also look for maximal P under given V and t_{cr}, or for maximal t_{cr} under given V and P.

In some cases, e.g. if creep buckling of columns made of Maxwell's material is discussed, the critical time does not exist. Then the logarithmic velocity of creep buckling may be used as a similar criterion, and also three interchanged formulations of optimal design are possible.

2. CONTOURS OF COMPLETE NONUNIQUENESS IN THE CASE OF STABILITY CONSTRAINTS

2.1. The concept of contours of complete nonuniqueness

The solution of an optimization problem may be nonunique: there may exist two, three... or even infinitely many equally optimal solutions. An interesting example of complete nonuniqueness of solution was given by W.Prager [41]. He studied transmission of a vertical concentrated force to a certain foundation contour by a two-bar truss. Assuming just strength constraints he stated that in the case of a circular foundation contour any truss is equally optimal: the shorter the bars, the larger the necessary cross-sectional area and the volume is constant.

M.Markiewicz and M.Życzkowski [42,43], noted that Prager's example is not just a curiosity, but makes it possible to develop a method of optimal design of trusses transmitting a force to a given foundation contour. Namely, if we construct a family of contours of complete nonuniqueness, then the points of tangency of a contour from this family to the given foundation contour determine the optimal solution (Fig.1). Of course, tangency must be understood here in a generalized sense (as the first common

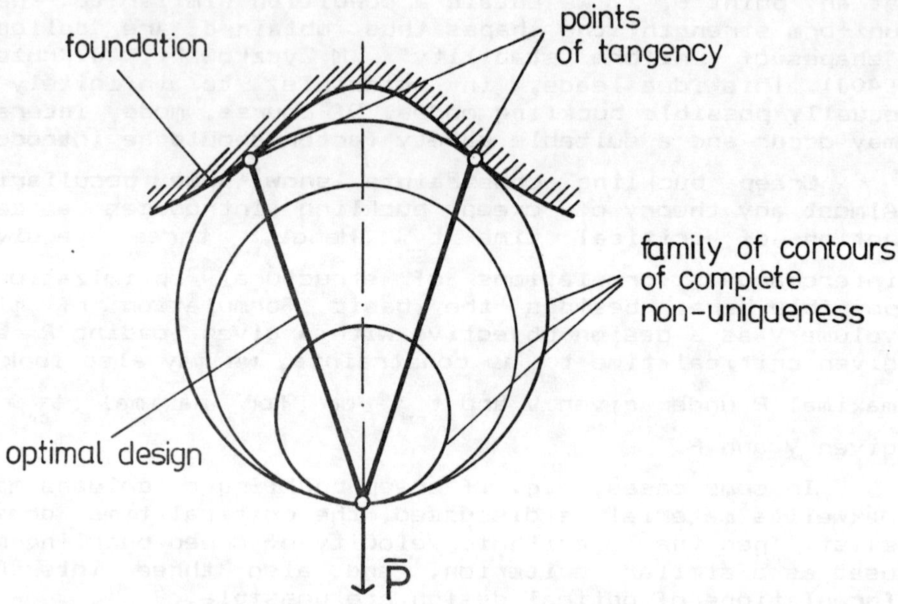

Fig. 1, Optimal truss determined by the method of contours of complete non-uniqueness for strength constraint

point), since both contours may have some corner points (cusps). Indeed, any other design corresponds to a larger circle, and hence to a larger volume.

If the bars are subject to compression and stability constraints are allowed for, the contours of complete nonuniqueness are no longer geometrically similar to each other, but the method holds. Following [43] we determine now the contours of complete nonuniqueness for symmetric two-bar trusses subject to stability constraints in elastic range. Such contours solve the problem of optimal transmission of a concentrated force to any symmetric foundation contour.

The volume of the truss, assumed as the design objective, is given by

$$V = 2Al, \qquad (2.1)$$

where A denotes the common cross-sectional area and l-common length of the bars. Consider now the constraints of the form

$$g_i(l,\ \varphi,\ A;\ P) \le 0, \qquad i = 1,2,\ldots n, \tag{2.2}$$

where φ is the angle shown in Fig.2. Regarding (2.2) as

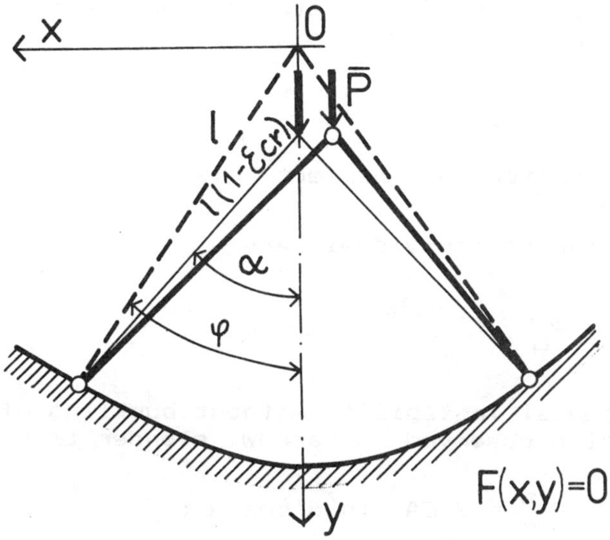

Fig. 2, Two-bar truss transmitting force P to given foundation contour

equations inside the range in which they are active we may eliminate the cross-sectional area A from (2.1) and (2.2) and obtain

$$f_i(l,\ \varphi;\ V,\ P) = 0, \qquad i = 1,2,\ldots n. \tag{2.3}$$

These equations specify individual segments of a family of curves in implicit form in the polar system of coordinates l, φ with the parameters V and P; in most cases these two parameters may be reduced to a single parameter. Practically, the active segments f_i are determined by the shortest distance l for a given φ. Each curve determines a contour of complete nonuniqueness, since the volume is constant here, independent of the design variables φ and A.

2.2. Elastic range

For a two-bar truss shown in Fig. 2 we consider three possible forms of instability (lateral buckling is disregarded). The dashed line shows the undeformed truss, whose apex is taken as the origin of polar coordinates l, φ. In the deformed precritical state the angle φ changes into α; normal force in the bars equals

$$N = \frac{P}{2\cos\alpha} . \tag{2.4}$$

Under the assumption of perfect elasticity, the critical forces are equal

(1) for buckling of individual bars

$$P_{cr}^{(1)} = \frac{2\pi^2 EI}{l^2}\cos\alpha ; \tag{2.5}$$

(2) for horizontal instability without buckling of the bars (bifurcation observed in narrow, slender trusses)

$$P_{cr}^{(2)} = 2\,EA\sin^2\alpha\cos\alpha ; \tag{2.6}$$

(3) for snap-through of the truss (wide, shallow trusses)

$$P_{cr}^{(3)} = 2\,EA\cos^3\alpha . \tag{2.7}$$

Equation (2.5) results directly from Euler's formula; the symbol I denotes here the moment of inertia corresponding to in-plane buckling. Eqs. (2.6) and (2.7) follow from the Mises theory; details are given e.g. by G.Bürgermeister, H.Steup and H.Kretzschmar [44].

In order to express the angle α in terms of the initial angle φ (which determines the shape of the contour in polar coordinates) we denote the critical (compressive) strain in the bars by ε_{cr} and write a simple geometric relation and Hooke's law in the form

$$l\sin\varphi = l\,(1-\varepsilon_{cr})\sin\alpha ,$$
$$\varepsilon_{cr} = P/2EA\cos\alpha . \tag{2.8}$$

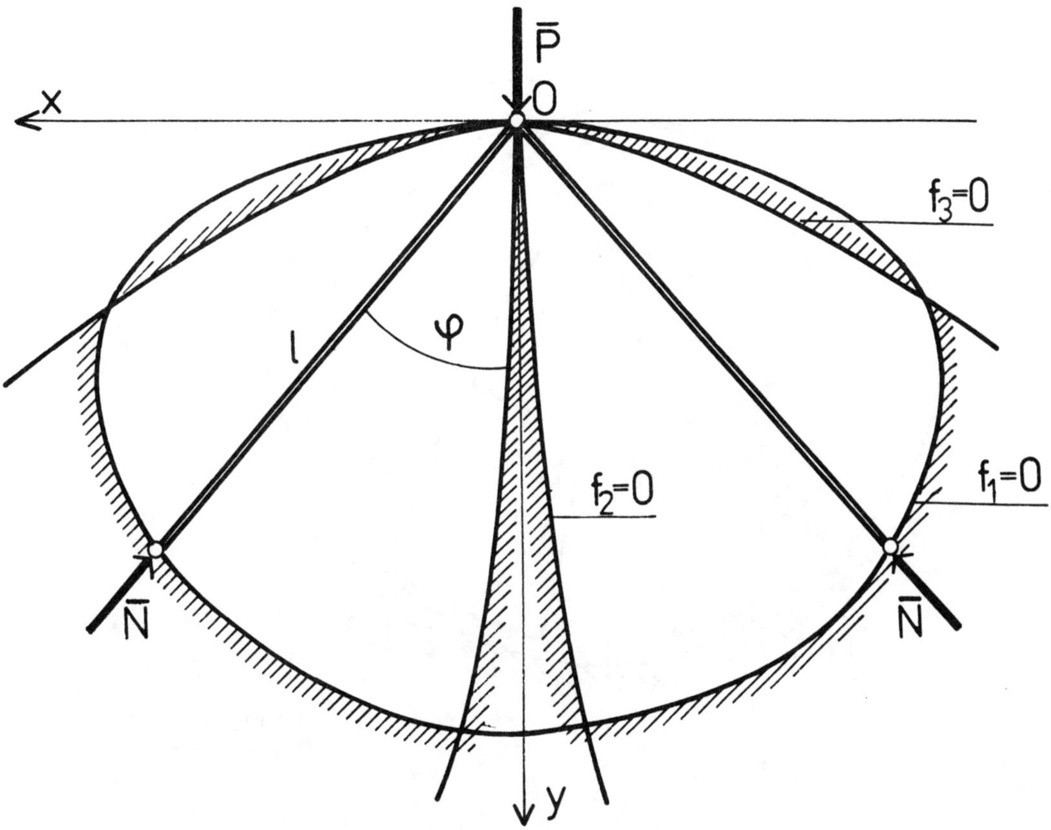

Fig. 3, Contour of complete non-uniqueness for truss design for stability constraints

Combining (2.8) successively with (2.5) - (2.7) we obtain for particular forms of the loss of stability

(1) $$\sin \varphi = \frac{1}{4}\left(3 + \sqrt{1 - 8\,\pi^2/\lambda_c^2}\right)\sin \alpha\,, \tag{2.9}$$

(2) $$\sin \varphi = \cos^2\alpha \;\sin \alpha\,, \tag{2.10}$$

(3) $$\sin \varphi = \sin^3\alpha\,, \tag{2.11}$$

where λ_c denotes the slenderness ratio of the bars. In the

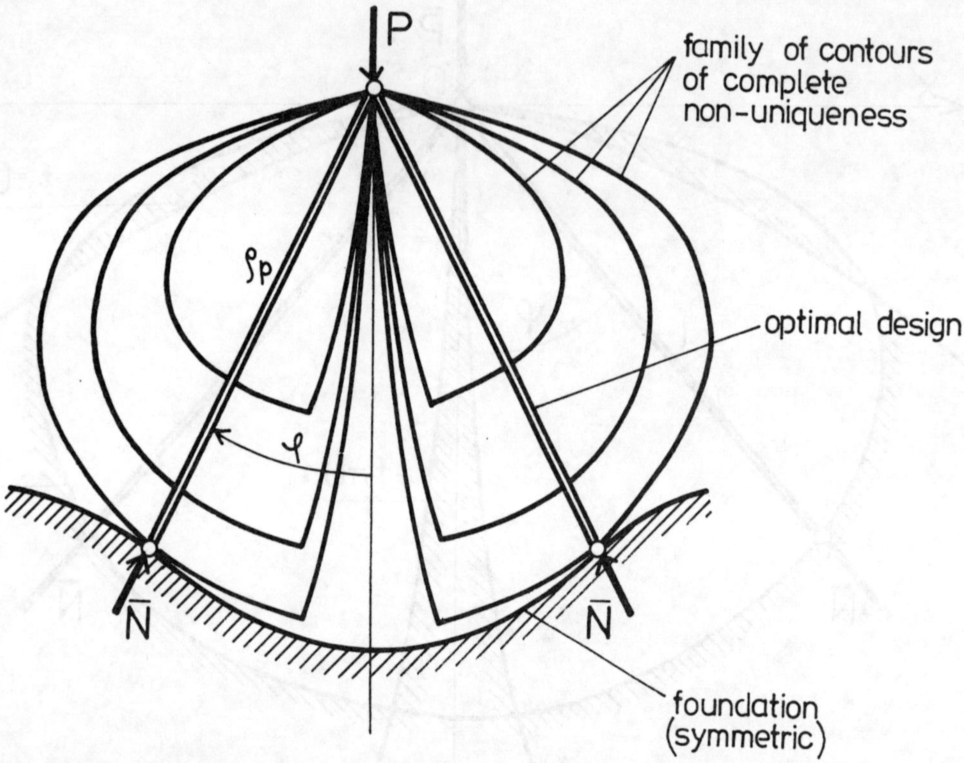

Fig. 4, Optimal truss determined by the method of contours of complete non-uniqueness for stability constrains

first case the difference between α and φ is very small and will be neglected, [43]. Introducing a certain safety factor against instability j, eliminating α from (2.9), (2.10) or (2.11), and A from (2.1), we obtain for assumed circular shape of cross-sections

$$f_1(l,\varphi;V,P) = \frac{8Pj}{\pi EV^2}\, l^4 - \cos\varphi = 0\ , \tag{2.12}$$

$$f_2(l,\varphi;V,P) = \frac{Pjl}{E\ V}\sin\varphi - \left(\sin^2\varphi + \frac{P^2j^2l^2}{E^2V^2}\right)^{3/2} = 0, \tag{2.13}$$

$$f_3(l,\varphi;V,P) = \frac{Pjl}{E\,V} - (1 - \sin^{2/3}\varphi\,)^{3/2} = 0\;. \tag{2.14}$$

A contour of complete nonuniqueness, described by Eqs. (2.12) - (2.14), is shown in Fig.3. The number of parameters may be reduced to one by introducing suitable dimensionless quantities. A one-parameter family of contours of complete nonuniqueness and its application to optimal design for a given foundation contour is shown in Fig.4.

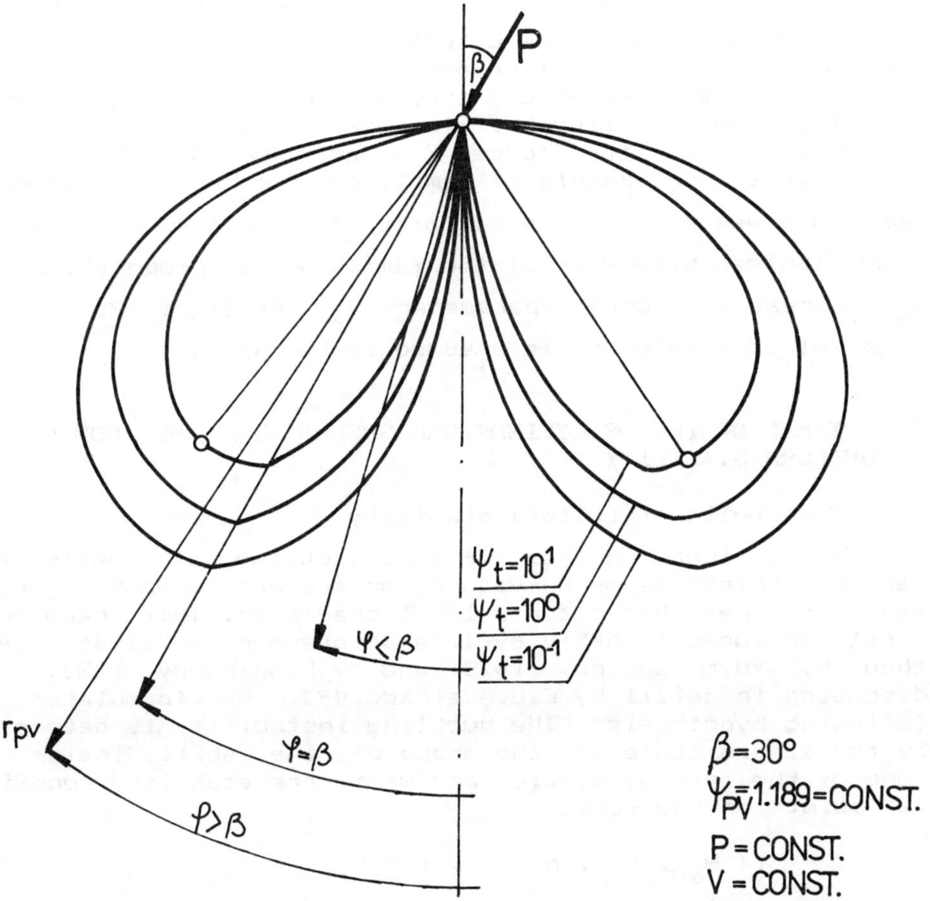

Fig. 5, Contours of complete non-uniqueness for creep buckling design

2.3. Inelastic range

M.Markiewicz generalized in [42] the above solution for the elastic-plastic range, adding to (2.12) - (2.14) three new equations corresponding to particular modes of plastic instability as described by Johnson-Ostenfeld's law. So, the contours of complete nonuniqueness might then consist of up to six segments. R.Wojdanowska and M.Życzkowski [45] considered a similar problem in creep conditions with some further generalization, namely oblique direction of the concentrated force to be transmitted. Cross-sectional areas are then different from each other, but the layout was assumed to be symmetric, $\varphi_1 = \varphi_2 = \varphi$. For bars in compression Rabotov-Shesterikov's creep buckling theory was assumed, and for bars in tension - Kachanov's theory of brittle creep rupture. Three formulations typical for optimization in creep conditions were considered: minimization of the volume V, maximization of the force P, and maximization of the life-time t_*. An example of family of contours of complete nonuniqueness adjusted to maximize t_* is shown in Fig.5; dimensionless parameter of the family ψ_t is proportional to t_*, whereas the other parameter ψ_{PV} contains P, V, and material constants and is assumed to be fixed.

3. OPTIMAL DESIGN OF CYLINDRICAL SHELLS VIA THE CONCEPT OF UNIFORM STABILITY

3.1. The shells of uniform stability

We mentioned in Sec. 1.6 that buckling of shells with variable thickness or curvatures or subject to nonhomogeneous state of stress has often a local character. This phenomenon, first mentioned probably by I.Ya.Shtayerman in 1936, [46], then by Yu.N.Rabotnov [47] and V.P.Shirshov [48], was discussed in detail by E.L.Axelrad [49]. He formulated the following hypothesis: "The buckling instability is determined by the stress state and the shape of the shell inside the zone of the initial buckle" and wrote the stability condition "at point" in the form

$$f(N_{\alpha\beta}, R_{\alpha\beta}, h, E, \nu) = 0, \tag{3.1}$$

where $R_{\alpha\beta}$ denotes the tensor of curvatures of the middle surface. Instead of membrane forces in the prebuckling state, $N_{\alpha\beta}$, we may also use the stresses $\sigma_{\alpha\beta}$, since the thickness h appears in (3.1) in any case.

Much before Axelrad's paper [49] the local stability condition was employed to shell optimization problems, first by M.Życzkowski and J.Krużelecki [40], 1973, published 1975. We give now a short presentation of the results achieved.

3.2. Cylindrical shell under overall bending

Consider a noncircular elastic cylindrical shell subject to pure overall bending, Fig.6. We look for optimal middle

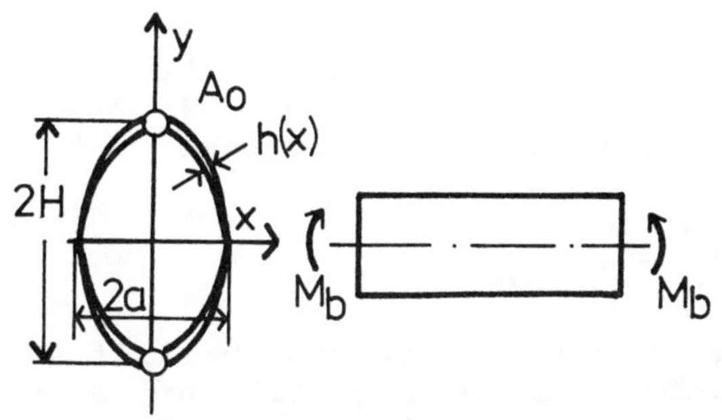

Fig. 6, Cylindrical shell under overall bending

surface $y = y(x)$ and optimal thickness distribution $h = h(x)$ of the shell as to minimize the overall cross-sectional area

$$A = 2\int_{-b}^{b} h(x)\sqrt{1+y'^2}\,dx + 2A_o \longrightarrow \min, \qquad (3.2)$$

where A_o denote concentrated areas in outer fibres; existence of such areas is typical for optimized cross-sections subject to bending and may be realized in the form of longitudinal ribs.We admit any sense of bending moment M_b, and hence the cross-section is assumed to be bisymmetric. The width 2b is regarded as unknown. According to the concept of local buckling, the cross-section is constant along the longitudinal axis, since the stress state is also constant.

The constraints are imposed on strength and local wall stability; other stability constraints (lateral buckling, Brazier's effect) are disregarded. The first constraint gives simply $\sigma_z = \sigma_o$ in outer fibres, where σ_o denotes the admissible normal stress. Hence (3.2) is complemented by an isoperimetric condition of a given bending moment

$$M_b = 2\int_{-b}^{b} \sigma_z(x)\, y(x)\, h(x)\, \sqrt{1+y'^2}\, dx + 2A_o H\sigma_o, \tag{3.3}$$

where $2H = 2y(0)$ is the height (depth) of the cross-section. The stress distribution may be presented in the form (positive in compression)

$$\sigma_z(x) = \frac{y(x)}{H}\,\sigma_o\,. \tag{3.4}$$

Now, the wall stability condition will be formulated locally as follows. For a circular cylindrical shell of constant thickness the critical stress is determined by a formula of the type

$$\sigma_{cr} = \beta E \frac{h}{R}\,, \tag{3.5}$$

where R denotes the radius of the shell; the coefficient β equals $\beta = 0.605$ according to the linear theory of stability of perfect shells (for $\nu = 0.3$), but may also be evaluated in some other way when allowing for imperfections or when using lower critical stress as decisive. Now we assume that (3.5) is also valid for nonhomogeneous state of stress, variable thickness $h = h(x)$ and variable radius of curvature $\rho = \rho(x)$. Combining (3.4) with (3.5) and introducing a suitable safety factor against buckling j we obtain

$$h(x) = \frac{j\sigma_o}{\beta EH}\, y(x)\, \rho(x) = \frac{\psi\, y(1+y'^2)^{3/2}}{H\,|y''|}\,, \tag{3.6}$$

where ψ denotes the dimensionless parameter

$$\psi = \frac{j\sigma_o}{\beta E}\,, \tag{3.7}$$

(in [40] a different definition of ψ is introduced).

Eliminating A_o from (3.3) (under the assumption $A_o > 0$) and substituting (3.6) into (3.2) we obtain the following expression to be minimized without any additional constraints:

$$A = \frac{M_b}{H\sigma_o} + \frac{4\psi}{H^3}\int_0^b \frac{y\,(1 + y'^2)^2}{(-y'')}(H^2 - y^2)\,dx \longrightarrow \min, \qquad (3.8)$$

since y'' is negative inside the interval under consideration. Introducing dimensionless area $a = (\sigma_o^2/\psi\, M_b^2)^{1/3}A$, dimensionless height $\chi = (\psi\,\sigma_o/M_b)^{1/3}H$, dimensionless variables $\xi = x/H$, $\eta = y/H$, and regarding η as independent variable, we rewrite (3.8) in the form

$$a = \frac{1}{\chi} + \chi^2\,\omega \quad \longrightarrow \quad \min, \qquad (3.9)$$

where

$$\omega = 4\int_0^1 \frac{\eta}{(-\xi'')}\,(1 - \eta^2)\,(1 + \xi'^2)^2\,d\eta\,. \qquad (3.10)$$

The expression (3.9) may be regarded as a function of the parameter χ (height) and as functional depending on the function $\xi = \xi\,(\eta)$, determining middle surface of the shell. Simple differentiation with respect to χ gives

$$\chi_{opt} = \sqrt[3]{\frac{1}{2\,\omega}}\,, \qquad a = \frac{3}{2}\sqrt[3]{2\,\omega} \longrightarrow \min. \qquad (3.11)$$

Hence, further optimization is reduced to minimization of the functional ω. In the form (3.10) it does not contain the dependent variable $\xi = \xi(\eta)$ explicitely, and hence the Euler-Lagrange equation is readily reduced to a third-order equation, [40]:

$$\xi'' = \frac{C_1\xi''^3}{2\eta(1-\eta^2)(1+\xi'^2)^2} + \frac{4\xi'\xi''}{1+\xi'^2} + \frac{(1-3\eta^2)\xi''}{2\eta(1-\eta^2)}\,. \qquad (3.12)$$

The transversality condition at $\eta = 0$ gives $C_1 = 0$. Details of numerical integration of (3.12) are given in [40].

Paper [40] considers also some restricted variants of optimization. First, if we assume constant wall thickness h, then (3.6) determines directly the function $y = y(x)$. An analogy with Euler's problem of finite deflections of an elastic column is seen, and hence the well-known solution is expressed by elliptic integrals. In dimensionless form we obtain

$$\xi = \frac{1}{\sqrt{\delta}} \left[2E(\phi_o, k_e^2) - 2E(\phi, k_e^2) - F(\phi_o, k_e^2) + F(\phi, k_e^2) \right] -$$

$$- \sqrt{\frac{1-z}{1+z}} + \eta \sqrt{\frac{2(1-z) + \delta(1-\eta^2)}{2(1+z) - \delta(1-\eta^2)}} , \tag{3.13}$$

where

$$\phi = \text{arc}\sin\left\{ \eta \sqrt{\frac{2\delta}{[2(1-z)+\delta]\ [2(1+z)-\delta(1-\eta^2)]}} \right\},$$

$$k_e^2 = \frac{2-2z+\delta}{4} , \qquad \delta = \frac{\psi H}{h} , \qquad z = \frac{1}{\sqrt{1+\eta_o'^2}} , \tag{3.14}$$

and ϕ_o is taken at $\xi = 0$, $\eta = 1$. If we require continuity of slope at $\xi = 0$, then $\eta_o' = 0$ and $z = 1$; moreover, a similar conditionat $\eta = 0$ leads to $\delta = 2$. If no such requirements are introduced, then both parameters z and δ are regarded as free and their optimal values are $z_{opt} = 0.981$, $\delta_{opt} = 3.430$.

Other restricted variants of optimization refer to prescribed middle surface and variable thickness, given by (3.6). Fig.7 shows the results of variational optimization (3.12), constant thickness (3.13) (both with additional requirements of continuity of slope), elliptic and parabolic profiles (with optimal parameters of ellipse or parabola, [40]). It is seen from (3.11) that A is inversely proportional to the height H, hence variational optimization is the best and prescribed parabolic profile the worst among the variants under discussion. This conclusion holds only for $A_o \neq 0$, since only then the functional can be reduced to (3.8) and parametric optimization to (3.11). For comparison, Fig.7 presents also the optimal constant-thickness shell without a rib ($A_o = 0$); the profile is higher, but by no means better than that with a rib.

3.3. Cylindrical shell under bending with axial force

Additional axial force N does not change essentially the solution given in Sec. 3.2, provided we admit any possible combination of senses of bending moment and of axial force. Then we arrive at bisymmetric cross-section with each half designed for the most inconvenient case (compression due to

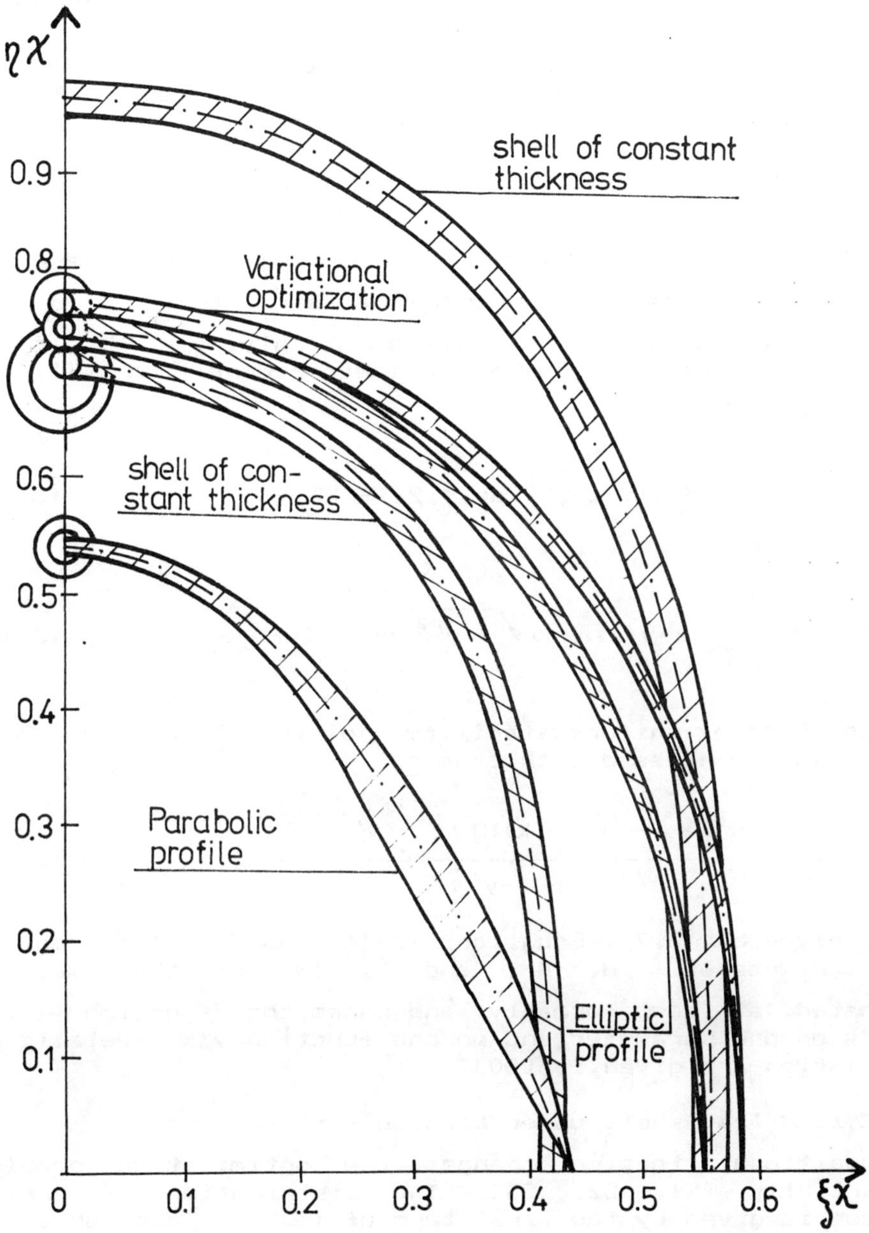

Fig. 7, Optimal cross - sectional shapes of a cylindrical shell under overall bending

bending combined with compressive axial force, J.Krużelecki [50]).

The objective function remains in the same form (3.2), but the stress distribution consist now of two terms, due to bending moment and due to axial force, respectively:

$$\sigma_z(x) = \sigma_{zM} + \sigma_{zN} = \frac{y(x)}{H} k \sigma_o + (1 - k) \sigma_o , \tag{3.15}$$

where $k = \sigma_{Mmax}/\sigma_o$ denotes the share of maximal stress σ_o due to bending moment. The strength constraint $\sigma_z \le \sigma_o$ is here satisfied as equality at one point only, $y = H$, as in the preceding case. The two isoperimetric conditions will be written in the form

$$N = 2 \int_{-b}^{b} \sigma_{zN} h(x) \sqrt{1+y'^2}\, dx + 2(1-k) A_o \sigma_o , \tag{3.16}$$

$$M_b = 2 \int_{-b}^{b} \sigma_{zM}(x) y(x) h(x) \sqrt{1+y'^2}\, dx + 2kHA_o \sigma_o . \tag{3.17}$$

The condition of uniform stability, joining (3.15) and (3.5) determines the necessary thickness

$$h(x) = \frac{\psi\, [ky + (1 - k)H]\, (1 + y'^2)^{3/2}}{H\, (-y'')} \tag{3.18}$$

with ψ given by (3.7). Equations (3.16) and (3.17) contain three parameters: H, A_o, and k. Two of them may be eliminated, at least formally, and hence the function (3.2) depends on one parameter and on one function $y(x)$. Details of optimization are given in [50].

3.4. Cylindrical shell under bending with torsion

Additional torsion changes the optimization problem substantially, [51, 52, 53]. The distribution of normal stresses is given by the first term of (3.15), and shearing stresses due to torsion may be found using the hydrodynamic analogy

$$\tau = \frac{c}{h} , \tag{3.19}$$

where the constant c depends on the torque M_t. Shearing stresses are no longer described by a monotonically increasing function of variable y, and hence it is no longer obvious that the strength constraint should be satisfied at and only at the point y = H. On the contrary, we have to write both constraints, related to strength and stability, in the form of weak inequalities and consider individual intervals in which any of them is active or passive, respectively. Free de Saint-Venant torsion will be considered and no stresses due to restricted warping will be allowed for.

As the stability constraint in local form we assume

$$\frac{j\sigma}{\sigma_{cr}} + \left[\frac{j\tau}{\tau_{cr}} \right]^2 \le 1 , \tag{3.20}$$

where σ_{cr} and τ_{cr} are critical stresses in pure compression and in pure torsion of a circular cylindrical shell, respectively, and j is the safety factor. The above form is suggested for circular cylindrical shells e.g by A.Pflüger [54] and A.S.Volmir [39]. According to the simplest, linear theory of shell stability, the critical stresses are given by the formulae

$$\sigma_{cr} = \frac{E}{[3(1-\nu^2)]^{1/2}} \frac{h}{R} , \quad \tau_{cr} = \frac{E}{3\sqrt{2}(1-\nu^2)^{3/4}} \left(\frac{h}{R}\right)^{3/2} , \tag{3.21}$$

in wich we replace now the radius R by the current radius of curvature ρ. Further, substituting the expressions for stresses, namely (3.4) for σ and (3.19) for τ, we rewrite (3.20) in the form

$$h^5 - k_1 y\rho h^4 - k_2\rho^3 \ge 0 , \tag{3.22}$$

where k_1 and k_2 depend on material constants, c, j, k, and H.

The strength constraint will be based on the Huber-Mises-Hencky failure hypothesis,

$$\sigma^2 + 3\tau^2 \le \sigma_0^2 , \tag{3.23}$$

and after substituton of the first term of (3.15) and (3.19)

$$\frac{k^2 y^2}{H^2} + \frac{3c^2}{h^2 \sigma_o} \le 1 . \tag{3.24}$$

As we mentioned above, the constraints (3.22) and (3.24) may be active or passive. In the general case we have two functions as design variables, namely $y = y(x)$ and $h = h(x)$. Possible combinations of active (a) and passive (p) constraints are gathered in Table 2, indicating also the type of relevant mathematical problems:

TABLE 2

		STABILITY CONSTRAINT a	STABILITY CONSTRAINT p
STRENGTH CONSTRAINT	a	differential equation	single variational optimization
STRENGTH CONSTRAINT	p	single variational optimization	double variational optimization

In the case of pure torsion we obtain as the optimal section a circular annulus and hence both strength and stability constraints are active. On the contrary, in the cases of pure bending only the stability constraint is active (except outer fibres). Hence, under combined loadings both these combinations may also be expected. Moreover, if the admissible stress σ_o is relatively low, a contour of uniform strength (without simultaneous uniform stability) may appear. Finally, in an optimal structure both constraints cannot be simultaneously passive and this case should be excluded.

The optimization problem is now stated as follows. We look for the function $y(x)$ and $h(x)$ and for the parameters A_o, H, k, minimizing the cross-sectional area (3.2) under the isoperimetric constraints (3.3) and

$$M_t = 2 \int_{-b}^{b} \tau h(y - xy')\, dx \tag{3.25}$$

and under the local stability and strength constraints, (3.22) and (3.24), respectively.

Consider first the profile of boat uniform strength and uniform stability (both constraints active). From (3.24) regarded as an equation we may easily determine the thickness,

$$h = \frac{cH\sqrt{3}}{\sigma_o \sqrt{H^2 - k^2 y^2}} . \tag{3.26}$$

Substituting (3.26) into (3.22), regarded also as equation, we solve the resulting cubic equation with respect to the radius of curvature ρ and hence derive a differential equation for $y = y(x)$. In dimensionless coordinates

$$\eta'' = - C(1+\eta'^2)^{3/2}(1-k^2\eta^2)^{5/6}\left[\left(1 + \sqrt{1 + \frac{D\eta^3}{1-k^2\eta^2}}\,\right)^{1/3} + \left(1 - \sqrt{1 + \frac{D\eta^3}{1-k^2\eta^2}}\,\right)^{1/3}\right]^{-1} , \tag{3.27}$$

where C and D are certain constants.

Consider now the profile of uniform strength only. Then the thickness is also given by (3.26), but the function $y = y(x)$ is to be determined from the Euler-Lagrange equation for the functional. One obtains, [52],

$$\frac{\xi' + \lambda_1 \xi' \eta^2}{\sqrt{(1+\xi'^2)(1-k^2\eta^2)}} - \frac{2}{\sqrt{3}}\lambda_2 \eta = B_1 , \tag{3.28}$$

where η is treated as the independent variable, λ_1 and λ_2 are Lagrangian multipliers.

The profile of uniform stability only is the most difficult to be determined. The thickness h should be evaluated from (3.22) regarded as an equation. This is an algebraic equation of the fifth degree with respect to h. Paper [52] derives for h a convenienet approximate formula. Substituting into the functional we derive the relevant Euler-Lagrange equation for the function $y = y(x)$. This equation, rather complicated, will not be quoted here but final shapes are shown in Fig. 8.

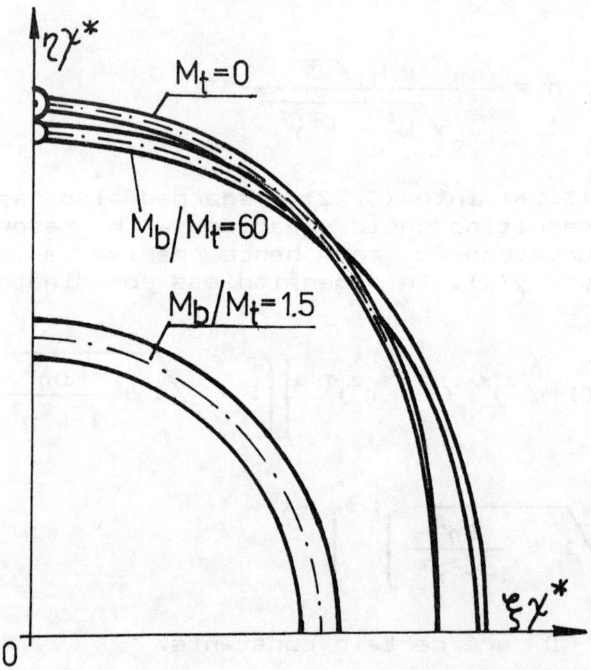

Fig. 8, Optimal cross-sectional shapes of a cylindrical shell under overall bending with torsion

Ranges of validity of individual solutions and the optimal dimensionless cross-sectional areas $a = A\sqrt[9]{E^2/(M_b^2 + M_t^2)}$ in terms of $M_t/\sqrt{M_b^2 + M_t^2}$ are shown in Fig.9.

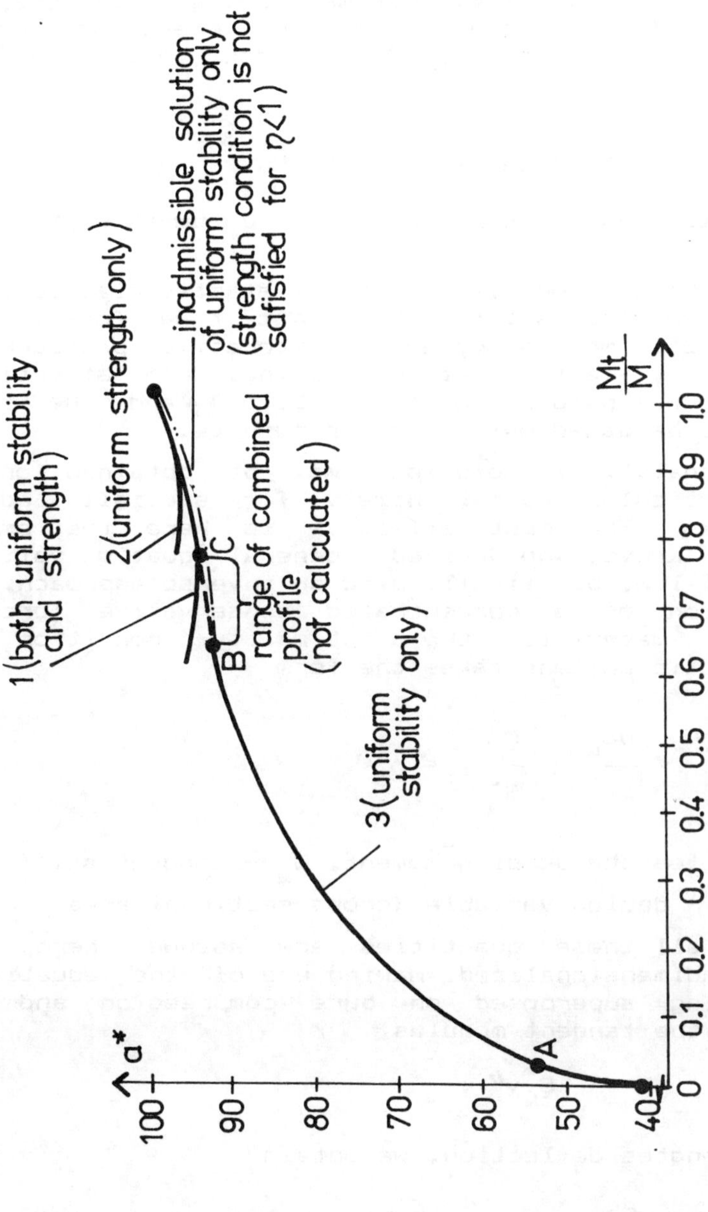

Fig. 9, Minimal cross-sectional arrea and ranges of validity of individual optimal solution for bending with torsion

Restricted variants of optimal design for prescribed shapes of middle surface and also for prescribed constant thickness are discussed by J.Krużelecki and M.Życzkowski [51,52].

4. OPTIMAL DESIGN OF INELASTIC COLUMNS

4.1. Elastic-plastic solutions for columns with solid sections

The problem of optimal design of elastic-plastic columns under Eulerian force was for the first time formulated by P.Laasonen [55] who proposed an iterative procedure for Ylinen's law of elastic-plastic buckling. Much attention to this problem was paid by W.Krzyś [56,57] and the present section will be based mainly on his results.

Exact analytical solutions may be obtained only in certain particular cases, namely for specially adjusted physical laws. The most effective is here the proposal suggested by Krzyś, who derived a general equation valid for any physical law, but finally used an inverse approach. Under the assumption of a concentrated conservative force and homogeneous material, the optimality condition for elastic-plastic columns takes the form

$$\frac{dS_M}{d\Phi} \frac{M^2}{S_M^2} = \Lambda_o , \tag{4.1}$$

where M denotes the bending moment, S_M – tangent stiffness in bending, Φ – design variable (cross-sectional area) and Λ_o – a constant; all these quantities are assumed here to be suitably nondimensionalized. Making use of the equation of small bending superposed on pure compression and hence governed by the tangent modulus

$$M = - S_M v'' , \tag{4.2}$$

where v denotes deflection, we obtain

$$\frac{dS_M}{d\Phi} v''^2 = \Lambda_o . \tag{4.3}$$

Krzyś noted that if S_M is a linear function of Φ, then the deflection line of an optimal column is always a parabola of the second degree or consists of segments of such parabolas. The curvature is constant, independently of the exponent in the geometrical relation $I = c_1 A^{\nu}$ ($\nu = 2$ for geometrically similar sections, $\nu = 1$ or $\nu = 3$ for plane-tapered bars); conversely, this exponent appears in resulting physical laws.

Indeed, let us assume S_M as a linear function of the cross-sectional area (control variable) Φ,

$$S_M = K_1 + K_2 \Phi \, , \tag{4.4}$$

where K_1 and K_2 are certain constants to be determined later. Combining (4.3) and (4.4) we obtain

$$v'' = \pm \, (\Lambda_0 / K_2)^{1/2} = \text{const.} \tag{4.5}$$

Further, since (4.2) yields $S_M = -M/v''$, and v'' is constant, we may write $S''_M = -M''/v'' = -P$, and after integration

$$S_M = B_1 + B_2 x - \frac{1}{2} P \, x^2 \, . \tag{4.6}$$

Equating (4.4) and (4.6) we determine the optimal shape

$$\Phi(x) = \frac{1}{K_2} \left(B_1 - K_1 + B_2 x - \frac{1}{2} P \, x^2 \right) \tag{4.7}$$

and from the state equations (4.2)

$$M(x) = \pm \, (\Lambda_0 / K_2)^{1/2} \left(B_1 + B_2 x - \frac{1}{2} P \, x^2 \right) . \tag{4.8}$$

Now we have to derive the physical law corresponding to the assumption (4.4). Suppose the whole of the column is in the elastic-plastic range before buckling, then

$$S_M = I \left. \frac{d\sigma}{d\varepsilon} \right|_{\sigma = \sigma_{cr}} . \tag{4.9}$$

Comparing (4.9) to (4.6), requiring $d\sigma/d\varepsilon = E$ for $\sigma = \sigma_H$ (proportional limit) and $d\sigma/d\varepsilon = 0$ for $\sigma = \sigma_0$ (yield-point stress) we finally determine K_1 and K_2 and arrive at

$$\frac{d\sigma}{d\varepsilon} = \frac{\sigma_0 - \sigma}{\sigma_0 - \sigma_H} \left(\frac{\sigma}{\sigma_H}\right)^{\nu - 1} E \, . \qquad (4.10)$$

In the particular case $\nu = 2$ (spatially uniformly affine columns) (4.10) turns into the Johnson-Ostenfeld law (F.Bleich [58]). In the case $\nu = 1$ and $\nu = 3$ it is a new law; accuracy of approximation of the real material behaviour is similar to that of other laws (Johnson- Ostenfeld, Ylinen) and was investigated in detail by W.Krzyś [56,57].

4.2. Optimization of imperfect columns under linear creep buckling constraints

Theory of creep buckling of imperfect, physically linear columns initiated by W.Siegfried [59], was developed by A.M.Freudenthal [60], A.R.Rzhanitsyn [61] and A.D.Ross [62]. For Maxwell's type body - under the assumption of geometrical linearity - an exponential increase of deflections in time was observed. The coefficient of time in the exponent is called the "logarithmic velocity of buckling". Hence a new kind of optimization problem may be formulated here: find the shape of the column of minimal volume for a given logarithmic velocity of buckling under a given force. This constraint is rather of stiffness type than of stability type, nevertheless it is typical for a group of creep buckling theories.

The above problem was solved for Maxwell's law by R.Wojdanowska [63], who found an analogy with classical optimization of an elastic column under the constraints of a given critical force. Here, following a paper by R.Wojdanowska and M.Życzkowski [64] we quote an extension of that analogy for much broader class of materials.

Assume the following linear creep law

$$f_1(t)\,\dot{\varepsilon} + f_2(t)\,\varepsilon = f_3(t)\,\dot{\sigma} + f_4(t)\,\sigma \, , \qquad (4.11)$$

where $f_i(t)$ are non-negative functions of time. This law generalizes the standard material (Prager - Hohenemser), since time - dependent coefficients take possible ageing effects into account. Using Bernoulli's hypothesis and integrating (4.11) multiplied by the coordinate y over the cross-sectional area we derive basic equation of creep bending and buckling. Further, substituting M = Pv we obtain

$$f_1(t) I \dot{v}'' + f_2(t) I (v'' - v''_-) + f_3(t)\, P\dot{v} + f_4(t)\, Pv = 0 \qquad (4.12)$$

where $v_- = v_-(x)$ denotes the initial deflection line. Now confine our considerations to such initial deflection lines which preserve their shape during buckling:

$$v(x,t) = v_-(x)\ T(t) \tag{4.13}$$

and separate the variables as follows

$$-\frac{I(x)v''_-(x)}{v_-(x)} = \frac{Pf_3(t)\dot{T}(t) + Pf_4(t)\ T(t)}{f_1(t)\ \dot{T}(t) + f_2(t)[T(t)-1]} = k^2 > 0. \tag{4.14}$$

It was proved in [64] that $\dot{T}/T$ (logarithmic velocity of buckling) is a monotonically decreasing function of k^2, hence k^2 may be regarded as a parameter responsible for the column stiffness. Hence, according to the constraint adopted, k^2 should be kept constant in the expression

$$I(x) = -k^2 \frac{v_-(x)}{v''_-(x)}\ . \tag{4.15}$$

Finally, using Chentsov's approach we arrive at the functional analogous to that for elastic columns, and so the analogy is proved. It is valid only for certain specific initial deflection lines $v_-(x)$; they were determined in [64].

4.3. Optimization of columns under nonlinear creep buckling constraints

From among numerous theories of creep buckling for physically non-linear creep we distinguish two basic groups: the first considers a critical time for imperfect columns defined by infinite increase of deflections or velocities (J.Kempner [65], N.J.Hoff [66]), whereas the second considers a quite different critical time, defined by loss of stability of perfect columns (F.R.Shanley [34], G.Gerard [67], Yu.N.Rabotnov and S.A.Shesterikov [68]).

Optimal design of columns simply supported or clamped at one end with a constraint based on Rabotnov-Shesterikov creep buckling theory was initiated by M.Życzkowski and R.Wojdanowska-Zając [69]. Minimal volume was the design objective, whereas the constraint used the strain-hardening creep law

$$\bar{\Phi} = \dot{\varepsilon}^c(\varepsilon^c)^{\mu} - \Gamma\sigma^n = 0\ , \tag{4.16}$$

where ε^c denotes inelastic strain,

$$\varepsilon^c = \varepsilon - \frac{\sigma}{E} \quad , \tag{4.17}$$

Γ, μ and n are constants, and dot denotes differentiation with respect to the time t. In order to describe instability we consider small variations superposed on the precritical state. First, if follows from (4.16) that these variations satisfy the following equation

$$\frac{\partial \bar{\Phi}}{\partial \dot{\varepsilon}^c} \delta \dot{\varepsilon}^c + \frac{\partial \bar{\Phi}}{\partial \varepsilon^c} \delta \varepsilon^c + \frac{\partial \bar{\Phi}}{\partial \sigma} \delta \sigma = 0 \ . \tag{4.18}$$

Now, the Rabotnov-Shesterikov theory defines the critical time t_{cr} as corresponding to quasi-equilibrium in the neighbouring state, and hence instability is determined by

$$\delta \dot{\varepsilon} = \delta \dot{\sigma} = 0 \tag{4.19}$$

for $t = t_{cr}$. In precritical state of pure compression we may integrate (4.16) with σ constant in time, obtaining

$$\varepsilon = \frac{\sigma}{E} + \left[\Gamma(\mu + 1)\sigma^n \ t\right]^{\frac{1}{\mu+1}} . \tag{4.20}$$

Buckling is characterized by $\delta\varepsilon = y\delta\varkappa$; substituting this relation into (4.18), making use of (4.19) and (4.20), multiplying by y and integrating over the cross-sectional area $A = A(x)$ we derive the following governing equation of creep buckling of nonprismatic bars:

$$v''(x) + \left\{\frac{P}{EI(x)} + \frac{n}{\mu}\left[\Gamma(\mu + 1)t_{cr}P^n\right]^{\frac{1}{\mu+1}}\left[I(x)\right]^{-1} \left[A(x)\right]^{-\frac{n-\mu-1}{\mu+1}}\right\} v(x) = 0 \ . \tag{4.21}$$

Paper [69] discussed minimization of the volume V with (4.21) as a constraint (subsidiary condition). Partly analytical and partly numerical solutions were given; optimal shapes are shown in Fig.10.

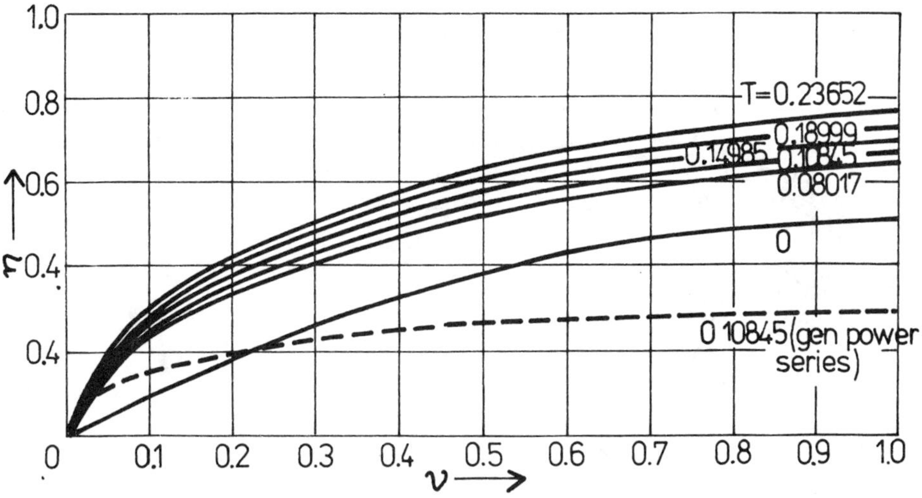

Fig. 10, Shapes of optimal columns for various creep buckling

The case of clamped-clamped columns leads, of course, to bimodal solutions; they were given by J.Błachut and M.Życzkowski in [70]. Optimal design of columns with respect to Kempner-Hoff creep buckling time was initiated by A.Wróblewski [71], and some related problems were investigated by W.Świsterski, A.Wróblewski and M.Życzkowski [72], and by M.Życzkowski [73].

5. MULTIMODAL OPTIMIZATION OF CIRCULAR ARCHES AGAINST CREEP BUCKLING

5.1. Formulation of the problem

Consider a circular arch under uniform external pressure subject to creep as described by strain-hardening creep law(4.16), Fig.11. Following the paper [74] we are looking for optimal distribution of the cross-sectional area along the axis of the arch as to minimize the volume under given critical loading in creep conditions. Two in-plane and two out-of-plane buckling modes will be considered.

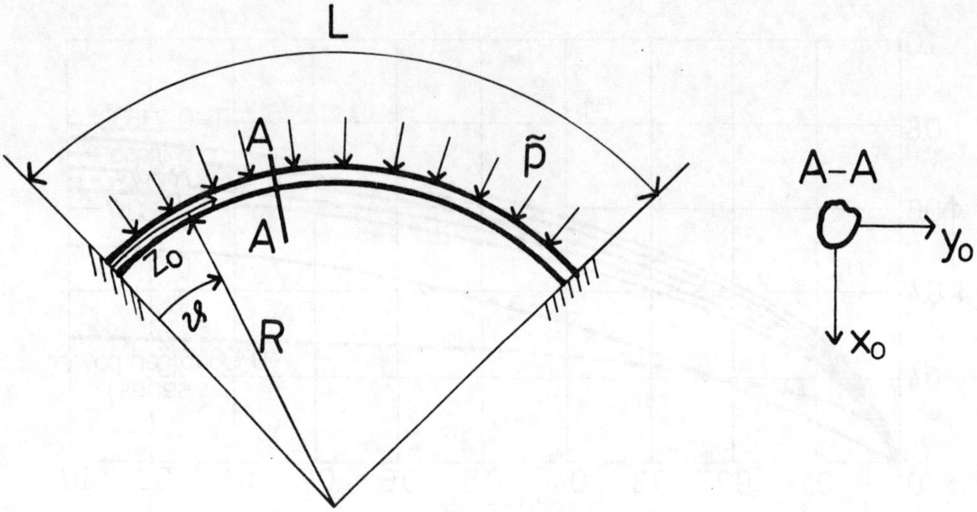

Fig. 11, Arch under consideration

5.2. Constitutive equations of creep buckling

In the precritical state $\tilde{\sigma}_z = -\tilde{p}\tilde{R}/\tilde{A} = \mathrm{const}(\tilde{t})$ and compressive strains due to creep are described by (4.20). Tilda over a symbol denotes dimensional quantities. Variations in creep buckling state are governed by (4.18) expressed in terms of effective stress σ_e and effective strain ε_e and by the similarity of deviators which gives:

$$\delta\tilde{s}_{ij} = \frac{\tilde{s}_{ij}}{\tilde{\sigma}_e^2}\left(\frac{\delta\tilde{\sigma}_e}{\delta\varepsilon_e} - \frac{\tilde{\sigma}_e}{\varepsilon_e}\right)\tilde{s}_{kl}\,\delta\varepsilon_{kl} + \frac{2}{3}\,\frac{\tilde{\sigma}_e}{\varepsilon_e}\,\delta\varepsilon_{ij}\,. \qquad (5.1)$$

The Rabotnov-Shesterikov theory requires (4.19) for effective stresses and strains.

Under the assumption of small displacements and rotations the buckling state is described by the following vector Y of state variables:

$$Y = (\tilde{w}_i,\ \alpha_i,\ \tilde{Q}_i,\ \tilde{M}_i) \qquad i=1,2,3 \qquad (5.2)$$

where:

$\tilde{w}_i = (\tilde{u}, \tilde{v}, \tilde{w})$ – displacements,

$\alpha_i = (\alpha, \beta, \gamma)$ – angles of rotation,

$\tilde{M}_i = (\tilde{M}_x, \tilde{M}_y, \tilde{M}_z)$ – bending and twisting moments,

$\tilde{Q}_i = (\tilde{Q}_x, \tilde{Q}_y, \tilde{N}_z)$ – shear and normal forces.

All the above quantities denote increments of state variables from the prebuckling state. They are determined from the equilibrium equations and geometrical relations of the buckling state (see [9]). In particular, under assumption of Bernoulli's and de Saint-Venant's hypotheses, moments are expressed as follows:

$$\tilde{M}_x = \iint_A \delta\tilde{\sigma}_z \tilde{y}\, d\tilde{A} = \iint_A \frac{\delta\tilde{\sigma}_e}{\delta\varepsilon_e} \delta\varepsilon_z \tilde{y}\, d\tilde{A} = \mathfrak{I}_x \frac{\delta\tilde{\sigma}_e}{\delta\varepsilon_e} \delta\tilde{\varkappa}_x = \tilde{B}_x\, \delta\tilde{\varkappa}_x$$

$$\tilde{M}_y = \iint_A \delta\tilde{\sigma}_z \tilde{x}\, d\tilde{A} = \mathfrak{I}_y \frac{\delta\tilde{\sigma}_e}{\delta\varepsilon_e} \delta\tilde{\varkappa}_y = \tilde{B}_y\, \delta\tilde{\varkappa}_y \tag{5.3}$$

$$\tilde{M}_z = \iint_A (\delta\tilde{\tau}_{zy}\tilde{x} - \delta\tilde{\tau}_{zx}\tilde{y})\, d\tilde{A} = \frac{1}{3} \frac{\tilde{\sigma}_e}{\varepsilon_e} \mathfrak{I}_t \delta\tilde{\varkappa}_z = \tilde{B}_t\, \delta\tilde{\varkappa}_z$$

where $\delta\tilde{\varkappa}_{x,y,z}$ denotes increments of curvatures (in fact $\delta\varkappa_z$ denotes unit angle of twist), $\mathfrak{I}_{x,y,t}$ and $\tilde{B}_{x,y,t}$ are the moments of inertia and rigidities under creep conditions of the cross-sectional area, respectively. These formulae are similar to corresponding elastic ones, but $\tilde{B}_x$, $\tilde{B}_y$, $\tilde{B}_t$ depend on time(are calculated for the critical time $\tilde{t}_{cr}$) which is "hidden"in ε_e and $\delta\varepsilon_e$ (eqs. 4.20). Other equilibrium and geometrical equations are the same as in elasticity.

5.3. The problem of behaviour of loading in the course of buckling

Critical state of arches depends essentially on the behaviour of loading. Various particular cases of such behaviour have been studied, but a more general theory has not been given as yet. Such an approach was proposed for columns by Kordas [75] who introduced four parameters as to describe linear relations between two loading components arising during buckling and two geometrical quantities. We are now going to give a similar, general approach describing the behaviour of loading by uniformly distributed radial pressure in arches during buckling.

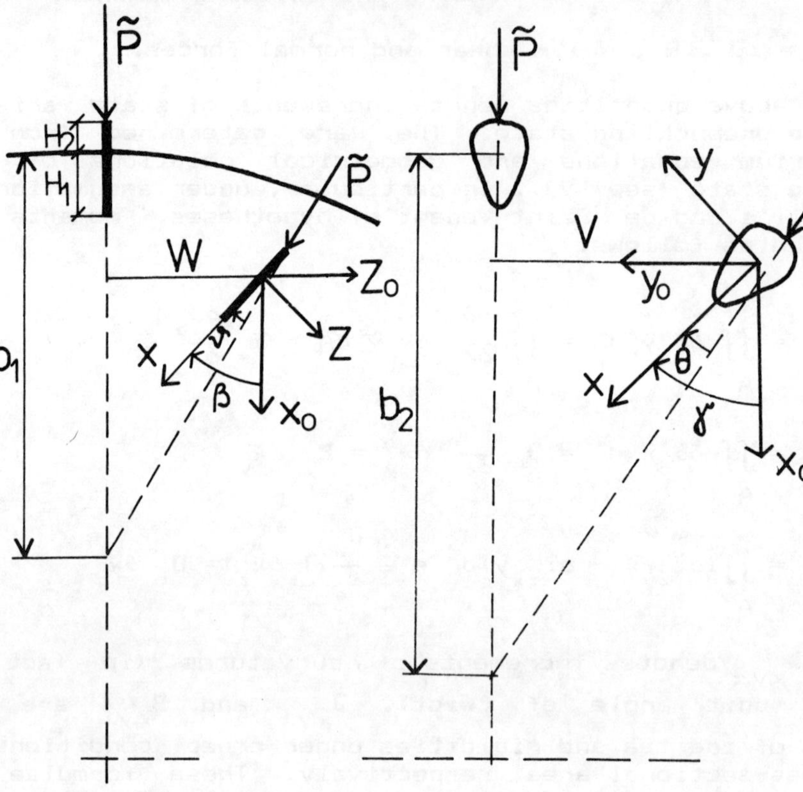

Fig. 12, Behaviour of loading in the course of arch buckling

In general, in the course of buckling the arch axis is loaded by forces and moments distributed over the element. They constitute a certain generalized vector of external forces $\mathfrak{U}_i = (\tilde{f}_x, \tilde{f}_y, \tilde{f}_z, \tilde{m}_x, \tilde{m}_y, \tilde{m}_z)^T$. The vector components are defined in the coordinate system (x,y,z) connected with the buckled state - see Figure 12a,b. They depend on the displacements w_i and rotations α_i of the arch axis which constitute the vector of generalized displacements

$\tilde{u}_i = (\tilde{u}, \tilde{v}, \tilde{w}, \tilde{\alpha}, \tilde{\beta}, \tilde{\gamma})^T$:

$$\tilde{U}_i = F_i(\tilde{u}_j) , \qquad i,j=1,\ldots,6 \qquad (5.4)$$

Since the values of critical loading depend just on linear terms of (5.4), we may perform relevant expansions and retain linear terms only:

$$\tilde{U}_i = \tilde{p}\, \tilde{a}_{ij}\, \tilde{u}_j \qquad (5.5)$$

where the summation convention holds and i,j=1,...,6. So, it may seem that the behaviour of loading during buckling is described by 36 parameters $\tilde{a}_{ij}$. However, with the accuracy of the first-order terms,

$$\tilde{f}_x = \tilde{p} , \quad \tilde{m}_x = 0 \qquad (5.6)$$

and the dependence on deflection $\tilde{u}$ and on rotation α about the normal axis usually does not take place, so we may retain (5.5) just with 16 parameters $\tilde{a}_{ij}$. A particular, very important case of load behaviour is that resulting in uncoupling of in-plane and out-of-plane buckling. Changing the order of remaining components in $\tilde{U}$ and $\tilde{u}$:

$$\tilde{U}_i = \left(\tilde{f}_z, \tilde{m}_y, \tilde{f}_y, \tilde{m}_z \right)^T ,$$
$$\tilde{u}_i = \left(\tilde{w}, \beta, \tilde{v}, \gamma \right)^T , \qquad (5.7)$$

we can write the matrix $\tilde{a}_{ij}$ in the form:

$$\tilde{a}_{ij} = \begin{pmatrix} \tilde{a}_{11} & \tilde{a}_{12} & 0 & 0 \\ \tilde{a}_{21} & \tilde{a}_{22} & 0 & 0 \\ 0 & 0 & \tilde{a}_{33} & \tilde{a}_{34} \\ 0 & 0 & \tilde{a}_{43} & \tilde{a}_{44} \end{pmatrix} \qquad (5.8)$$

with 8 parameters.

Practically, the components $\tilde{a}_{ij}$ are not independent. Figure 12a,b presents behaviour of loading during buckling which is described by three parameters $\tilde{a}$, $\tilde{b}_1, \tilde{b}_2$. For linearized relations the components $\tilde{a}_{ij}$ can be written:

$$\tilde{a}_{11} = -\frac{1}{\tilde{b}_1}\left(1 + \frac{\tilde{a}}{\tilde{b}_1}\right), \qquad \tilde{a}_{12} = \left(1 + \frac{\tilde{a}}{\tilde{b}_1}\right)$$

$$\tilde{a}_{21} = -\frac{\tilde{a}}{\tilde{b}_1}\left(1 + \frac{\tilde{a}}{\tilde{b}_1}\right), \qquad \tilde{a}_{22} = \tilde{a}\left(1 + \frac{\tilde{a}}{\tilde{b}_1}\right)$$

$$\tilde{a}_{33} = -\frac{1}{\tilde{b}_2}\left(1 + \frac{\tilde{a}}{\tilde{b}_2}\right), \qquad \tilde{a}_{34} = -\left(1 + \frac{\tilde{a}}{\tilde{b}_2}\right)$$

$$\tilde{a}_{43} = \frac{\tilde{a}}{\tilde{b}_2}\left(1 + \frac{\tilde{a}}{\tilde{b}_2}\right), \qquad \tilde{a}_{44} = \tilde{a}\left(1 + \frac{\tilde{a}}{\tilde{b}_2}\right) \tag{5.9}$$

The parameter $\tilde{a}$ describes the point of pressure acting. For example $\tilde{a}=\tilde{A}_1$ denotes internal surface of arch; $\tilde{a}=-\tilde{A}_2$ external one; $\tilde{a}=0$ is typical case of loading (pressure acts on arch axis). Parameters $\tilde{b}_1, \tilde{b}_2$ describe directions of pressure during buckling (in-plane and out-of-plane respectively). Typical cases are: $\tilde{b}_i=\infty$ - direction fixed in space; $\tilde{b}_i=\tilde{R}$ -direction to the centre of initial curvature; $\tilde{b}_i=\tilde{a}$ -direction normal to the deformed axis. A more general behaviour is also possible, but seldom in engineering applications.

5.3. Final form of state equations

In the present chapter, considerations are confined to the case of uncoupled in-plane and out-of-plane buckling. We introduce dimensionless variables:

$$z=\frac{\tilde{z}}{L}, \quad u_i=\frac{\tilde{u}_i}{L}, \quad Q_i=\frac{\tilde{Q}_i L^4}{E V^2}, \quad M_i=\frac{\tilde{M}_i L^3}{EV^2}, \quad f_i=\frac{\tilde{f}_i R^3 L^2}{EV^2},$$

$$m_i=\frac{\tilde{m}_i L^4}{EV^2}, \quad J_x=\frac{\mathfrak{J}_x L^2}{V^2}, \quad J_y=\frac{\mathfrak{J}_y L^2}{V^2}, \quad J_t=\frac{\mathfrak{J}_t L^2}{V^2}, \quad S(z)=\frac{\tilde{A}(z)}{\tilde{A}^*}, \tag{5.10}$$

and parameters:

$$p=\frac{\tilde{p}R^3L^2}{EV^2},\quad T=E\left[\Gamma(\mu+1)\tilde{t}_{cr}\right]^{1/(1+\mu)}\left(\frac{EV^2}{R^2\tilde{A}^*L^2}\right)^{\nu},\quad \nu=\frac{n}{\mu+1}-1,$$

$$\zeta=\frac{L}{R},\quad \omega_x=\frac{V^2}{\mathfrak{J}_x^*L^2},\quad \omega_y=\frac{V^2}{\mathfrak{J}_y^*L^2},\quad \omega_z=\frac{V^2}{\mathfrak{J}_t^*L^2},$$

$$a=\frac{\tilde{a}}{R},\quad b_i=\frac{\tilde{b}_i}{R}. \tag{5.11}$$

In the equations (5.11) constants marked with star are quantities defined for a prismatic arch of the same volume as the optimal arch:

$$V=\tilde{A}^*L=\tilde{A}^*L\int_0^1 S(z)\,dz \tag{5.12}$$

After introducing the quantities K_x, K_y instead of the shear forces Q_x, Q_y:

$$K_x=Q_x-p\zeta^3\beta,$$
$$K_y=Q_y+p\zeta^3\alpha, \tag{5.13}$$

the in-plane and out-of-plane buckling are described by the state vector $Y_{IN}=(u,w,\beta,K_x,N_z,M_y)$ and $Y_{OUT}=(v,\alpha,\gamma,K_y,M_x,M_z)$ respectively. The state equations take the following form:

IN-PLANE BUCKLING

$$u'=\beta-\zeta w,$$
$$w'=\zeta u,$$
$$\beta'=\frac{M_y}{J_y}\omega_y\left[1+\frac{n}{\mu}T\left(\frac{p}{S}\right)^{\nu}\right],$$
$$K_x'=-\zeta N_z, \tag{5.14}$$
$$N_z'=\zeta K_x+p\zeta\beta(1-a_{12})-p\zeta^4a_{11}w,$$
$$M_y'=-K_x-p\zeta^2\beta(1+a_{22})-p\zeta^3a_{21}w,$$

OUT-OF-PLANE BUCKLING

$$
\begin{aligned}
v' &= -\alpha ,\\
\alpha' &= \frac{M_x}{J_x}\,\omega_x\left[1+\frac{n}{\mu}T\left(\frac{p}{S}\right)^{\nu}\right]-\zeta\gamma ,\\
\gamma' &= \frac{M_z}{J_t}\,\omega_z\left[1+T\left(\frac{p}{S}\right)^{\nu}\right]+\zeta\alpha ,\\
K_y' &= -p\zeta^3\gamma(1+a_{34}) - p\zeta^4 a_{33} v ,\\
M_x' &= -\zeta M_z - p\zeta^2\alpha + K_y ,\\
M_z' &= \zeta M_x - p\zeta^2\gamma a_{44} - p\zeta^3 a_{43} v .
\end{aligned}
\tag{5.15}
$$

Considering symmetric (index S) and antisymmetric (index A) forms of in-plane and out-of-plane buckling the boundary conditions for the state equations may be written as follows:

IN-PLANE	OUT-OF-PLANE
$u^{S,A}(0)=w^{S,A}(0)=\beta^{S,A}(0) = 0,$	$v^{S,A}(0)=\alpha^{S,A}(0)=\gamma^{S,A}(0) = 0,$
$\beta^S(\frac{1}{2})=K_x^S(\frac{1}{2})=w^S(\frac{1}{2}) = 0,$	$\alpha^S(\frac{1}{2})=M_z^S(\frac{1}{2})=K_y^S(\frac{1}{2}) = 0,$
$u^A(\frac{1}{2})=N_z^A(\frac{1}{2})=M_y^A(\frac{1}{2}) = 0,$	$\gamma^A(\frac{1}{2})=v^A(\frac{1}{2})=M_x^A(\frac{1}{2}) = 0,$ (5.16)

Boundary conditions that distinguish symmetric and antisymmetric buckling modes may be defined for z=0 and $z=\frac{1}{2}$, due to symmetry of the structure in the prebuckling state.

5.4. The optimal design problem

In the present paper consideration are confined to the case of a rectangular cross-section with ratio χ=B/H=const. The problem of optimization discussed here is to determine the design variable $\tilde{A}(z)$ (cross-sectional area function) that satisfies the state equations (5.14,5.15) condition (5.12)and minimizes total volume of the arches for given external load and critical time:

$$
V = \int_0^L \tilde{A}(\tilde{z})\, d\tilde{z} \longrightarrow \min ,
$$

$$
\tilde{p},\ \tilde{t}_{cr} = \text{const.} \tag{5.17}
$$

The geometrical parameters of the considered cross-section have the form:

$$J_x = J_y = J_t = S^2(z) \ , \quad \omega_x = 12\chi \ ,$$
$$\omega_y = \frac{12}{\chi} \ , \quad \frac{1}{\omega_z} = \frac{\chi}{3}\left[\frac{1}{3} - \frac{64}{\pi^5}\chi \ \mathrm{th}\left(\frac{\pi}{2\chi}\right)\right] \tag{5.18}$$

Other cases of change of dimensions of the rectangular cross-section for elastic arch are presented in [9].

The necessary optimality condition is derived by use of Pontryagin's maximum principle. It is easy to prove that if the conditions:

$$a_{43} = a_{34} + 1 \ , \qquad a_{21} = a_{12} - 1 \tag{5.19}$$

are satisfied, the problem is selfadjoint. It means that it is sufficient to solve the set of state equations only.

The conditions (5.19) are satisfied in two cases: (i) $a=0$ i.e. the pressure acts on arch axis (except the case of follower pressure); (ii) $b_1=b_2=\infty$ i.e. the direction of pressure acting is fixed in space.

In the present paper, considerations are confined to a clamped-clamped arch and three cases of load for which the problem is selfadjoint:

case A - direction fixed in space and pressure acts on arch axis:

$$a=0, \quad b_1=b_2=\infty \tag{5.20}$$

case B - direction fixed in space and pressure acts on external upper surface of the arch:

$$a= -\frac{H}{2R} = -\frac{H^*}{2R}\sqrt{S} \ , \quad b_1=b_2=\infty \tag{5.21}$$

case C - direction to the center of initial curvature and pressure acts on arch axis:

$$a=0, \quad b_1=b_2=R \tag{5.22}$$

The Hamiltonian H may be written in the form:

$$H = \lambda_1 H^S_{IN} + \lambda_2 H^A_{IN} + \lambda_3 H^S_{OUT} + \lambda_4 H^A_{OUT} + \lambda_o S =$$

$$= \lambda_1 \left\{ \left(\frac{M^S_y}{S}\right)^2 \omega_y \left[1 + \frac{n}{\mu} T \left(\frac{p_1}{S}\right)^{\nu}\right] + p_1 \zeta^2 (1+a_{22}) \left(\beta^S\right)^2 \right\} +$$

$$+ \lambda_2 \left\{ \left(\frac{M^A_y}{S}\right)^2 \omega_y \left[1 + \frac{n}{\mu} T \left(\frac{p_2}{S}\right)^{\nu}\right] + p_2 \zeta^2 (1+a_{22}) \left(\beta^A\right)^2 \right\} +$$

$$+ \lambda_3 \left\{ \left(\frac{M^A_x}{S}\right)^2 \omega_x \left[1 + T \left(\frac{p_3}{S}\right)^{\nu}\right] + \left(\frac{M^A_z}{S}\right)^2 \omega_z \left[1 + \frac{n}{\mu} T \left(\frac{p_3}{S}\right)^{\nu}\right] + p_3 \zeta^2 a_{44} \left(\gamma^A\right)^2 \right\} +$$

$$+ \lambda_4 \left\{ \left(\frac{M^S_x}{S}\right)^2 \omega_x \left[1 + T \left(\frac{p_4}{S}\right)^{\nu}\right] + \left(\frac{M^S_z}{S}\right)^2 \omega_z \left[1 + \frac{n}{\mu} T \left(\frac{p_4}{S}\right)^{\nu}\right] + p_4 \zeta^2 a_{44} \left(\gamma^S\right)^2 \right\} +$$

$$+ \lambda_o S \,, \tag{5.23}$$

$$p_1 = p^S_{IN}, \quad p_2 = p^A_{IN}, \quad p_3 = p^S_{OUT}, \quad p_4 = p^A_{OUT}$$

where terms independent of S are omitted; $\lambda_o, \lambda_1, \lambda_2, \lambda_3, \lambda_4$ are nonnegative constants that are to be determined. The necessary optimality condition has the form:

$$\frac{\partial H}{\partial S} = 0 \tag{5.24}$$

5.5. Numerical examples

The problem is solved by used the numerical iterative method presented in [76]. The results are obtained for arch steepnes $\zeta = \frac{\pi}{2}$. The dependence of critical loading p on ratio χ for same values of the critical time T are presented in the Fig. 13 for the loading cases A.

The range of application of suitable optimization modalities are marked (short dashed line) in this figure. In the problem under consideration there exist the ranges of uni-, bi- and trimodal formulations.

For each critical time the optimal ratio χ_{opt} (for which the critical pressure is the largest) exists. These points are connected by long dashed line. It is interesting that the optimal ratio χ_{opt} is clearly dependent on the critical time whereas the ranges of modalities are practically independent of time. The optimal ratio χ is determined with respect to uni- or bimodal formulation of the problem.

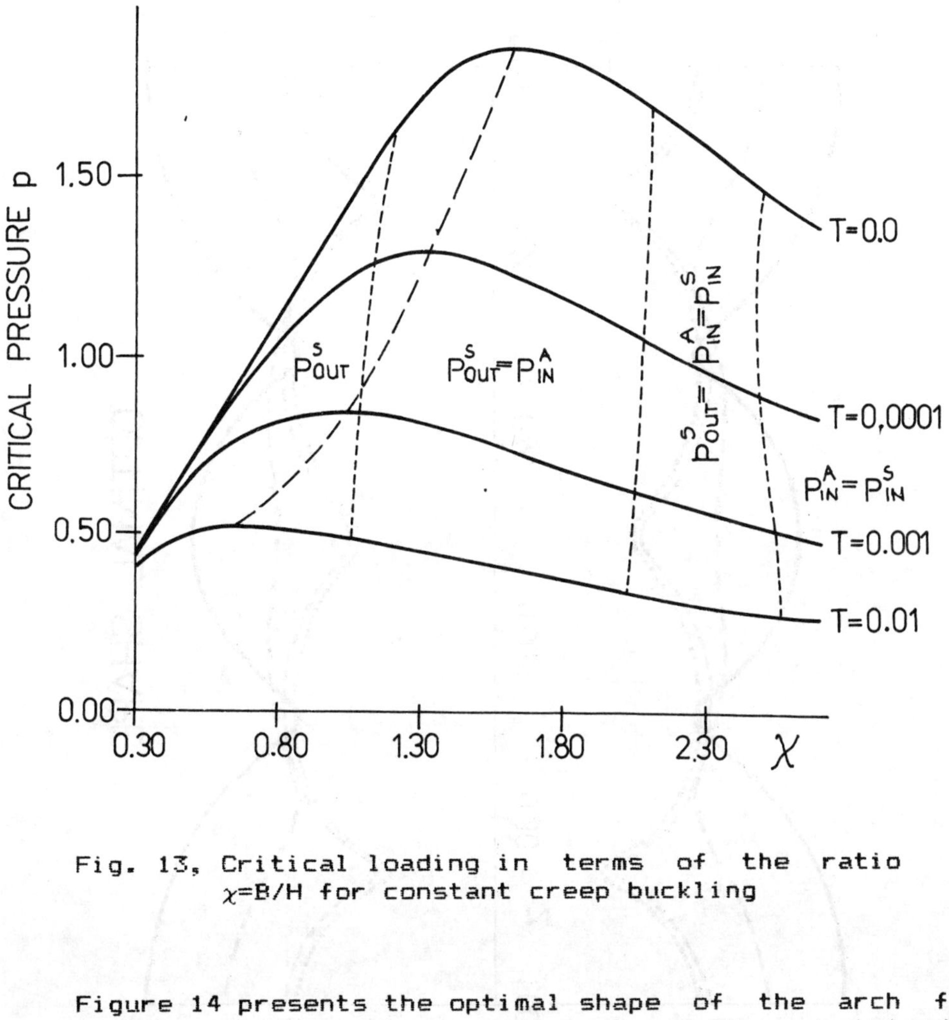

Fig. 13, Critical loading in terms of the ratio χ=B/H for constant creep buckling

Figure 14 presents the optimal shape of the arch for unimodal (χ=1.0), bimodal (χ=1.6 and χ=2.7) and trimodal (χ=2.6) formulation of the problem with respect to first case of loading and the critical time T=0.

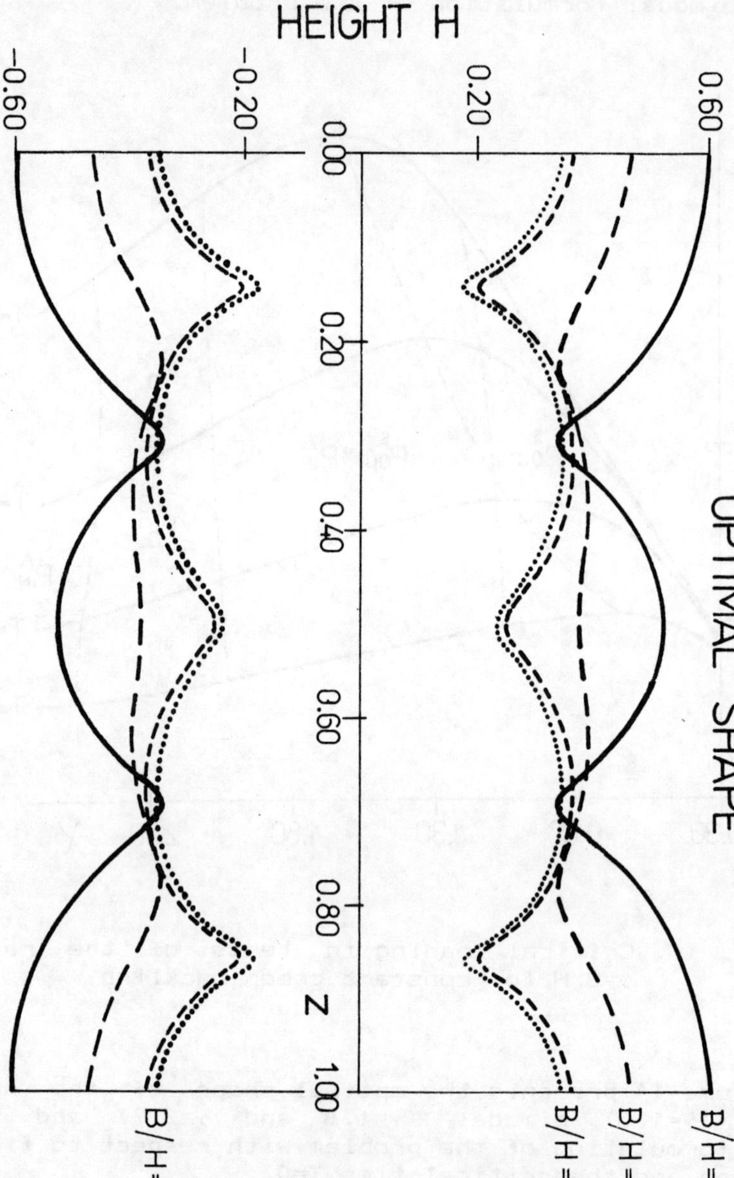

Fig. 14, Optimal shapes of the arch for various B/H

6. OPTIMAL DESIGN OF PLATES AND SHELLS UNDER CREEP BUCKLING CONSTRAINS

6.1. Formulation of the problem for circular plates

Consider a circular plate loaded by in-plane uniformly distributed pressure subject to creep as described by strain

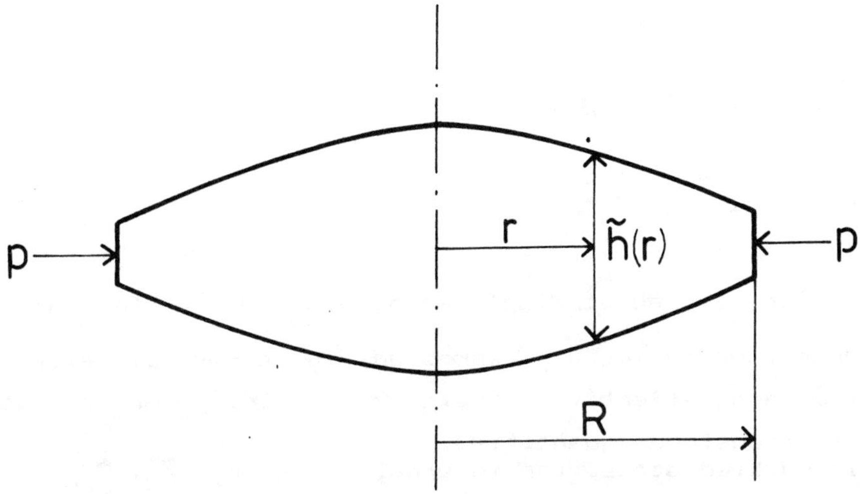

Fig. 15, Circular plate under consideration

hardening creep law (4.16) - Fig. 15. Following the paper [77] by Wróblewski we minimize the total volume of the plate under given critical loading in creep conditions (the critical time as well as the critical pressure), assuming the thickness as the design variable. A circularly symmetric thickness distribution $\tilde{h}=\tilde{h}(r)$ will be analyzed, butunsymmetric buckling modes $\tilde{w}=\tilde{w}(r,\theta)$ will also be admitted.

6.2. Constitutive equations of creep buckling

The stability analysis of plates of variable stiffness,loaded by in-plane forces requires at first, the solution of membrane stress distribution problem (precritical state), and just then application of stability criterion. The assumption of small deflections during buckling makes it possible to disregard mutual interaction between membrane state and bending state.

In precritical state two components of the internal forces are different from zero, namely normal forces $\tilde{N}_r$, $\tilde{N}_\theta$. Assuming the law of the similarity of stress and strain deviators and of incompressibility of material one can derive the governing equations of the precritical state:

$$\tilde{N}_\theta = \frac{1}{2}\tilde{N}_r + \beta\frac{\tilde{u}}{r}$$

$$\frac{d(r\tilde{N}_r)}{dr} = \frac{1}{2}\tilde{N}_r + \beta\frac{\tilde{u}}{r}$$

$$\frac{d\tilde{u}}{dr} = \frac{3}{4}\frac{1}{\beta}\tilde{N}_r - \frac{1}{2}\frac{\tilde{u}}{r} \qquad (6.1)$$

where $\tilde{u}=\tilde{u}(r)$ is radial displacement and $\beta = \frac{\tilde{\sigma}_e}{\varepsilon_e}$. The variable β is described by (4.20) expressed in terms of effective stress $\tilde{\sigma}_e$ and effective strain ε_e. Tilda over a symbol denotes dimensional quantities.

In buckled state, the internal forces $\tilde{M}_r$, $\tilde{M}_\theta$, $\tilde{M}_{r\theta}$ arise. Under the assumption of the Rabotnov – Shesterikov creep stability theory, making use of equation (5.1), moments are expressed as follows:

$$\tilde{M}_r = -\frac{\tilde{h}^3}{9}\beta\left[\tilde{a}_{11}\tilde{w}''_{rr} + \tilde{a}_{12}\left(\frac{1}{r}\tilde{w}'_r + \frac{1}{r^2}\tilde{w}''_{\theta\theta}\right)\right]$$

$$\tilde{M}_\theta = -\frac{\tilde{h}^3}{9}\beta\left[\tilde{a}_{21}\tilde{w}''_{rr} + \tilde{a}_{22}\left(\frac{1}{r}\tilde{w}'_r + \frac{1}{r^2}\tilde{w}''_{\theta\theta}\right)\right]$$

$$\tilde{M}_{r\theta} = -\frac{\tilde{h}^3}{9}\beta\,\tilde{a}_{33}\left(\frac{1}{r}\tilde{w}'_\theta\right)'_r \qquad (6.2)$$

where:

$$\tilde{a}_{11} = \frac{\tilde{\sigma}_r^2}{\tilde{\sigma}_e^2}(a+1), \quad \tilde{a}_{12} = \tilde{a}_{21} = \frac{\tilde{\sigma}_r\tilde{\sigma}_\theta}{\tilde{\sigma}_e^2}\left(a+\frac{1}{2}\right), \quad \tilde{a}_{22} = \frac{\tilde{\sigma}_\theta^2}{\tilde{\sigma}_e^2}(a+1)$$

$$\tilde{a}_{33} = \frac{1}{2}, \qquad a = \frac{3}{4}\left(\frac{\varepsilon_e}{\tilde{\sigma}_e}\frac{\delta\tilde{\sigma}_e}{\delta\varepsilon_e} - 1\right), \qquad (6.3)$$

and $\tilde{w}=\tilde{w}(r,\theta)$ is deflection of plate.

The internal forces in buckled state have to satisfy the equilibrium equation:

$$\left(r\tilde{M}_r\right)''_{rr} + \frac{2}{r}\left(r\tilde{M}_{r\theta}\right)''_{r\theta} + \frac{1}{r}\left(\tilde{M}_\theta\right)''_{\theta\theta} - \left(\tilde{M}_\theta\right)'_r + r\tilde{N}_r\left(\tilde{w}\right)''_{rr} + \\ + \tilde{N}_\theta\left[\frac{1}{r}\left(\tilde{w}\right)''_{\theta\theta} - \left(\tilde{w}\right)'_r\right] = 0 \tag{6.4}$$

Now, we introduce the following dimensionless variables:

$$\xi=\frac{r}{R}, \quad \Phi=\frac{\tilde{h}}{\tilde{H}_o}, \quad w=\frac{\tilde{w}}{R}, \quad u=\frac{\tilde{u}}{R}, \quad \beta=\frac{\tilde{\beta}}{E},$$

$$N_i=\frac{9R^2\tilde{N}_i}{E\tilde{H}_o^3}, \quad M_j=\frac{9R\tilde{M}_j}{E\tilde{H}_o^3}, \quad \sigma_k=\frac{\tilde{\sigma}_k}{E}, \quad \begin{matrix} i=r,\theta \\ j=r,\theta,r\theta \\ k=r,\theta,e \end{matrix} \tag{6.5}$$

and parameters:

$$p=\frac{9R^2}{E\tilde{H}_o^2}\tilde{p}, \quad \chi=\frac{9R^2}{\tilde{H}_o^2}, \quad T=\left[\Gamma(\mu+1)E^n t_{cr}\right]^{1/(\mu+1)}\left(\frac{1}{\chi}\right)^{\nu},$$

$$\nu=\frac{n-\mu-1}{\mu+1}, \tag{6.6}$$

where $\tilde{H}_o=\text{const}(\xi)$ is the plate thickness of the same volume as the optimal plate:

$$V = \pi R^2\tilde{H}_o = 2\pi R^2\tilde{H}_o\int_0^1 \xi\,\Phi(\xi)\,d\xi \tag{6.7}$$

Assuming, that variables in the solution may be separated as follows:

$$\begin{aligned} w(\xi,\theta) &= w(\xi)\cos m\theta \\ M_r(\xi,\theta) &= M_r(\xi)\cos m\theta \\ M_\theta(\xi,\theta) &= M_\theta(\xi)\cos m\theta \\ M_{r\theta}(\xi,\theta) &= M_{r\theta}(\xi)\sin m\theta \end{aligned} \tag{6.8}$$

and introducing new state variables:

$$M = \xi M_r \,, \qquad N = \xi N_r \,, \qquad \varphi = \frac{dw}{d\xi} \,,$$
$$Q = \frac{dM}{d\xi} + 2mM_{r\theta} - M_\theta \,, \tag{6.9}$$

we may write the state equations in the following form:

1) the precritical state (membrane state)

$$N_\theta = \frac{1}{2}\frac{N}{\xi} + \beta \frac{u}{\xi}\frac{\Phi}{\chi} \,,$$
$$N' = \frac{1}{2}\frac{N}{\xi} + \beta \frac{u}{\xi}\frac{\Phi}{\chi} \,,$$
$$u' = \frac{3}{4}\frac{1}{\beta}\frac{\chi}{\Phi}\frac{N}{\xi} - \frac{1}{2}\frac{u}{\xi} \,, \tag{6.10}$$

2) the critical state

$$w' = \varphi \,,$$
$$\varphi' = -\frac{M}{\beta\Phi^3 a_{11}\xi} - \frac{a_{12}}{a_{11}}\frac{\varphi}{\xi} + m^2 \frac{a_{12}}{a_{11}}\frac{w}{\xi^2} \,,$$
$$M' = Q + \frac{a_{12}}{a_{11}}\frac{M}{\xi} + \beta\Phi^3 \frac{\varphi}{\xi}\left(\frac{a_{12}^2}{a_{11}} - a_{22} - m^2\right) +$$
$$+ \beta\Phi^3 \frac{w}{\xi^2}\left(1 + a_{22} - \frac{a_{12}^2}{a_{11}}\right)$$
$$Q' = \frac{M}{\xi^2}\left(\frac{1}{\beta\Phi^3 a_{11}}\frac{N}{\xi} + m^2 \frac{a_{12}}{a_{11}}\right) + \beta\Phi^3 \frac{\varphi}{\xi^2}\left(\frac{a_{12}^2}{a_{11}} - a_{22} - 1\right) +$$
$$+ \left[\frac{N}{\xi}\left(\frac{1}{2} - \frac{a_{12}}{a_{11}}\right) + \beta \frac{\Phi}{\chi}\frac{u}{\xi}\right]\left(m^2 \frac{w}{\xi} - \varphi\right) +$$
$$+ \beta\Phi^3 \frac{w}{\xi^3}\left[1 - m^2\left(\frac{a_{12}^2}{a_{11}} - a_{22}\right)\right] \tag{6.11}$$

The boundary conditions take the forms:

$$u(0) = 0 \,, \quad N(1) = -p \,, \quad w(1) = 0 \,, \quad \varphi(1) = 0 \,,$$

$$\begin{array}{lll} \varphi(0) = 0 \,, & Q(0) = 0 & \text{for } m=0 \,, \\ w(0) = 0 \,, & M(0) = 0 & \text{for } m=1,3,5,\ldots \\ w(0) = 0 \,, & \varphi(0) = 0 & \text{for } m=2,4,6,\ldots \end{array} \tag{6.12}$$

6.3. The optimal design problem

The problem of optimization is to determine the design variable $\Phi(\xi)$ (thickness function) that satisfies the state equations (6.10, 6.11), condition (6.7) and minimizes the total volume of the plate for given external load and the critical time:

$$V = \int_0^R r\, h(\tilde{r})\, dr \longrightarrow \min \,,$$

$$\tilde{p},\ t_{cr} = \text{const} \,. \tag{6.13}$$

The necessary optimality condition is derived by use of Pontryagin's maximum principle. The Hamiltonian H may be written in the form:

$$H = \bar{u}u' + \bar{N}N' + \bar{w}w' + \bar{\varphi}\varphi' + \bar{M}M' + \bar{Q}Q' + \lambda\Phi \tag{6.14}$$

where barred symbols denote the adjoint functions and λ is a constant which is determined by the relation (6.7). The adjoint functions are defined as:

$$\bar{f}' = -\frac{\partial H}{\partial f} \,, \qquad f=u,N,w,\varphi,M,Q \tag{6.15}$$

and we assume here that the terms $\frac{\partial\beta}{\partial u}$ and $\frac{\partial\beta}{\partial N}$ may be neglected. The boundary conditions one can obtain from the transversality conditions.

Finally, the Hamiltonian takes the form:

$$\frac{1}{\xi} H = \frac{1}{\beta\Phi^3 a_{11}} \left[\frac{\bar{Q}}{\xi}\frac{N}{\xi}\frac{M}{\xi^2} - \frac{\bar{\varphi}}{\xi}\frac{M}{\xi} \right] + \beta\Phi^3 \left\{ \left(\frac{a_{12}^2}{a_{11}} - a_{22} - m^2 \right) \frac{\bar{M}}{\xi}\frac{\varphi}{\xi} + \right.$$

$$\left. + m^2 \left[1 - m^2 \left(\frac{a_{12}^2}{a_{11}} - a_{22} \right) \right] \frac{\bar{Q}}{\xi}\frac{w}{\xi^3} + m^2 \left(\frac{a_{12}^2}{a_{11}} - a_{22} - 1 \right) \frac{\bar{Q}}{\xi}\frac{\varphi}{\xi^2} \right\} +$$

$$+ \beta \frac{\Phi}{\chi} \left[\frac{\bar{Q}}{\xi} \frac{u}{\xi} \left(m^2 \frac{w}{\xi} - \varphi \right) + \frac{\bar{N}}{\xi} \frac{u}{\xi} \right] + \frac{3}{4} \frac{1}{\beta} \frac{\chi}{\Phi} \frac{\bar{u}}{\xi} \frac{N}{\xi} + \lambda\Phi \tag{6.16}$$

where terms without design variable Φ are omitted. The necessary optimality condition:

$$\frac{\partial H}{\partial \Phi} = 0 , \tag{6.17}$$

determines the shape of plate.

6.4. Numerical solution and examples

The state equations (6.10, 6.11) as well as the Hamiltonian (6.16) and condition (6.15) show singularities when ξ tends to zero (there exist terms of the type $\frac{1}{\xi}$, $\frac{1}{\xi^2}$, $\frac{1}{\xi^3}$). At the vicinity $\xi=0$, there is a necessity of expansion of the state variables into power series of the radial variable ξ:

$$u = u_o + u_1\xi + \ldots ,$$
$$N = N_o + N_1\xi + \ldots ,$$
$$w = \xi^m \left(w_o + w_2\xi^2 + \ldots \right) ,$$
$$\varphi = \xi^{m-1} \left(\varphi_o + \varphi_2\xi^2 + \ldots \right) ,$$
$$M = \xi^{m-1} \left(M_o + M_2\xi^2 + \ldots \right) ,$$
$$Q = \xi^{m-2} \left(Q_o + Q_2\xi^2 + \ldots \right) . \tag{6.18}$$

Substituting these series into state equations and boundary conditions we can calculate the coefficients:

$$u_o = N_o = \varphi_o = M_o = Q_o = 0 ,$$

$$u_1 = \frac{1}{2} \frac{1}{\beta} \frac{\chi}{\Phi(0)} N_1$$

$$\varphi_2 = (m+2) w_2 , \quad M_2 = \beta\Phi^3(0)(m+1) \left[(m-2) a_{12} - (m+2) a_{11} \right] w_2$$

$$Q_2 = \beta\Phi^3(0) \left\{ (m+1) \left[(m-2) a_{12} - (m+2) a_{11} \right] \left(m+1 - \frac{a_{12}}{a_{11}} \right) + \right.$$

$$- \left(a_{22} - \frac{a_{12}}{a_{11}}\right)(m-2)(m+1) + m^2(m+1)\Big\} \tag{6.19}$$

One can notice that terms of the type, for example, $\frac{w}{\xi}$, $\frac{\varphi}{\xi}$, $\frac{w}{\xi^2}$, are not singular. For adjoint functions similar a approach is applied.

The numerical iterative method presented in [76] is used here. Figure 16 presents the values of critical pressure p_{01}, p_{02}, p_{11} in dependence on constraint Φ_{min}. The first index corresponds to the number of circumferential wave lengths along the perimeter, whereas the second index to the number of radial half-wave lengths along the radius. Dashed line denotes the range of inactive constraint. For circular plate it is shown that critical pressure p_{01} is always less then other; however, for annular plates the situation may be quite different.

Figure 17 presents the optimal shape of the plate for T=0 and T=0.001.

6.5. An example of optimal design of cylindrical shells

Optimal design of shells under stability constrains is usually much more complicated than that of plates. Hence,following the paper by Rysz [78] we make use here of the simplest creep buckling theory, namely that proposed by Gerard [67] and based on critical strain which is supposed to be independent of inelastic properties of the material. Though this theory is often criticized, nevertheless is gives a certain estimate of creep buckling state. Moreover, we make use of the local approach to shell buckling problems as described in Sec. 3 for the elastic range.

Consider a cylindrical shell under overall bending due to concentrated moments M_b and tension-compression due to axial forces N. Both senses of M_b and N are admitted, and hence the cross-section of the shell is assumed bisymmetric.We look for the minimal volume equivalent here to minimal cross-sectional area:

$$A = 2\int_{-H}^{+H} h\sqrt{1 + x'^2}\, dy + 2A_o \longrightarrow \min \tag{6.20}$$

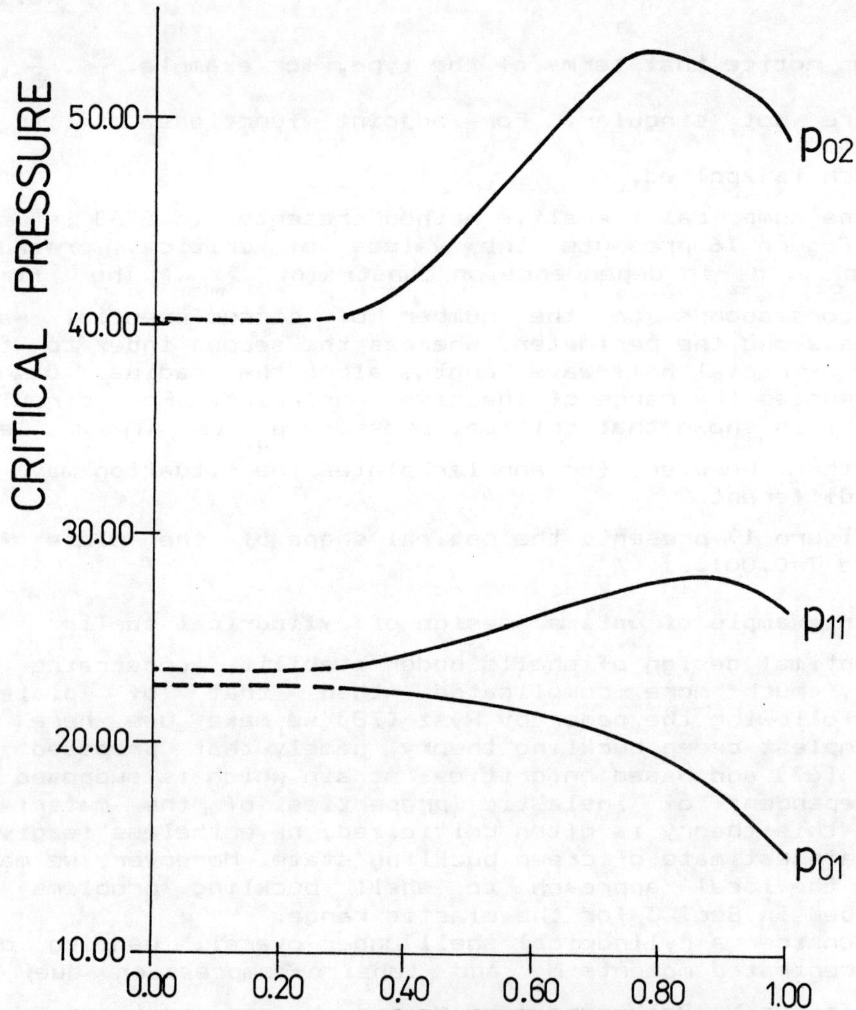

Fig. 16, Critical pressures for a circular plate in trms of Φ_{min}

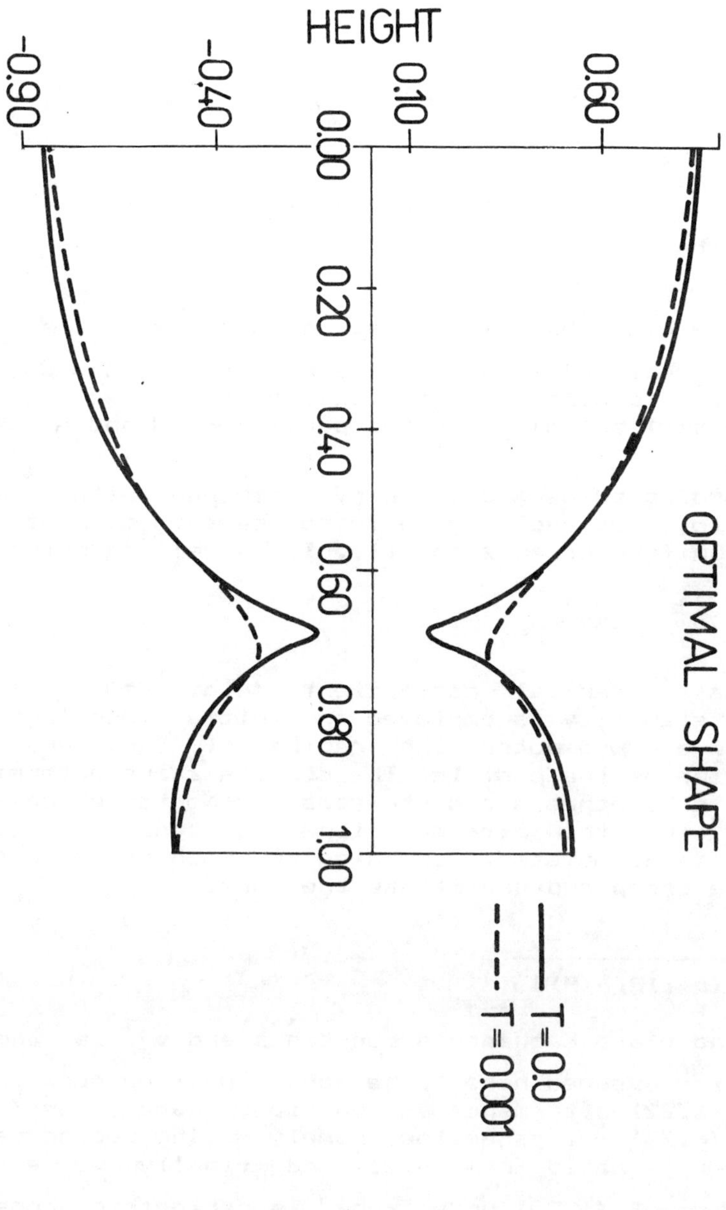

Fig. 17, Optimal shapes of a circular plate for various creep buckling times

under the constraints of the given bending moment and normal forces:

$$M_b = 2\int_{-H}^{+H} \sigma_z yh \sqrt{1 + x'^2}\, dy + A_o H\left(\sigma_{r1} - \sigma_{r2}\right) = \text{const}$$

$$N = 2\int_{-H}^{+H} \sigma_z h \sqrt{1 + x'^2}\, dy + A_o\left(\sigma_{r1} + \sigma_{r2}\right) = \text{const}$$

(6.21)

where A_o denotes the rib area, σ_{r1} and σ_{r2} are stresses carried by ribs, H is the overall height (depth). So, in the optimization problem under consideration we have two functions x(y) and h(y) and two parameters H and A_o as design variables.

According to Gerard's theory combined with the local approach to creep buckling we introduce the concept of the shape of uniform creep stability. It is defined by:

$$\varepsilon_{cr} = \beta \frac{h}{\rho} = \text{const.} \tag{6.22}$$

where β is a numerical coefficient (e.g. β=0.6 if linear theory of stability is employed and initial imperfections are neglected) and ρ denotes the radius of curvature of the central line of the profile. The strains ε are determined by Bernoulli's hypothesis and stresses by Norton's creep law. Moreover, for the extreme fibre in tension a strength condition is formulated. According to Kachanov's hypothesis of brittle creep rupture it has the form:

$$t_R = \frac{1}{(\nu+1)R[\sigma(H)]^{\nu}}, \tag{6.23}$$

where R and ν are Kachanov's constants and t_R is the creep rupture time assumed here to be equal to creep buckling time. Equation (6.22) determines a function, namely h=h(y), and equation (6.23) - a parameter, namely H. The second parameter A_o may be eliminated from (6.21) and finally we arrive at minimization of (6.20) under one isoperimetric constraint. Optimal shape x=x(y) is then determined by Euler-Lagrange equation which is integrated numerically. An example is shown in Figure 18.

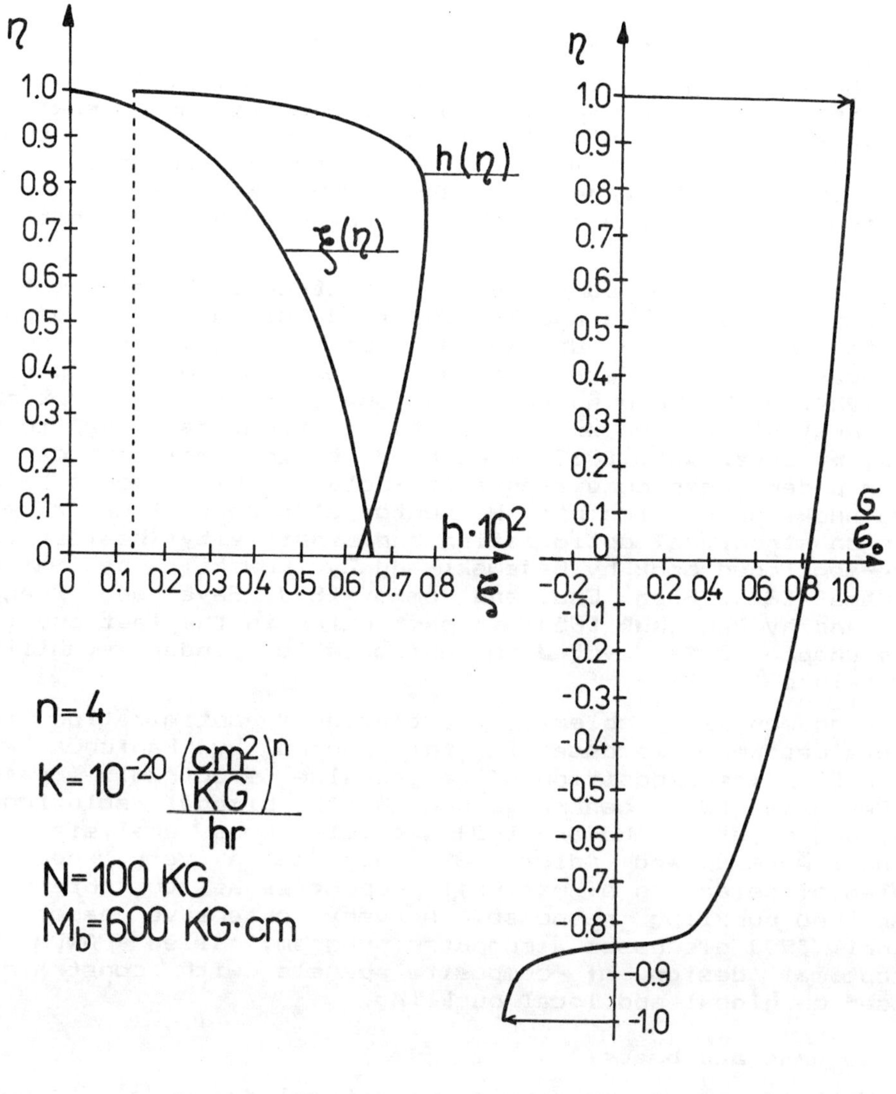

Fig. 18, Optimal shape and the relevant stress distribution for a cylindrical shell under bending with normal force for creep buckling and creep rupture

7. RECENT RESULTS ON OPTIMAL STRUCTURAL DESIGN UNDER STABILITY CONSTRAINTS

7.1. Surveys and papers of general character

The monograph by Gajewski and Życzkowski [9] gives a relatively comprehensive list of references on structural optimization under stability constraints. However, these references were closed with the years 1985-86 and more recent papers could not have been included. Hence we quote here some new results, regarding this short chapter as a supplement to [9].

Recently, three survey papers on structural optimization were published: Haftka and Grandhi [79] discuss 124 papers from the period 1973-1985, whereas Levy and Lev [80] quote 194 papers, 39 books and 7 computer programs from the period 1980-1984. Both above surveys pay some attention to optimal structural design under stability constraints. The third survey by Życzkowski [81] is devoted to optimal structural design under creep conditions; it includes some papers with creep buckling constraints. We menton also here four recent books on structural optimization and sensitivity (besides the above-mentioned book by Gajewski and Życzkowski): by Haftka and Kamat [82], Haug, Choi and Komkov [83], Save and Prager [84], and by Banichuk [85]; in particular in the last one the whole chapter 8 is devoted to optimization under stability constraints.

Fundamental problems of structural optimization for eigenvalues were discussed in the papers by Banichuk and Barsuk [86] (decomposition of eigenvalue spectrum), Bratus and Seyranian [87] (general approach to bimodal solutions) Shin, and Plaut and Haftka [88] (simultaneous analysis and design): Antman and Adler [89] discuss a very specific problem of selection of material properties as to obtain a prescribed buckling response. A very extensive paper by Bushnell [90] discusses a computer program, called PANDA 2, for optimal design of composite panels with constraints imposed on global and local buckling.

7.2. Columns and beams

Most recent papers devoted to optimal design of columns discuss multimodal problems. Seyranian [91] gives an analytical solution to bimodal optimization of clamped-clamped columns; Madsen [92] uses also analytical approach to evaluate higher buckling modes for unimodal optimal columns; Olhoff [93] discusses variational approach to bimodal design; Efremov and Matveev [94] consider conservative and non-conservative loading behaviour.

An additional gain may be achieved if we replace solid cross-sections of columns by thin-walled ones. Optimal design of thin-walled columns with closed profile was discussed by Pfefferkorn [95] (annular sections), whereas open profiles were considered by Larichev [96], Mikulski and Szymczak [97] (I-section constant along the axis). Hasegawa, Abo, Mauroof and Nishino [98] considered optimal thin-walled columns allowing for initial imperfections.

Optimal design of beams with respect to lateral buckling was discussed by Kartvelishvili [99] (closed thin-walled sections, sensitivity analysis), Wang, Thevendran, Teo and Kitipornchai [100] (rectangular section, parametric optimal design).

7.3. Plates and shells

Parametric optimal design of sandwich plates was discussed by Vinson in [101] (honeycomb core, uniaxial compression) and in [102] (compression with shear). Nakagiri and Takabatake [103] evaluated optimal ply angles in composite plates. A general sensitivity analysis for buckling of plates is due to Z.Mróz [104]; however, no effective application to optimal design was given.

Parametric optimal design of stiffened cylindrical shells under biaxial compression was discussed by McGrattan [105]: he assumed thickness, and dimensions of ribs as design variables. Błachut [106] determined optimal barrel-shaped shells with respect to their elastic stability. Rysz [78] considered cylindrical shells under overall bending with compression under creep conditions; determined optimal middle surface and optimal thickness under creep buckling and creep rupture constraints.

Finally, it should be noted that some other recent papers on optimal structural design under stability constraints, quoted by the co-authors of the present lecture notes, have not been included in this chapter.

REFERENCES

1. Prager, W.: Optimality criteria in structural design, Proc. Nat. Acad. Sci. USA 61 (1968), 3, 794 - 796.
2. Prager, W.: Conditions for structural optimality, Computers and Structures 2 (1972), 5, 833 - 840.
3. Prager, W. and J.E. Taylor: Problems of optimal structural design, Trans. ASME, J. Appl. Mech. 35 (1968), 1, 102 - 106.
4. Berke, L. and V.B. Venkayya: Reviev of optimality criteria approaches in structural optimization, Proc. Struct. Optimiz. Symp. ASME, AMD 7 (1974), 23 -34.
5. Save, M.A.: A general criterion for optimal structural design, J. Optimiz. Theory and Appl. 15 (1975), 1, 119 - 129.
6. Fleury, C. and M. Geradin: Optimality criteria and mathematical programming in structural weight optimization, Comput. and Struct. 8 (1978), 7 - 17.
7. Mróz, Z. and A. Mironov: Optimal design for global mechanical constraints, Arch. Mech. Stos. 32 (1980), 4, 505 - 516.
8. Galileo Galilei Linceo: Discorsi e dimostrazioni matematiche, Leiden 1638.
9. Gajewski, A. and M. Życzkowski: Optimal structural design under stability constraints, Kluwer - Nijhoff, Dordrecht 1988.
10. Krzyś, W. and M. Życzkowski: Klasyfikacja problemów kształtowania wytrzymałościowego, Czasopismo Techniczne 68 (1963), 2, 1 - 4.
11. Życzkowski, M.: Optimal structural design in rheology, J. Appl. Mech. 38 (1971), 1, 39 - 46.
12. Życzkowski, M. and A. Gajewski: Optimal structural design under stability constraints, Proc. IUTAM Symp. Collapse - the Buckling of Structures, London 1982, Cambridge Univ. Press 1983, 299 - 332.
13. Razani, R.: The behavior of the fully stressed design of structures and its relationship to minimum weight design, AIAA Journal 3 (1965), 12, 2262 - 2268.
14. Kicher, T.P.: Optimum design - minimum weight versus fully stressed, Proc. ASCE, J. Struct. Div. 92 (1966), 6, 265 - 279.
15. Reinschmidt, K., C.A. Cornell and J.F. Brotchie: Iterative design and structural optimization, Proc. ASCE, J. Struct. Div. 92 (1966), ST6, 281 - 318.
16. Malkov, V.P. and R.G. Strongin: Minimum weight design based on strength constraints (in Russian), Mietody Reshenya Zadach Uprugosti i Plastichnosti 4, Gorky 1971, 138 - 149.

17. Gallagher, R.H.: Fully stressed design, Optimum structural design: theory and applications, Wiley, New York 1973, 19 - 32.
18. Nemirovsky, Yu.V. and B.S. Reznikov: Beams and plates of uniform strength in creep conditions (in Russian), Mashinovedenye (1969), 2, 58 - 64.
19. Życzkowski, M. and W. Świsterski: Optimal structural design of flexible beams with respect to creep rupture time, Proc. IUTAM Symp. Structural Control, Waterloo 1979, North - Holland 1980, 795 - 810.
20. Drucker, D.C. and R.T. Shield: Design for minimum weight, Proc. 9th Int. Congr. Appl. Mech., Brussels 1956, vol. 5 (1957), 212 - 222.
21. Drucker, D.C. and R.T. Shield: Bounds on minimum weight design, Quart. Appl. Math. 15 (1957), 269 -281.
22. Zavelani - Rossi, A.: Minimum - weight design for two - dimensional bodies, Meccanica 4 (1969), 4, 445 - 452.
23. Kordas, Z. and M. Życzkowski: Investigations of the shape of thick-walled non-circular cylinders showing full plasticization at the collapse, Bull. Acad. Pol., Ser. Sci. Techn. 18 (1970), 10, 839 - 847 (English extensive summary); Rozpr. Inż. 18 (1970), 3, 371 - 390 (Polish full text).
24. Kordas, Z.: Problematyka określania kształtów ciał wykazujących całkowite uplastycznienie w stadium zniszczenia, Zeszyty Naukowe Politechniki Krakowskiej, Podstawowe Nauki Techniczne 15 (1977).
25. Bochenek, B., Z. Kordas and M. Życzkowski: Optimal plastic design of a cross section under torsion with small bending, J. Struct. Mech. 11 (1983), 3, 383 - 400.
26. Skrzypek, J. and M. Życzkowski: Termination of processes of finite plastic deformations of incomplete toroidal shells, Solid Mech. Arch. 8 (1983), 1, 39 - 98.
27. Szuwalski, K. and M. Życzkowski: On the phenomenon of decohesion in perfect plasticity, Int. J. Solid Struct. 9 (1973), 1, 85 - 98.
28. Wasiutyński, Z.: O kształtowaniu wytrzymałościowym, Akademia Nauk Technicznych, Warszawa 1939.
29. Prager, W.: Optimal structural design for given stiffness in stationary creep, Z. angew Math. Physik 19 (1968), 252 - 256.
30. Niordson, F.: On the optimal design of a vibrating beam, Quart. Appl. Math. 23 (1965), 1, 47 - 53.
31. Olhoff, N.: A survey of the optimal design of vibrating structural elements, Shock and Vibration Digest 8 (1976), 8, 3 - 10; 9, 3 - 10.

32. Troitsky, V.A.: Optimization of elastic bars in the presence of free vibrations (in Russian), Izv. AN SSSR, Mekh. Tverd. Tela (1976), 3, 145 -152.
33. Shanley, F.R.: Principles of structural design for minimum weight, J. Aero. Sci. 16 (1949), 3.
34. Shanley, F.R.: Weight - strength analysis of aircraft structures, McGraw-Hill, New York - Toronto - London 1952.
35. Spunt, L.: Optimum structural design, Prentice - Hall, Englewood Cliffs, N.J., 1971.
36. Neut, A., van der : The interaction of local buckling and column failure of thin-walled compression members, Proc. 12th Int. Congr. Appl. Mech. Stanford 1968, Springer 1969, 389 - 399.
37. Thompson, J.M.T.: Optimization as a generator of structural instability, Int. J. Mech. Sci. 14 (1972), 9, 627 - 629.
38. Thompson, J.M.T. and W.J. Supple: Erosion of optimum designs by compound branching phenomena, J. Mech. Phys. Solids 21 (1973), 3, 135 - 144.
39. Volmir, A.S.: Stability of elastic systems (in Russian), Fizmatgiz, Moskva 1963; Stability of deformable systems (in Russian), Nauka, Moskva 1967.
40. Życzkowski, M. and J. Krużelecki: Optimal design of shells with respect to their stability, Proc. IUTAM Symp. Optimization in Structural Design, Warsaw 1973, Springer 1975, 229 - 247.
41. Prager, W.: Introduction to structural optimization, CISM Courses 212, Springer, Wien - New York 1974.
42. Markiewicz, M.: Kształtowanie prostych ustrojów kratowych przy warunkach stateczności sprężysto - -plastycznej metodą wyznaczania konturu całkowitej niejednoznaczności, Rozpr. Inż. 28 (1980), 4, 569 - 584.
43. Markiewicz, M. and M. Życzkowski: Contour of complete non-uniqueness as a method of structural optimization with stability constraints, J. Optimiz. Theory Appl. 35 (1981), 1, 23 - 30.
44. Bürgermeister, G. and H. Steup: Stabilitätstheorie, Teil 1, Akademie - Verlag, Berlin 1957; Teil 2 (with H. Kretzschmar), Berlin 1963.
45. Wojdanowska, R. and M. Życzkowski: Optimal trusses transmitting a force to a given contour in creep conditions, Int. J. Mech. Sci. 26 (1984), 1, 21 - 28.
46. Shtaerman, I.Ya.: Stability of shells (in Russian), Trudy Kievsk. Aviats. Instituta 1 (1936).
47. Rabotnov, Yu.N.: Local stability of shells (in Russian), Dokl. AN SSSR, Novaya Seria 52 (1946), 2, 111 - 112.

48. Shirshov, V.P.: Local stability of shells (in Russian), Trudy II Vsesoy. Konf. po Teorii Plastin i Obolochek, Lvov 1961, Kiev 1962, 314 - 317.
49. Axelrad, E.L.: On local buckling of thin shells, Int. J. Non-Linear Mech. 20 (1985), 4, 249 - 259.
50. Krużelecki, J.: Optimal design of a cylindrical shell under overall bending with axial force, Bull. Acad. Pol., Ser. Sci. Techn. 35 (1987) (English extensive summary); Rozpr. Inż. 33 (1985), 1/2, 135 - 149, (Polish full text).
51. Krużelecki, J.: Optimization of shells under combined loadings via the concept of uniform stability, Optimization of distributed parameter structures, Ed. by E.J. Haug and J. Cea, Nijhoff, Vol. II, 929 — 950 (1981).
52. Krużelecki, J. and M. Życzkowski: Optimal design of an elastic cylindrical shell under overal bending with torsion, Solid Mechanics Archives 9 (1984), 3, 269 - 306.
53. Krużelecki, J. and M. Życzkowski: Optimal structural design of shells - a survey, Solid Mechanics Archives 10 (1985), 2, 101 -170.
54. Pflüger, A.: Stabilitätsprobleme der Elastostatik, Springer, Berlin - Göttingen - Heidelberg 1950 (1964, 1975).
55. Laasonen, P.: Nurjahdustuen edullisimmasta poikipinnanvalinnasta, Tekn. Aikakauslehti 38 (1948), 2, 49.
56. Krzyś, W.: Optimale Formen Gedrückter dünnwandiger Stützen im elastisch-plastischen Bereich, Wiss. Z. TU Dresden 17 (1968), 2, 407 - 410.
57. Krzyś, W.: Optimum design of thin - walled closed cross - section columns, Bull. Acad. Pol., Ser. Sci. Techn. 21 (1973), 8, 409 - 420.
58. Bleich, F.: Buckling strength of metal structures, McGraw-Hill, New York 1952.
59. Siegfried, W.: Failure from creep as influenced by the state of stress, J. Appl. Mech. 10 (1943), 4, 202 - 212.
60. Freudenthal, A.M.: Some time effects in structural analysis, Rep. 6th Int. Congr. Appl. Mech., Paris 1946.
61. Rzhanitsyn, A.R.: Deformation processes of structures consisting of viscoelastic elements, (in Russian), Dokl. AN SSSR 52 (1946), 25, 1.
62. Ross, A.D.: The effect of creep on instability and indeterminacy investigated by plastic models, Struct. Eng. 24 (1946), 413.
63. Wojdanowska, R.: Optimal design of weakly curved compressed bars with Maxwell type creep effects, Arch. Mech. Stos. 30 (1978), 6, 845 - 851.

64. Wojdanowska, R. and M. Życzkowski: On optimal imperfect columns subject to linear creep buckling, J.Appl. Mech. 47 (1980), 2, 438 - 439.
65. Kempner, J.: Creep bending and buckling of non-linearly viscoelastic columns, PIBAL Rep. No. 200, Brooklyn 1952; NACA TN 3137, Jan. 1954.
66. Hoff, N.J.: Buckling and stability, J. Roy. Aero. Sci. 58 (1954), 3 - 52.
67. Gerard, G.: A creep buckling hypothesis, J. Aero. Sci. 23 (1956), 9, 879 - 882.
68. Rabotnov, Yu.N. (G.N.) and S.A. Shesterikov: Creep stability of columns and plates, Prikl. Mat. Mekh. 21 (1957), 3, 406 -412 (Russian version); J. Mech. Phys. Solids 6 (1957), 1, 27 - 34 (English version).
69. Życzkowski, M. and R. Wojdanowska-Zając: Optimal structural design with respect to creep buckling, Proc. IUTAM Symp. Creep in Structures 2, Göteborg 1970, Springer 1972, 371 - 387.
70. Błachut, J. and M. Życzkowski: Bimodal optimal design of clamped-clamped columns under creep conditions, Int. J. Solids Struct. 20 (1984), 6, 571 - 577.
71. Wróblewski, A.: Parametryczna optymalizacja prętów mimośrodowo ściskanych w nawiązaniu do teorii wyboczenia pełzającego Kempnera - Hoffa (in print).
72. Świsterski, W., A. Wróblewski and M. Życzkowski: Geometrically non-linear eccentrically compressed columns of uniform creep strength vs. optimal columns, Int. J. Non-Linear Mech. 18 (1983), 4, 287 - 296.
73. Życzkowski, M.: Recent results on optimal design in creep conditions, Euromech Coll. 164 on Optimization Methods in Structural Design, Siegen 1982, Bibliograph. Inst. Zürich 1983, 444 - 449.
74. Wróblewski, A. and M. Życzkowski: On multimodal optimization of circular arches against plane and spatial creep buckling, Structural Optimization 1 (1989), 2.
75. Kordas, Z.: Stability of the elastically clamped compressed bar in the general case of behaviour of the loading, Bull. Acad. Pol., Ser. Sci. Techn. 11 (1963), 419 - 428 (English extensive summary); Rozpr. Inż. 11 (1963), 3, 435 - 448 (Polish full text).
76. Bochenek, B. and A. Gajewski: Multimodal optimal design of a circular funicular arch with respect to in-plane and out-of-plane buckling, J. Struct. Mech. 14 (1986), 3, 257 - 274.
77. Wróblewski, A.: Optimal design of a circular plate with respect to creep buckling (in print).

78. Rysz, M. and M. Życzkowski: Optimal design of a cylindrical shell under overall bending and axial force with respect to creep stability, Structural Optimization 1 (1989), 1.
79. Haftka, R.T. and R.V. Grandhi: Structural shape optimization - a survey, 26th Struct. Dyn. and Mat. Conf., Part I, New York 1985, 617 - 628.
80. Levy, R. and O.E. Lev: Recent developments in structural optimization, Proc. ASCE, J. Struct. Engng. 113 (1987), 9, 1939 - 1962.
81. Życzkowski, M.: Optimal structural design under creep conditions, Mech. Teor. Stos. 24 (1986), 3, 243 - 258 (Polish version), Appl. Mech. Rev., 41 (1988), 12, 453 - 461 (English extended version).
82. Haftka, R.T. and M.P. Kamat: Elements of structural optimization, Nijhoff, Dordrecht 1985.
83. Haug, E.J., K.K. Choi and V. Komkov: Design sensitivity analysis of structural systems, Academic Press, Orlando - San Diego - New York 1986.
84. Save, M. and W. Prager: Structural optimization, Vol.1, Optimality criteria, Plenum Press, New York 1985.
85. Banichuk, N.V.: Introduction to structural optimization (in Russian), Nauka, Moskva 1986.
86. Banichuk, N.V. and A.A. Barsuk: Application of spectral decomposition of eigenvalues in structural optimization under stability constraints (in Russian), Problemy Ustoych. i Pred. Nes. Sposobn. Konstruktsiy, Leningrad 1983, 17 - 24.
87. Bratus, A.S. and A.P. Seyranian: Bimodal solutions in optimization of eigenvalues (in Russian), Prikl. Mat. Mekh. 47 (1983), 4, 546 - 554.
88. Shin, Y.S., R.H. Plaut and R.T. Haftka: Simultaneous analysis and design for eigenvalue maximization, AIAA/ ASME/ASCE/AHS 28th Struct., Struct. Dyn. and Mat. Conf., Monterey, 1987, New York 1987, 334 - 342.
89. Antman, S.S. and C.L. Adler: Design of material properties that yield a prescribed global buckling response, Trans. ASME, J. Appl. Mech. 54 (1987), 2, 263 - 266.
90. Bushnell,D.: PANDA 2 - program for minimum weight design of stiffened, composite, locally buckled panels, Comp. and Struct. 25 (1987), 4, 469 - 605.
91. Seyranian, A.P.: On a certain problem of Lagrange (in Russian), Izv. AN SSSR, Mekh. Tverd. Tela (1984), 2, 101 - 111.
92. Madsen, N.: Analytical determination of higher buckling modes for unimodal optimal columns, J. Struct. Mech. 11 (1984), 4, 545 - 560.

93. Olhoff, N.: Structural optimization by variational methods, Computer Aided Optimal Design, Proc. NATO Adv. Study Institute, Tróia 1986, Springer 1987, 87 -164.
94. Efremov, A.Yu. and K.A. Matveev: Shape optimization of bars in stability problems (in Russian), Dinam. i Prochnost Aviats. Konstr., Novosibirsk 1986, 89 - 92.
95. Pfefferkorn, W.: Der dünnwandige Knickstab mit minimierter Masse, IfL - Mitteilungen 25 (1986), 3, 65 - 67.
96. Larichev, A.D.: Optimization of stability of thin-walled bars of open cross-section, (in Russian), Prikl. Probl. Prochn. Plastichn. (Gorky), 1986, No. 34, 97 - 103.
97. Mikulski, T. and C. Szymczak: Optymalne kształtowanie przekroju poprzecznego ściskanych prętów cienkościennych o przekroju otwartym, Zesz. Nauk. Polit. Gdańskiej 42 (1987), 73 - 92.
98. Hasegawa, A., H. Abo, M. Mauroof and F. Nishino: A simplifiield analysis and optimality on the steel column behavior with local buckling, Proc. Jap. Soc. Eng. 1986, No. 374, 195 - 204.
99. Kartvelishvili, V.M.: On optimal solutions to a Prandtl's problem (in Russian), Issled. po stroit. mekh. i nadezhn. konstr., Moskva 1986, 81 - 89.
100. Wang,C.M., V. Thevendran, K.L. Teo and S. Kitipornchai: Optimal design of tapered beams for maximum buckling strength, Eng. Struct. 8 (1986), 4, 276 - 284.
101. Vison, J.R.: Optimum design of composite honeycomb sandwich panels subjected to uniaxial compression, AIAA Journal 24 (1986), 10, 1690 - 1696.
102. Vinson, J.R.: Minimum weight web-core sandwich panels subjected to combined uniaxial compression and in-plane shear loads, AIAA/ASME/ASCE/AHS 28th Struct., Struct. Dyn. and Mat. Conf., Monterey 1987, New York 1987, 282 - 288.
103. Nakagiri, S. and H. Takabatake: Optimum design of FRP laminated plates under axial compression by use of the Hessian matrix, Proc. Int. Conf. Computer Mechanics '86, Tokyo, Vol. 2, Tokyo 1986, X/71 - X/76.
104. Mróz, Z.: Sensitivity analysis and optimal design with account for varying shape and support conditions, Computer Aided Optimal Design, Proc. NATO Adv. Study Institute, Tróia 1986, Springer 1987, 407 - 438.
105. McGrattan, R.J.: Weight optimization of stiffened cylindrical panels, Trans. ASME, J. Pressure Vessel Techn. 109 (1987), 1, 1 - 9.
106. Błachut, J.: Optimal barrel - shaped shells under buckling constraints, AIAA Journal 25 (1987), 1, 186 - 188.

PART II

A. Gajewski
Technical University of Cracow, Cracow, Poland

ABSTRACT

After a very short description of Pontryagin's maximum principle and sensitivity analysis as applied to eigenvalue problems, a unified approach to column optimization has been presented. Particular attention has been paid to multimodal solutions obtained for compressed columns in an elastic medium and elastically clamped columns for buckling in two planes. Next, a general statement of the optimization of arches has been formulated. The necessity of multimodal optimization was pointed out, especially if in-plane and out-of-plane buckling was taken into account. The sensitivity analysis has been applied to a new optimization problem of annular plates compressed by uniformly distributed non-conservative forces. Both the precritical membrane state and the small transverse vibration has been taken into account. Finally, the parametrical optimization of a visco-elastic column compressed by a follower force with respect to its dynamic stability, as well as the optimization of a plane bar system in conditions of internal resonance has been considered.

1. SELECTED METHODS OF STRUCTURAL OPTIMIZATION

1.1. General remarks

There is a fairly extensive literature on optimization methods: references to over 500 monographs and textbooks are given in the book by A. Gajewski and M. Zyczkowski [1]. The great majority of them are devoted to the calculus of variations. In these lectures however, we will consider practical aspects of two other methods of structural optimization, namely optimal control theory and sensitivity analysis. These modern methods may be treated as extensions of the classical calculus of variations and of late they have been frequently used. The results presented in our lectures have mostly been obtained on the basis of optimal control methods or sensitivity analysis.

1.2. Pontryagin's maximum principle

1.2.1. Equations of state and boundary conditions

Consider a "dynamic system" whose state is determined by n state variables $y_i = y_i(x)$, $x \in [x_o, x_1]$, governed by n quasilinear ordinary differential equations of the first order:

$$y_i' = f_i(x, y_i, \phi_j), \quad i=1..n; \quad j=1..r. \qquad (1.2.1)$$

Such equations can describe e.g. the stability of a deformable body. We may regard the state as a point in an n-dimensional state- (or phase-) space **Y**, $y_i \in \mathbf{Y}$. Functions f_i may depend on the state variables y_i, the independent variable x (in non-autonomous problems), and on some control variables (design variables) ϕ_j, assumed to be piece-wise continuous and belonging to an r-dimensional control space Φ, $\phi_j \in \Phi$. Moreover, **Y** and Φ may be restricted to certain subspaces by additional inequality constraints (cf. 1.2.4).

Further, we assume that the initial and final states $y_i(x_o)$ and $y_i(x_1)$ lie on certain prescribed manifolds

$$B_m[x_o, x_1, y_i(x_o), y_i(x_1)] = 0, \quad m=1..m_o<n. \qquad (1.2.2)$$

If x_o and x_1 are preassigned, as in most cases under consideration, then the problem will be referred to as a fixed time problem; if x_1 is free or subject only to a certain constraint, then the problem will be called a free-time problem. Free-time problems may also be encountered in structural optimization, e.g. when designing the tallest column under self-weight.

1.2.2. The objective functional

Here we consider the following objective functional (cost function)

$$J\,[y_i,\phi_j] = \int_{x_o}^{x_1} f_o(x,y_i,\phi_j)\,dx \tag{1.2.3}$$

which is to be minimized. If we introduce a new variable $y_o(x)$, defined by the differential equation and the initial condition

$$y_o' = f_o(x,\ y_i,\ \phi_j)\ , \qquad y_o(x_o) = 0\ , \tag{1.2.4}$$

then we may write the objective functional in the form

$$J = y_o(x_1). \tag{1.2.5}$$

1.2.3. The Hamiltonian and the maximum principle

We first reduce a general, non-autonomous problem to an autonomous one. This may easily be done by introducing a new state variable $y_{n+1} = x$ by the equation

$$y_{n+1}' = f_{n+1} = 1\ , \qquad y_{n+1}(x_o) = x_o\ . \tag{1.2.6}$$

Now, the space of state variables is extended to an (n+2) - dimentional space with the vector of state:

$$y = (y_o,\ y_1,\dots y_n,\ y_{n+1})^T. \tag{1.2.7}$$

We introduce a scalar function *H*, called the Hamiltonian,

$$\mathcal{H} = \psi_i f_i + \psi_{n+1} = H + \psi_{n+1}, \quad i=0,1,2..n, \tag{1.2.8}$$

where $\psi_i = \psi_i(x)$ is an adjoint vector of state, corresponding to the vector of Lagrangian multipliers; it is defined by the system of $n+2$ adjoint differential equations

$$\psi_i' = -\psi_\alpha f_{\alpha,i} \tag{1.2.9}$$

where $i,\alpha = 0,1,...n+1$, and the summation convention over α holds. From (1.2.9) it follows that $\psi_o' = 0$ (since y_o does notappear in any of the functions f_α), and hence ψ_o=const.; usually it is assumed that $\psi_o = -1$.

The maximum principle will now be stated as follows:

optimal control $\tilde{\phi}_j$, maximizing the functional(1.2.5) is reached if, for $x \in [x_o, x_1]$

(a) the Hamiltonian (1.2.8) is maximized

$$\sup_{\phi_j \in \Phi} \mathcal{H}(\psi_i, \tilde{y}_i, \phi_j) = \mathcal{H}(\psi_i, \tilde{y}_i, \tilde{\phi}_j) = M(\psi_i, \tilde{y}_i), \tag{1.2.10}$$

(b) $\quad M(\psi_i, \tilde{y}_i) = 0, \qquad (1.2.11)$

(c) the vector of state and the adjoint vector of state satisfy the canonical Hamiltonian equations:

$$\tilde{y}_i' = \frac{\partial \mathcal{H}}{\partial \psi_i}, \qquad \psi_i' = -\frac{\partial \mathcal{H}}{\partial y_i}, \tag{1.2.12}$$

(d) the vector of state satisfies the boundary conditions (1.2.2), whereas the adjoint vector of state satisfies the transversality conditions

$$\psi_i(x_o)\, \delta y_i(x_o) - \psi_i(x_o)\, \delta y_i(x_o) = 0 \tag{1.2.13}$$

where m_o variations may be eliminated by using (1.2.2), namely

$$\frac{\partial B_m}{\partial y_i(x_o)}\, \delta y_i(x_o) + \frac{\partial B_m}{\partial y_i(x_1)}\, \delta y_i(x_1) = 0, \tag{1.2.14}$$

and the coefficients of the remaining variations should vanish. Considering the simplest case, if a variation,- say $\delta y_i(x_o)$- is free, we obtain $\psi_i(x_o)=0$.

If H is an analytical function of ϕ_j then (1.2.10) requires

$$\frac{\partial H}{\partial \phi_j}\bigg|_{\phi_j=\tilde{\phi}_j} = 0 . \tag{1.2.15}$$

1.2.4. Inequality constraints

Control variables and state variables may be restricted by some inequality constraints. In the simplest case they are imposed on control variables only

$$\varphi_j(\phi_\alpha) \le 0, \qquad j = 1,2,\dots m. \tag{1.2.16}$$

Then, if the Hamiltonian -regarded as $H(\phi_\alpha)$- has a negative semi-definite Hessian matrix $\frac{\partial^2 H}{\partial\phi_\alpha\partial\phi_\beta}$ where $x\in[x_o,x_1]$ then the optimal configuration $\tilde{\phi}_\alpha$ can be determined either by (1.2.15) or by $\varphi_j = 0$. Conversely, if the Hessian matrix is positive semidefinite then the optimal configuration is always determined by any of $\varphi_j=0$ yielding a larger value of the Hamiltonian.

If more general inequalities are present

$$\varphi_j(y_i,\phi_\alpha) \le 0, \qquad j=1,2,\dots m, \tag{1.2.17}$$

then, in general, part of the trajectory lies inside (1.2.17) and is determined by (1.2.10) - (1.2.12), and part lies at the boundary (1.2.17). Conditions at boundary points are discussed in detail by L.S.Pontryagin et al.[2].

1.2.5. Additional parametric optimization

In many problems the functionals depend not only on the control variables $\phi_j(x)$, but also on the parameters a_i, where i = 1,2,..s (for example, the piece-wise constant functions $\phi_j(x) = a_j$). Besides (1.2.10) - (1.2.12) one determines optimal values of the parameters in most cases from the necessary conditions

$$\int_{x_0}^{x_1} \frac{\partial H}{\partial a_i}\, dx = 0, \qquad i=1,2,..s. \tag{1.2.18}$$

1.3. Sensitivity analysis

1.3.1. General remarks

According to the opinion of P.Pedersen [3], "sensitivity analysis is the "cornerstone" of any approach to optimal design". Indeed, it becomes a fruitful area of engineering research, applied mainly to optimization problems, but also to approximate analysis, analytical model improvement and assessment of design trends. An up-to-date survey of methods applicable to the calculation of structural sensitivity derivatives for finite element modeled structures has been presented by H.M.Adelman and R.T.Haftka [4]. The design sensitivity analysis of distributed-parameter structures (continuous models) is broadly discussed in the books of E.J.Haug and J.S.Arora [5] and E.J.Haug, K.K.Choi and V.Komkov [6].

In general, sensitivity analysis aims at determining the effects of perturbation control variables ϕ_j on the state of the system y(x) (including eigenvalues), on the objective functional $J_o(\phi_j, y_i)$ and on constraints. This aim is achieved via sensitivity operators, i.e. the mappings ascribing variations of the functions of state, objective functional and constraints to the variations of control variables.

1.3.2. Eigenvalue problems

For our purposes - optimization under stability constraints - the most important thing is sensitivity analysis as applied to eigenvalue problems. Such an analysis was developed by W.H.Wittrick [7], R.H.Plaut and K.Huseyin [8], M.Farshad [9], E.J.Haug and B.Rousselet [10], A.P.Seyranian and A.V.Sharanyuk [11], J.-L.Claudon and M.Sunakawa [12], P.Pedersen and A.P.Seyranian [13], G.Szefer [14].

Here we confine our analysis to ordinary differential equations. On the other hand, we introduce geometrical non-linearity of the system under consideration in order to discuss various forms of the loss of stability - bifurcation, snap-through and flutter. Equations of precritical state (independent of time) and of critical state will be assumed in the form:

$$Y_i'=G_i(x,Y_\alpha,\phi_\beta,P)\ , \qquad i=1..I, \qquad (1.3.1)$$

$$Z_j'=A_{j\alpha}(x,Y_i,\phi_\beta,P,\lambda)Z_\alpha \qquad j=1..J, \qquad (1.3.2)$$

$$B_k^{(0)}[Y_i(0),P]=0; \qquad B_l^{(1)}[Y_i(1),P]=0; \qquad (1.3.3)$$

$$k=1..K, \quad l=1..L, \quad K+L=I,$$

$$\mu_{m\alpha}[P,\lambda,Y_i(0)]Z_\alpha(0)=0; \qquad \nu_{n\alpha}[P,\lambda,Y_i(1)]Z_\alpha(1)=0 \qquad (1.3.4)$$

$$m=1..M, \quad n=1..N, \quad M+N=J.$$

It should be noted that the variables of the precritical state Y_i describe static deformations of the system and are real, whereas the variables of critical state Z_j may describe vibrations and, in general, are complex. Also, nonlinear boundary conditions (1.3.3) depend on the load parameter P whereas linear boundary conditions (1.3.4) may depend on both the load parameter and the frequency of vibrations (complex).

In order to determine the sensitivity of eigenvalues (P or λ) to variations of the control functions ϕ_β we introduce the variables of adjoint state χ_i and ψ_j as satisfying the following linear differential equations

$$\chi_i'=-\chi_\alpha\frac{\partial G_\alpha}{\partial Y_i}-\psi_\alpha\frac{\partial A_{\alpha\beta}}{\partial Y_i}Z_\beta \qquad (1.3.5)$$

$$\psi_j'=-A_{\alpha j}\psi_\alpha \qquad (1.3.6)$$

and the relevant transversality conditions

$$\chi_i(0)=-\lambda_\alpha^{(0)}\frac{\partial B_\alpha^{(0)}}{\partial Y_i(0)}-\Lambda_\alpha^{(0)}\frac{\partial \mu_{\alpha\beta}}{\partial Y_i(0)}Z_\beta(0) \qquad (1.3.7)$$

$$\chi_i(1)= \lambda_\alpha^{(1)} \frac{\partial B_\alpha^{(1)}}{\partial Y_i(1)} + \Lambda_\alpha^{(1)} \frac{\partial \nu_{\alpha\beta}}{\partial Y_i(1)} Z_\beta(1) \qquad (1.3.8)$$

$$\psi_j(0)=-\Lambda_\alpha^{(0)} \mu_{\alpha j} \qquad (1.3.9)$$

$$\psi_j(1)= \Lambda_\alpha^{(1)} \nu_{\alpha j} . \qquad (1.3.10)$$

The constant parameters λ_i and Λ_j can be expressed by the boundary values of state functions and then eliminated.

By calculating variations of (1.3.1) and (1.3.2), multiplying them by χ_i and ψ_j respectively, and integrating, we can eliminate δY_i and δZ_j and derive the following equation for sensitivity operators:

$$\int_0^1 \left(\chi_\alpha \frac{\partial G_\alpha}{\partial \phi_\beta} + \psi_\alpha \frac{\partial A_{\alpha\gamma}}{\partial \phi_\beta} Z_\gamma\right) \delta\phi_\beta \, dx +$$

$$+ \delta P \left[\lambda_\alpha^{(0)} \frac{\partial B_\alpha^{(0)}}{\partial P} + \lambda_\beta^{(1)} \frac{\partial B_\beta^{(1)}}{\partial P} + \Lambda_\alpha^{(0)} \frac{\partial \mu_{\alpha\beta}}{\partial P} Z_\beta(0) + \right.$$

$$\left. + \Lambda_\alpha^{(1)} \frac{\partial \nu_{\alpha\beta}}{\partial P} Z_\beta(1) + \int_0^1 \left(\chi_\alpha \frac{\partial G_\alpha}{\partial P} + \psi_\alpha \frac{\partial A_{\alpha\beta}}{\partial P} Z_\beta\right) dx \right] +$$

$$+ \delta\lambda \left[\Lambda_\alpha^{(0)} \frac{\partial \mu_{\alpha\beta}}{\partial \lambda} Z_\beta(0) + \Lambda_\alpha^{(1)} \frac{\partial \nu_{\alpha\beta}}{\partial \lambda} Z_\beta(1) + \right.$$

$$\left. + \int_0^1 \psi_\alpha \frac{\partial A_{\alpha\beta}}{\partial \lambda} Z_\beta \, dx \right] = 0 \qquad (1.3.11)$$

In the particular case of the static criterion of stability (λ=0) and one control function ϕ we obtain

$$\delta P = \int_0^1 g(x) \, \delta\phi \, dx \qquad (1.3.12)$$

where

$$g(x) = -\left(\chi_\alpha \frac{\partial G_\alpha}{\partial \phi} + \psi_\alpha \frac{\partial A_{\alpha\gamma}}{\partial \phi} Z_\gamma\right) \frac{1}{C}. \qquad (1.3.13)$$

The constant C denotes the constant factor associated with

δP in the equation (1.3.11).

Moreover, if the volume V is constant and given by

$$V = \int_0^1 m_0(x,\phi)\, dx = \text{const}. \qquad (1.3.14)$$

then, using the theory of minimax (V.F.Demyanov and V.N. Malozemov [15]), we obtain the following optimality condition:

$$g(x) + \Lambda_0 \frac{\partial m_0}{\partial \phi} = 0, \qquad (1.3.15)$$

where Λ_0 denotes a constant Lagrangian multiplier.

1.3.3. Simple examples

Consider a cantilever elastic column compressed by a concentrated force P at its free end. The behaviour of the force during buckling is shown in Fig.1; it is a polar force in case "a" (conservative) and an anti-tangential one in case "b" (non-conservative).

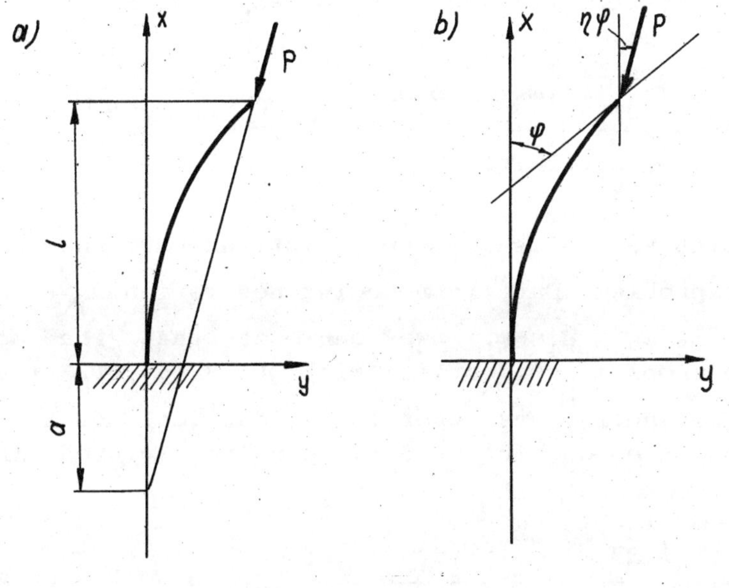

Fig.1. Conservative (a) and non-conservative (b) loadings

Assuming that the column axis is inextensible, the equations of the precritical state may be omitted, while non-dimensional equations of the critical state (as far as the static stability criterion is concerned) are of the form:

$$Z_1'=Z_2; \quad Z_2'=-\frac{Z_3}{F(\phi)}; \quad Z_3'=Z_4+PZ_2; \quad Z_4'=0; \tag{1.3.16}$$

where

$$P=\bar{P}l^2/E_oI_o, \quad F(\phi)=EI/E_oI_o, \quad \phi=A/A_o, \quad \alpha=a/l,$$

$$(Z_1=y,\ Z_2=\varphi,\ Z_3=M,\ Z_4=Q).$$

In case "a" (polar force) the boundary conditions and coefficients μ_{ij} and ν_{ij} are as follows:

$$Z_1(0)=Z_2(0)=Z_3(1)=0; \quad Z_4(1)+\frac{P}{1+\alpha}Z_1(1)=0 \tag{1.3.17}$$

$$\mu_{11}=1,\ \mu_{12}=0,\ \mu_{13}=0,\ \mu_{14}=0,\ \mu_{21}=0,\ \mu_{22}=1,\ \mu_{23}=0,\ \mu_{24}=0,$$

$$\nu_{11}=0,\ \nu_{12}=0,\ \nu_{13}=1,\ \nu_{14}=0,\ \nu_{21}=\frac{P}{1+\alpha},\ \nu_{22}=0,\ \nu_{23}=0,\ \nu_{24}=1.$$

Thus, equation (1.3.11) leads to the expression

$$\delta P = \frac{\int_o^1 \frac{d}{d\phi}\left(\frac{1}{F}\right)\psi_2 Z_3\ \delta\phi\ dx}{\frac{1}{1+\alpha}\psi_4(1)Z_1(1)+\int_o^1 \psi_3 Z_2 dx}, \tag{1.3.18}$$

where adjoint variables ψ_i are solutions of the adjoint eigenvalue problem. But, introducing new variables $\bar{Z}_1=\psi_4$, $\bar{Z}_2=-\psi_3$, $\bar{Z}_3=\psi_2$, $\bar{Z}_4=-\psi_1$ we can see that the adjoint eigenvalue problem is identical with (1.3.16),(1.3.17). The variables Z_i and $\bar{Z}_i$ are therefore identical (selfadjoint problem) and the equation (1.3.19) can be written in the form:

$$\delta P = \frac{\int_o^1 \frac{d}{d\phi}\left(\frac{1}{F}\right) M^2\ \delta\phi\ dx}{\frac{1}{1+\alpha}[y(1)]^2-\int_o^1 \varphi^2 dx}, \tag{1.3.19}$$

In case "b" (anti-tangential force) the boundary conditions

$$Z_1(0)=Z_2(0)=Z_3(1)=0, \quad Z_4(1)+\eta P Z_2(1)=0 \qquad (1.3.20)$$

imply $\mu_{11}=1$, $\mu_{22}=1$, $\nu_{13}=1$, $\nu_{22}=\eta P$, $\nu_{24}=1$ and other coefficients equal to zero.

Although the adjoint equations of state are here identical with (1.3.16), the adjoint boundary conditions are quite different from (1.3.20) ($Z_3(1)-\eta P Z_1(1)=0$). The variation of critical force is here expressed by the quotient:

$$\delta P = \frac{\int_0^1 \frac{d}{d\phi}(\frac{1}{F})\ M\bar{M}\ \delta\phi\ dx}{\eta\bar{y}(1)\varphi(1) - \int_0^1 \varphi\bar{\varphi}\ dx}, \qquad (1.3.21)$$

A more complicated - non-conservative - problem will be presented in Sec.4, where the precritical state will be also taken into account.

2. THE OPTIMAL STRUCTURAL DESIGN OF COLUMNS

2.1. General remarks

The great majority of papers devoted to the optimal structural design of columns with respect to stability constraints have been discussed in several survey papers by: J.Błachut and A.Gajewski [16] (a unified approach based on Pontryagin's maximum principle), E.J.Haug [17] (continuous parameter structures with multiple eigenvalues), N.Olhoff [18] (optimal design with respect to structural eigenvalues), T.A.Weishaar and R.H.Plaut [19] (non-conservative problems).

The most general formulation of the column optimization problem has been presented in the monograph by A.Gajewski and M.Życzkowski [1], where nearly 250 papers concerning columns are cited. More than 90% of them deal with single eigenvalues (buckling load or frequency of vibration) and only one control function (the cross-sectional area) and assume that buckling or vibrations occur in one plane only.

In 1976 N.Olhoff and S.H.Rasmussen [20] criticized the classical, but incorrect solution of I.Tadjbakhsh and J.B.Keller [21] obtained for a clamped-clamped column and arrived at the correct one on the basis of optimization with respect to a double eigenvalue connected with two fundamental forms of buckling (still in one plane). Since then some similar bimodal problems have been solved, e.g. for vibrating compressed columns and vibrating compressed columns in an elastic medium.

The determination of two independent control functions, e.g. the dimensions of a rectangular cross-section is an important problem which so far has not been discussed. As will be shown, certain solutions can be obtained here even in unimodal formulation and for buckling in one plane. However, it is more natural to introduce the multimodal (bi-, tri-, quadri-) formulation of the problem, i.e. optimization with respect to the critical load associated with two, three or four buckling modes arising in two principal planes. In such a case we may speak about both multimodal optimization and simultaneous mode design.

In this section we present a fairly general approach to the column optimization problem and many very well-known solutions can be treated as particular cases of that formulation.

2.2. The stability of non-prismatic columns

We base our considerations on the equations of motion of a compressed bar; in this way some problems of optimization with respect to vibrations and the application of the kinetic criterion of stability can be included. Making use of the kinetic criterion of stability however, we shall assume the material of the columns to be linearly-elastic or linearly-viscoelastic. On the other hand, if the static criterion of stability is sufficient, we shall also be able to consider the material to be non-linearly elastic, elastic-plastic, or non-linearly creeping.

In what follows, we present a theory of stability, confined to the discussion of motion in one plane, without torsion, and to linearized constitutive equations; this theory will be sufficient to provide a wide formulation of optimization problems. Stretchability of axis and the effect of shear will also be allowed for. Despite the partial mutual cancellation of these two effects when calculating the critical force, they are sometimes quite considerable. The remaining assumptions conform to Kirchhoff's theory.

2.2.1. Nonlinear equations of motion

A straight undeformed bar is subjected to a system of loadings originally acting in the direction ξ (Fig.2). In order to investigate its stability we consider its vibrations in the plane (ξ, η).

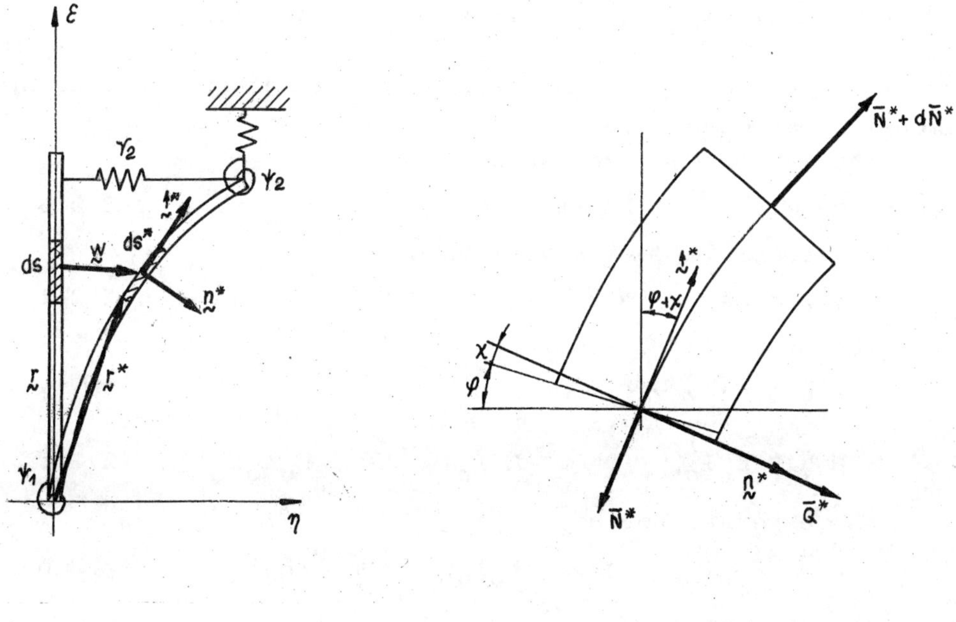

Fig.2. Column before and after buckling and its element

Such a motion may be described in a spatial system, determined by the initial configuration, using six first-order nonlinear partial differential equations:

$$u' = -1 + (1+\varepsilon_o)\cos(\varphi+\chi)$$

$$v' = (1+\varepsilon_o)\sin(\varphi+\chi)$$

$$\varphi' = -\hat{S}_M^{-1}\ M \qquad (2.2.1)$$

$$M' = -m + (1+\varepsilon_o)[Q\cos(\varphi+\chi) - N\sin(\varphi+\chi)] - \alpha\rho\Psi\ \ddot{\varphi}$$

$$Q' = -q + \rho\phi\ \ddot{v}$$

$$N' = -p + \rho\phi\ \ddot{u}$$

where

$$\varepsilon_o = \alpha\hat{S}_N^{-1}[N\cos(\varphi+\chi) + Q\sin(\varphi+\chi)]$$

$$\chi = \alpha\hat{S}_Q^{-1}\left\{(1+\varepsilon_o)[Q\cos(\varphi+\chi) - N\sin(\varphi+\chi)]\right\} \qquad (2.2.2)$$

In these equations we have introduced the following dimensionless quantities:

the independent variables:

$$x = s/l, \qquad t = \bar{t}/t_o \qquad (2.2.3)$$

the possible design variables:

$$\phi(x) = A/A_o, \quad \Psi(x) = I/I_o, \quad \Xi = \bar{E}/E_o, \quad \rho = \bar{\rho}/\rho_o \qquad (2.2.4)$$

the state variables:

$$u = \bar{u}/l, \quad v = \bar{v}/l,$$

$$N = \bar{N}l^2/E_oI_o, \quad Q = \bar{Q}l^2/E_oI_o, \quad M = \bar{M}l/E_oI_o, \qquad (2.2.5)$$

the external forces:

$$p = \bar{p}l^3/E_oI_o, \quad q = \bar{q}l^3/E_oI_o, \quad m = \bar{m}l^2/E_oI_o \qquad (2.2.6)$$

the stiffnesses

$$\hat{S}_M = \hat{\bar{S}}_M/E_oI_o, \quad \hat{S}_N = \hat{\bar{S}}_N/E_oA_o, \quad \hat{S}_Q = \hat{\bar{S}}_Q/E_oA_o, \qquad (2.2.7)$$

the parameters

$$\alpha=I_o/A_o l^2, \qquad t_o=(\rho_o A_o l^4/E_o I_o)^{1/2}, \qquad (2.2.8)$$

where A_o and I_o denote the cross-sectional area and the moment of inertia of the cross-section of a prismatic bar of the same length and volume as the given non-prismatic bar, where E_o and ρ_o are certain constants of stress and density dimensions, and where t_o denotes a constant of the dimension of time which may be treated as a unit of time. The angle φ represents the rotation of the cross-section and the additional angle χ is due to shear, $\hat{S}_N$, $\hat{S}_Q$ and $\hat{S}_M$ denote stiffnesses in tension, shear and bending respectively. In creep problems they may be trated as certain operators of time.

It should be noted that the distributed external loadings m,q,p may depend on state variables and on the coordinate x

$$\mathbf{p} = (m,q,p)^T = \mathbf{p}(x,u,v,\varphi,M,Q,N). \qquad (2.2.9)$$

Such a dependence is particularly important when discussing the stability of a bar, since it describes the behaviour of the loading in the course of buckling.

The system of governing equations (2.2.1),(2.2.2) should be complemented by appropriate boundary conditions.

2.2.2. The momentless precritical state and relevant conditions of loss of stability

The stability of equilibrium of a bar (or of motion in the case of creep) will be investigated by introducing a small vibration superposed on the basic motion (precritical state) of the bar. In most cases, an initially straight bar remains straight in the precritical state:

$$u_o'=\alpha f_o(N_o,\phi,t^*), \qquad N_o'=-p_o(x,\phi,P),$$
$$u_o(0)=0, \qquad N_o(1)=N_{o1}[P,u_o(1)]. \qquad (2.2.10)$$

The linear vibrations are described by the following linear boundary value problem (cf.(1.3.2),(1.3.4)):

$$Z'_j=A_{j\alpha}Z_\alpha, \qquad \mu_{m\alpha}Z_\alpha(0)=0, \qquad \nu_{n\alpha}Z_\alpha(1)=0 \tag{2.2.11}$$

$$j=1..6, \quad m=1..M, \quad n=1..N, \quad \alpha=1..6, \quad M+N=6$$

where

$$\mathbb{Z} = (u,v,\varphi,M,Q,N)^T$$

$$\mathbb{A} = \begin{bmatrix} 0 & 0 & 0 & 0 & 0 & \alpha/S_N \\ 0 & 0 & a & 0 & (1+\varepsilon_{oo})\alpha/S_Q & 0 \\ 0 & 0 & 0 & -1/S_M & 0 & 0 \\ -\alpha_{11} & -\alpha_{12} & -\tilde{\alpha}_{13} & -\alpha_{14} & -\alpha_{15}+a & -\alpha_{16} \\ -\alpha_{21} & -\alpha_{22}+\rho\lambda^2\phi & -\alpha_{23} & -\alpha_{24} & -\alpha_{25} & -\alpha_{26} \\ -\alpha_{31}+\rho\lambda^2\phi & -\alpha_{32} & -\alpha_{33} & -\alpha_{34} & -\alpha_{35} & -\alpha_{36} \end{bmatrix} \tag{2.2.12}$$

$$a=(1+\varepsilon_{oo})(1-\alpha N_o/S_Q), \qquad \tilde{\alpha}_{13}=\alpha_{13}+N_o a+\alpha\rho\lambda^2\Psi. \tag{2.2.13}$$

In general, the matrices $\mathbb{A}$, μ_{ij}, ν_{ij} and the vector of the state variables are complex.

The coefficients α_{ij} characterize the behaviour of distributed loadings during vibration (or buckling), whereas μ_{ij} and ν_{ij} characterize the behaviour of concentrated end forces, modes of support, values of concentrated masses and the moments of inertia at the ends, etc. They may also depend on the critical time t^* (visco-elastic columns). In some cases, the parameters μ_{ij} and ν_{ij} may be complex: for example, if a concentrated mass is attached to the free end of a vibrating visco-elastic column, the end shear force and moment depend on the square of the complex frequency of vibration, i.e. $Q(1)=-\alpha\lambda^2y(1)$, $M(1)=\beta\lambda^2\varphi(1)$ (α,β-are certain constants).

Extensibility of axis and shear deformation may be neglected by making the substitutions $\alpha S_N^{-1}=0$ and $\alpha S_Q^{-1}=0$, respectively. If both these substitutions are accepted, then $\varepsilon_o=0$ and $\chi=0$.

The nonlinear initial boundary value problem (2.2.10) (prebuckling state) and the linear homogeneous boundary value problem (2.2.11) determine the relation between the loading parameter P, the frequency of vibration, and possibly also the time parameter t (in creep problems)

$$f(P,\lambda,t) = 0. \qquad (2.2.14)$$

If the frequency λ vanishes for a certain value of loading parameter P, then the bar looses its stability by bifurcation or by snap-through (the static criterion of stability),

$$f(P,t^{*}) = 0, \qquad (2.2.15)$$

and if only the real part of the frequency passes from negative to positive values, then the loss of stability is due to flutter

$$f(P,\lambda_{c}^{I},t^{*}) = 0, \qquad (2.2.16)$$

where $\lambda^{I}=\mathrm{Im}\,\lambda$ denotes an imaginary part of λ.

2.3. A unified approach to the optimization of columns

2.3.1. A general statement of the problem

The functional representing the volume of the column will be assumed to be a design objective (cost function, the criterion of optimization):

$$V = A_{o}l\int_{o}^{1}\phi(x)dx \longrightarrow \min. \qquad (2.3.1)$$

The basic design variable is the dimensionless cross-sectional area $\phi=\phi(x)$, for a specific shape of cross-section. In practice, for the important cases one can use the relation

$$S_{M}=c_{1}\phi^{\nu}+c_{2}\phi+c_{3}, \qquad (2.3.2)$$

where c_{1},c_{2},c_{3} are constants. In the particular case $c_{2}=c_{3}=0$, $c_{1}=1$ we obtain the family of affine bars; for $\nu=2$ we have spatially uniformly affine bars (similar cross-sections, pyramids, cones, etc.).

In some problems of multimodal optimization, where loss of stability may take place in two mutually perpendicular planes, we usually assume several independent design variables, forming a vector $\phi_j=(\phi_1,\phi_2,\ldots\phi_n)$. In this case the stiffnesses of bending in both planes are certain functions of these design variables

$$S_{M_1}=S_{M_1}(\phi_j), \qquad S_{M_2}=S_{M_2}(\phi_j). \tag{2.3.3}$$

The first and basic constraint is a given constant critical force (or system of loadings) P_{cr}=const. A dual approach $P_{cr} \longrightarrow$ max. at V=const. is also frequently used. In creep problems an additional constraint is t_*=const.

Further constraints may refer to limitation of stresses (strength constraint in an elastic range)

$$\frac{\bar{N}_o(x)}{A_o\sigma_2(x)} \le \phi(x) \le \frac{\bar{N}_o(x)}{A_o\sigma_1(x)} \tag{2.3.4}$$

Inequality (2.3.4) may formally be written in the form

$$\phi_1(x) \le \phi(x) \le \phi_2(x) \tag{2.3.5}$$

and regarded as a purely geometrical constraint. However, there is a slight practical difference between (2.3.5) and (2.3.4).

2.3.2. Solution by Pontryagin's maximum principle

Pontryagin's maximum principle makes use of the equations of state which, in the case under consideration, describe both precritical and critical states (2.2.10), (2.2.11).

In order to make use of Pontryagin's maximum principle, we rewrite the system (2.2.11) in the form of a system of real equations. If we introduce the notation for real and imaginary parts:

$$\lambda=\lambda^R + i\lambda^I , \qquad i=\sqrt{-1} , \tag{2.3.6}$$

and similarly for A_{ij}, Z_j, μ_{ij}, ν_{ij}, then the equations (2.2.16) take the form

$$Z'^R_j = A^R_{j\alpha}Z^R_\alpha - A^I_{j\alpha}Z^I_\alpha\,, \qquad Z'^I_j = A^R_{j\alpha}Z^I_\alpha + A^I_{j\alpha}Z^R_\alpha\,, \tag{2.3.7}$$

$$\begin{aligned} &\mu^R_{m\alpha}Z^R_\alpha(0)-\mu^I_{m\alpha}Z^I_\alpha(0)=0, \qquad \mu^R_{m\alpha}Z^I_\alpha(0)+\mu^I_{m\alpha}Z^R_\alpha(0)=0, \\ &\nu^R_{n\alpha}Z^R_\alpha(1)-\nu^I_{n\alpha}Z^I_\alpha(1)=0, \qquad \nu^R_{n\alpha}Z^I_\alpha(1)+\nu^I_{n\alpha}Z^R_\alpha(1)=0. \end{aligned} \tag{2.3.8}$$

In order to present the design objective in a form adequate to Pontryagin's maximum principle, we introduce a new function $Z_o(x)$ by the equations:

$$Z'_o = \phi\,, \qquad Z_o(0)=0, \tag{2.3.9}$$

and write (2.3.1) thus:

$$V=A_o l\; Z_o(1) \tag{2.3.10}$$

In what follows, we assume the volume $A_o l$ to be equal to the minimal volume for a given load parameter P and a given time t_*

$$A_o l=V_{min} \tag{2.3.11}$$

(defining A_o in this manner). Hence (2.3.10), for the optimal control function $\tilde{\phi}(x)$, takes the form

$$Z_o(1)=1, \tag{2.3.12}$$

or, in other words, that function is normalized as follows

$$\int_o^1 \tilde{\phi}(x)\, dx = 1. \tag{2.3.13}$$

The Hamiltonian H, related to the systems of equations of state (2.2.10),(2.2.11) takes the form

$$H = \alpha\, \chi_{u_o} f_o(N_o,\phi_j,t_*) - \chi_{N_o} p_o(x,\phi_j,P) + (\overset{*}{\psi}_\alpha A_{\alpha\beta}Z_\beta)^R + \psi_o\phi, \tag{2.3.14}$$

where the quantities with asterisks over their symbols denote complex conjugate values.

According to the general scheme described in Sec.1.2.

we form an adjoint system of differential equations:

$$\chi'_{u_o}=0, \qquad \chi'_{N_o} = -\alpha\, \chi_{u_o}\frac{\partial f_o}{\partial N_o} - \left(\overset{*}{\psi}_\alpha \frac{\partial A_{\alpha\beta}}{\partial N_o} Z_\beta\right)^R \qquad (2.3.15)$$

$$\psi'_o=0, \qquad \psi'_j=-\overset{*}{A}_{\alpha j}\psi_\alpha. \qquad (2.3.16)$$

Making use of the transversality conditions (1.2.13), (1.2.14) at the end-points of the interval (0,1), we obtain a relevant system of boundary conditions:

$$\chi_{N_o}(0)=0, \qquad \chi_{u_o}(1) + \frac{\partial N_{o1}}{\partial[u_o(1)]}\chi_{N_o}(1)=0 \qquad (2.3.17)$$

$$\overset{*}{\bar{\mu}}_{i\alpha}\psi_\alpha(0)=0, \qquad \overset{*}{\bar{\nu}}_{i\alpha}\psi_\alpha(1)=0, \qquad (2.3.18)$$

where the new constants $\bar{\mu}_{ij}$, $\bar{\nu}_{ij}$ are certain functions of $\mu_{i\alpha}$ and $\nu_{i\alpha}$ ($i=1,2,3$, $\alpha=1..6$).

In order to obtain simpler notation and easier physical interpretation, we introduce the new dependent variables ($\bar{N}_o, \bar{u}_o, \bar{Z}_j$) of the adjoint equations by means of the following substitutions:

$$\chi_{u_o}=-k\bar{N}_o, \quad \chi_{N_o}=k\bar{u}_o, \quad \overset{*}{\psi}_j=(-1)^j k\bar{Z}_{n+1-j} \qquad (2.3.19)$$

where k denotes an arbitrary real constant, different from zero, and $j=1..n$.

Substituting these new variables (2.3.19) into (2.3.14)-(2.3.18) we obtain

the Hamiltonian

$$H = k[-\alpha\bar{N}_o f_o(N_o,\phi,t_*)-p_o(x,\phi,P)\bar{u}_o+(-1)^\alpha(A_{\alpha\beta}\bar{Z}_{n+1-\alpha}Z_\beta)^R + $$
$$+\Lambda_o\phi] \qquad \Lambda_o=\psi_o/k. \qquad (2.3.20)$$

the adjoint equations of precritical state with boundary conditions

$$\bar{N}'_o=0, \qquad \bar{u}'_o=\alpha\bar{N}_o\frac{\partial f_o}{\partial N_o} + (-1)^\alpha\left(\frac{\partial A_{\alpha\beta}}{\partial N_o}\bar{Z}_{n+1-\alpha}Z_\beta\right)^R \qquad (2.3.21)$$

$$\bar{u}_o(0)=0, \qquad \bar{N}_o(1) - \frac{\partial N_{o1}}{\partial[u_o(1)]}\,\bar{u}_o(1)=0, \tag{2.3.22}$$

the adjoint equations of critical state with boundary conditions

$$\bar{z}'_j = (-1)^{j+\alpha+1} A_{n+1-\alpha,\,n+1-j}\bar{z}_\alpha, \qquad \alpha=1..n, \tag{2.3.23}$$

$$(-1)^{n+1-\alpha}\,\bar{\mu}_{i,n+1-\alpha}\bar{z}_\alpha(0)=0, \qquad (-1)^{n+1-\alpha}\,\bar{\nu}_{i,n+1-\alpha}\bar{z}_\alpha(1)=0, \tag{2.3.24}$$

$i=1,2,3, \qquad \alpha=1..n.$

The necessary conditions of optimality are expressed by the solutions of the boundary-value problems (2.2.10), (2.2.11), (2.3.21) - (2.3.24) assuming that the optimal control $\tilde{\phi}_j(x)$ is determined by the supremum of the Hamiltonian (2.3.20) with respect to $\phi_j(x)$.

In many cases, the optimal shape $\tilde{\phi}_j(x)$ is determined by the optimality condition (1.2.15), here taking the form:

$$\left[-\alpha\bar{N}_o\frac{\partial f_o}{\partial\phi_j} - \bar{u}_o\frac{\partial p_o}{\partial\phi_j} + (-1)^{\alpha}\left(\frac{\partial A_{\alpha\beta}}{\partial\phi_j}\,\bar{z}_{n+1-\alpha}\,z_\beta\right)^R\right]_{\tilde{\phi}_j} + \Lambda_0 = 0$$

$$\alpha,\beta=1..n. \tag{2.3.25}$$

2.3.3. Multimodal formulation

Denoting subsequent eigenvalues and relevant state variables by the subscript ι ($\iota=1..L$), we have to allow for simultaneous L systems of equations of state and L adjoint systems. The Hamiltonian (2.3.20) should then be generalized to the form

$$H = \sum_{\iota=1}^{L} c_{(\iota)}\Big[-\alpha\bar{N}_{o,(\iota)}f_{o,(\iota)} - p_{o,(\iota)}\bar{u}_{o,(\iota)} + (-1)^{\alpha}\left(A_{\alpha\beta}\bar{z}_{n+1-\alpha,(\iota)}z_{\beta,(\iota)}\right)^R\Big] + \psi_o\phi, \tag{2.3.26}$$

where the arguments of $f_{o,(\iota)}$ and $p_{o,(\iota)}$ are as in

(2.3.20), and the constants $c_{(\iota)}$ are different from zero as long as the mode ι intervenes in optimization.

The optimality condition determines the optimal design in terms of state variables. The solution of the corresponding boundary-value problems for the same loading parameters determines the critical value of those parameters, the values of the constants $c_{(\iota)}$ (one of them remaining arbitrary), and the constant ψ_0.

2.3.4. The self adjoint boundary value problem of the critical state

Comparing the equations of the critical state (2.2.11) - (2.2.13) with the adjoint equations (2.3.23), (2.3.24) one may observe that in the case of some loadings these equations and the boundary conditions become identical. Then the solutions of both boundary-value problems are also identical up to a certain multiplier; however, this multiplier may be fixed as +1. Such cases are called self-adjoint.

In fact, the self-adjointness of the boundary-value problem of the critical state is a necessary and sufficient condition of the conservative behaviour of loading; the critical loading parameters may then always be determined by the static criterion of stability, $\lambda=0$. The variables of the critical and precritical states are real.

Such a situation occurs if the loading of the column does not contain distributed follower forces and if the behaviour of concentrated forces is conservative (A. Gajewski, M. Zyczkowski [22], M. Farshad, I. Tadjbakhsh [23]). The number of equations to be solved is then reduced by half.

2.4. The unimodal and bimodal optimal design of columns with respect to buckling or vibration in one plane

2.4.1. General remarks

We now confine our considerations to the conservative behaviour of loadings, and also to linear elasticity (neglecting extensibility and shear effects). The conditions of self-adjointness of the equations of the critical state are satisfied not only for loadings which are independent of the state variables (such as eulerian force of spatially fixed direction and materially fixed point of application acting at the free end, the weight of the column, an arbitrarily distributed load of spatially fixed direction) but also for loading proportional to the deflection v(x) and the perpendicular dimension b(x) (Winkler's foundation), for various types of behaviour of the concentrated force P acting at the free end if the following condition is satisfied

$$\nu_{23}=-\nu_{32}. \qquad (2.4.1)$$

Moreover, assuming the axis of the column to be inextensible and neglecting shear effects, the matrix $\mathbb{A}$ (2.2.12) can be simplified by the following assumptions:

$$\alpha_{ij}=0, \quad \varepsilon_o=0, \quad \chi=0, \quad a=1, \quad \alpha S_N^{-1}=0, \quad \alpha S_Q^{-1}=0 \qquad (2.4.2)$$

In what follows we also omit the weight of the column and other distributed forces; hence the precritical state is confined to the relation $N_o=-P$. The first and sixth row in the matrix $\mathbb{A}$ (2.2.12) can also be dropped and further analysis may be confined to the remaining four equations.

2.4.2. The optimal design of vibrating compressed columns with respect to one control function

The bimodal optimization of columns with respect to the frequency of transverse vibrations under axial compression are considered by A. Gajewski [24] and B. Bochenek and A. Gajewski [25]. These papers generalize the classical problem of N. Olhoff and R.H. Rasmussen [20].

The ends of the column are assumed to be elastically clamped with different flexibilities at each end, characterized by two constants ξ_1 and ξ_2 (Fig. 3). If these flexibilities are equal we have the symmetric structure for which symmetric or antisymmetric modes of vibration (or buckling) can be distinguished by the appropriate boundary conditions at the points x=0 and x=1/2. In the case of a non-symmetric structure $\xi_1\neq\xi_2$ the boundary conditions are the same for all modes of vibration:

$$v_{(l)}(0)=0; \qquad M_{(l)}(0) + \frac{1}{\xi_1}\varphi_{(l)}(0)=0 \qquad (2.4.3)$$
$$v_{(l)}(1)=0; \qquad M_{(l)}(1) + \frac{1}{\xi_2}\varphi_{(l)}(1)=0$$

The modes themselves can be differentiated by the number of zeros of the deflection lines.

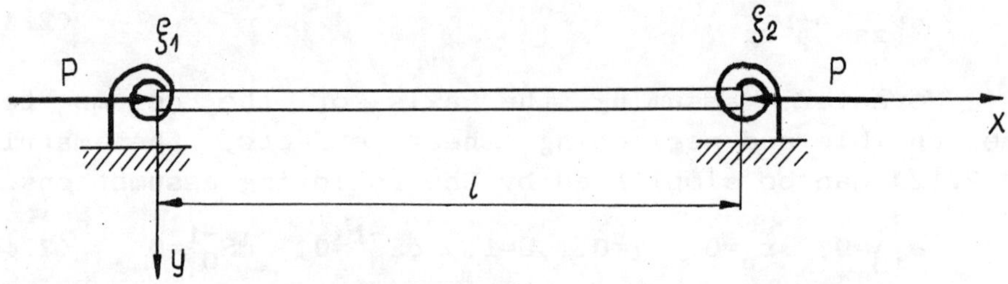

Fig.3. Elastically clamped compressed column

The optimality condition (2.3.25) is here very simple and in the case of bimodal formulation leads to the equation:

$$\phi(x) = \left\{\frac{\nu[(1-\mu)M^2_{(1)}+\mu M^2_{(2)}]}{\Lambda_o+(1-\mu)\Omega_{(1)}v^2_{(1)}+\mu\Omega_{(2)}v^2_{(2)}}\right\}^{\frac{1}{1+\nu}}, \qquad (2.4.4)$$

where the constants μ and Λ_o were introduced instead of $c_{(l)}$ and ψ_o, and $\Omega=-(\mathrm{Im}\lambda)^2$.

Some results of numerical calculations are presented in Fig.4. The dependence of the frequency of vibration on the values of lower geometrical constraint, for various values of compressive force shows that the bimodal formulation is necessary for $P\in(30,52.36)$.

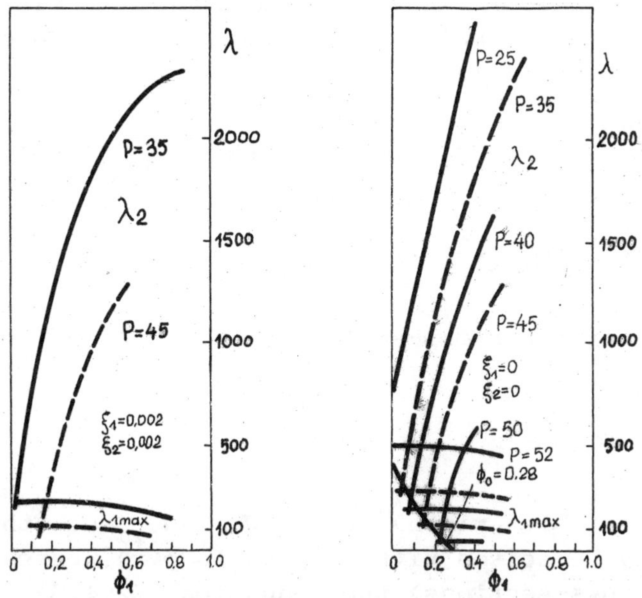

Fig. 4. Maximal (first) and second frequences of vibration versus lower geometrical constraint

2.4.3. The optimization of column in an elastic medium with respect to one control function

The necessity of multimodal optimization is even more visible for a column in an elastic medium (Fig.5): indeed, even for prismatic columns the number of half-wavelengths is unknown a priori and the occurence of two equally probable buckling modes may often be encountered.

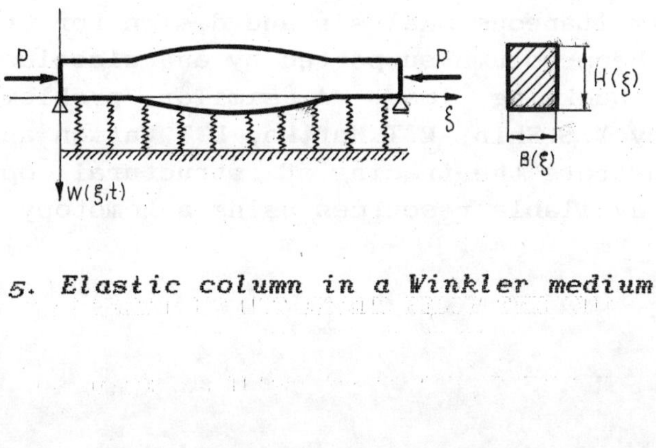

Fig. 5. Elastic column in a Winkler medium

If we consider vibrations (or buckling) in one plane only, then the optimality condition (2.3.25) takes the form, (generalizing (2.4.4)):

$$\frac{1}{S_M^2}\frac{\partial S_M}{\partial \phi}\left[(1-\mu)M^2_{(1)}+\mu M^2_{(2)}\right] + Kk(x)\frac{\partial b}{\partial \phi}\left[(1-\mu)v^2_{(1)}+\mu v^2_{(2)}\right]-$$

$$-\left[(1-\mu)\Omega_{(1)}v^2_{(1)}+\mu\Omega_{(2)}v^2_{(2)}\right] = \Lambda_0. \qquad (2.4.5)$$

For a rectangular cross-section of width b and depth h we have

$$S_M = \phi^{\nu}, \qquad b = \phi^{\frac{3-\nu}{2}}. \qquad (2.4.6)$$

If $\nu=1$ (variable width) and $\Omega_{(1)}=\Omega_{(2)}=\Omega$, the equation (2.4.5) may be solved with respect to $\phi(x)$, whereas for $\nu=2$ (similar cross-sections) the equation (2.4.5) is of the sixth degree with respect to $\phi^{1/2}$ and should be solved numerically.

Fig.6 shows the domains of unimodal and bimodal optimization for $\nu=1$, $k(x)\equiv1$ and $\Omega=0$ in the plane $K - \phi_1$ (the upper bound not being introduced, $\phi_2 \longrightarrow \infty$), while Fig.7 shows several optimal shapes of columns. Details and further results are given by A.Gajewski [26].

The bimodal optimization of columns in Winkler's medium was also investigated by J.Kiusalaas [27], S.I.Repin [28], and A.D.Larichev [29]. The problem has recently been studied by R.H.Plaut, L.W.Johnson and N.Olhoff [30] for columns of an idealized sandwich cross-section and various boundary conditions: pinned-pinned, clamped-clamped, and pinned-clamped. More recently, Y.S.Shin, R.H.Plaut and R.T.Haftka [31] have used simultaneous analysis and design for the optimum design of a beam-column supported by an elastic foundation for maximum buckling load. A similar problem has been considered by Y.S.Shin, R.T.Haftka, L.T.Watson and R.H.Plaut [32] to illustrate the tracing of structural optima as a function of available resources using a homotopy method.

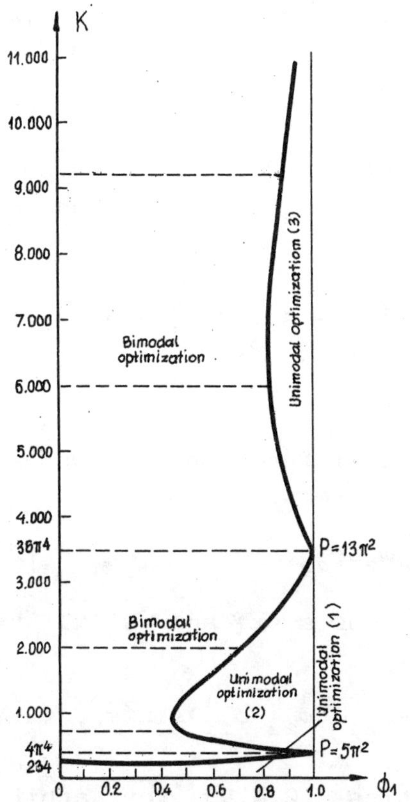

Fig.6. Validity ranges of unimodal and bimodal optimization

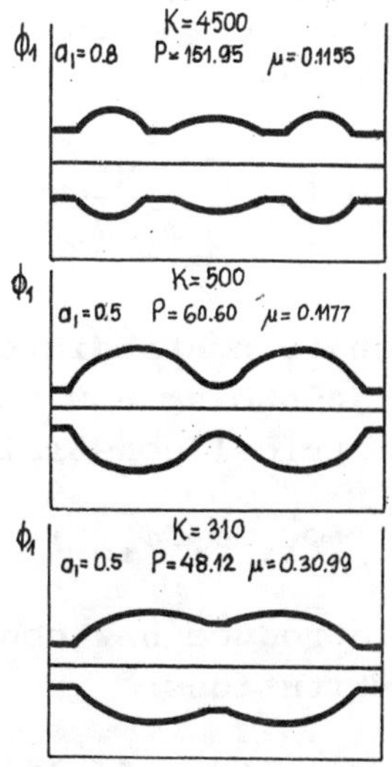

Fig.7. Optimal shapes of columns

2.4.4. The optimization of columns with respect to two control functions

The determination of two independent design variables, namely the dimensions of a rectangular cross-section is an important problem which so far has not been discussed. Indeed, it seems to be very unusual, especially in the case of the unimodal optimization of columns which are buckled or are vibrating in one plane only. The occurrence of such a possibility depends mainly on the existence of suitable geometrical constraints, but also on other factors such as the behaviour of loads, foundation resistance, etc.

Considering a cantilever column of rectangular cross-section -compressed by the eulerian force- which deflects in the (x,v) plane, we have:

$$S_M = \phi_1(x)\phi_2^3(x) \quad , \tag{2.4.7}$$

and the analytical optimality conditions (2.3.25) with respect to the functions ϕ_1 and ϕ_2 lead to the set of equations:

$$\phi_2\left(\Lambda_o - \frac{M^2}{\phi_1^2\ \phi_2^4}\right) = 0, \qquad \phi_1\left(\Lambda_o - 3\ \frac{M^2}{\phi_1^2\ \phi_2^4}\right) = 0, \tag{2.4.8}$$

which is **contradictory.**

Assuming now that the control functions are subject to geometrical constraints of the form:

$$\phi_{11} \le \phi_1 \le \phi_{12} \quad ; \qquad \phi_{21} \le \phi_2 \le \phi_{22}, \tag{2.4.9}$$

we introduce new control functions $s_1(x)$ and $s_2(x)$ by the substitutions:

$$\phi_1 = \phi_{11} + (\phi_{12} - \phi_{11})\sin^2 s_1; \quad \phi_2 = \phi_{21} + (\phi_{22} - \phi_{21})\sin^2 s_2, \tag{2.4.10}$$

which assure the inequalities (2.4.9) for arbitrarily chosen values of s_1 and s_2. After substitution of (2.4.10) into (2.3.20) the Hamiltonian becomes a function of the variables s_1 and s_2 and the optimality conditions (2.3.25) (with respect to s_1 and s_2) take the form:

$$\begin{aligned} &\frac{\partial H}{\partial \phi_1}(\phi_{12} - \phi_{11}) \sin s_1 \cos s_1 = 0, \\ &\frac{\partial H}{\partial \phi_2}(\phi_{22} - \phi_{21}) \sin s_2 \cos s_2 = 0, \end{aligned} \tag{2.4.11}$$

In each step of the numerical integration of the state equations (2.2.11) and the normalization equation (2.3.9) the control functions ϕ_1 and ϕ_2 should be determined from the supremum condition of the Hamiltonian (2.3.20). One of possible solutions is shown in Fig.8.

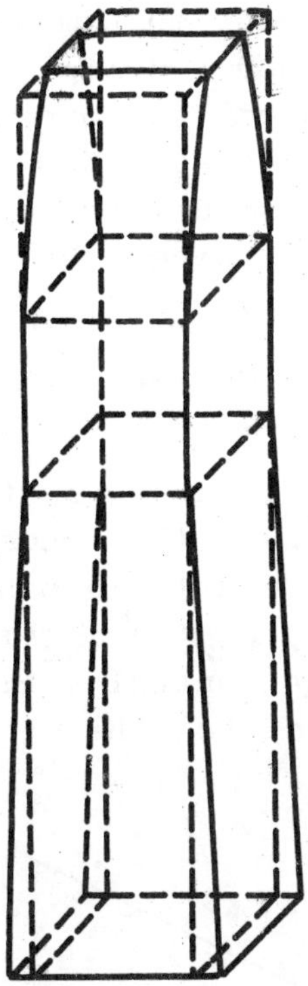

Fig.8. Optimal shape of a column optimized with respect to two control functions

2.5. The optimization of columns with respect to buckling in two planes

2.5.1. General remarks

A column designed optimally for plane buckling in the plane (x,y) may loose its stiffness in the perpendicular plane (x,z) and subsequently may buckle in the second plane. This situation is typical for different modes of support in two mutually perpendicular planes, for example double clamping in the plane (x,z) and simple support in the plane (x,y), as in connecting rods. In such cases optimization for buckling in two planes is necessary - it leads to multimodal optimization, with a number of simultaneous modes equal to 2,3, or 4, depending on the mode of support and additional geometric constraints.

Here we can speak about multimodal optimization in each plane and simultaneous mode design in two planes. Both optimal design with respect to one control function and optimal design with respect to two control functions is possible. The numerical solutions of the problems under discussion have been obtained by B. Bochenek [33] and B. Bochenek and M. Nowak [34].

2.5.2. Multimodal optimal design with respect to two control functions

Let us consider a column of rectangular cross-section b x h as shown in Fig. 9. The ends of the column are elastically clamped with equal clamping rigidities at both ends $\psi_{xy}(0)=\psi_{xy}(1)=\psi_{xy}$, $\psi_{xz}(0)=\psi_{xz}(1)=\psi_{xz}$, but in general $\psi_{xy}\neq\psi_{xz}$. The independent control variables $b(x)=\phi_1(x)$ and $h(x)=\phi_2(x)$ are subjected to the geometrical constraints (2.4.9).

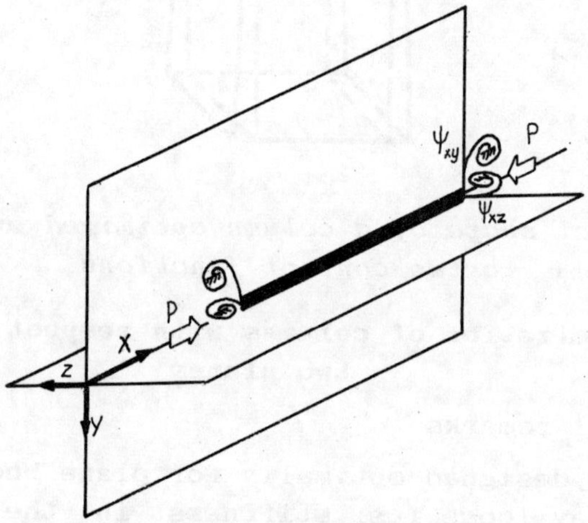

Fig. 9. The column with elastically clamped ends in two planes

Dimensionless bending stiffnesses in the planes (x,y) and (x,z) and the cross-sectional area are equal:

$$S_{M(1)}=S_{M(2)}=\phi_1\phi_2^3;\quad S_{M(3)}=S_{M(4)}=\phi_1^3\phi_2;\quad \phi=\phi_1\phi_2,\quad (2.5.1)$$

where the numeration of the modes allows for two buckling

modes in each plane.

The buckling state of the structure within the linear theory is described by two sets of four first-order differential equations of the type (2.2.16) with appropriate boundary conditions:

$$w'_{i(l)} = \bar{A}_{ij} w_{j(l)}; \qquad v'_{i(l)} = \tilde{A}_{ij} v_{j(l)}; \qquad (2.5.2)$$

$$l = 1,2; \qquad i,j = 1..4,$$

where:

$$\bar{A}_{ij} = \begin{Bmatrix} 0, & 1, & 0, & 0 \\ 0, & 0, & \dfrac{-1}{\phi_1 \phi_2^3}, & 0 \\ 0, & P, & 0, & 1 \\ 0, & 0, & 0, & 0 \end{Bmatrix}, \qquad \tilde{A}_{ij} = \begin{Bmatrix} 0, & 1, & 0, & 0 \\ 0, & 0, & \dfrac{-1}{\phi_1^3 \phi_2}, & 0 \\ 0, & P, & 0, & 1 \\ 0, & 0, & 0, & 0 \end{Bmatrix} \qquad (2.5.3)$$

The boundary conditions may be formulated for $x=0$ and $x=1/2$ because of the symmetry of column (in the prebuckling state) and distinguish symmetric and antisymmetric buckling modes. The boundary value problems are self-adjoint.

The Hamiltonian (2.3.26), constructed for the equations (2.5.2) takes the form:

$$\begin{aligned} H = {} & \bar{c}_{(1)} \left[2w_{4(1)} w_{2(1)} + \frac{w_{3(1)}^2}{\phi_1 \phi_2^3} + P\, w_{2(1)}^2 \right] + \\ & + \bar{c}_{(2)} \left[2w_{4(2)} w_{2(2)} + \frac{w_{3(2)}^2}{\phi_1 \phi_2^3} + P\, w_{2(2)}^2 \right] + \\ & + \tilde{c}_{(1)} \left[2v_{4(1)} v_{2(1)} + \frac{v_{3(1)}^2}{\phi_1^3 \phi_2} + P\, v_{2(1)}^2 \right] + \\ & + \tilde{c}_{(2)} \left[2v_{4(2)} v_{2(3)} + \frac{v_{3(2)}^2}{\phi_1^3 \phi_2} + P\, v_{2(2)}^2 \right] + \psi_0 \phi_1 \phi_2 , \end{aligned} \qquad (2.5.4)$$

where $\bar{c}_{(l)}$, $\tilde{c}_{(l)}$ are certain constants which have to be determined.

The extremum conditions of the Hamiltonian (2.5.4) with respect to the control variables ϕ_1 and ϕ_2 make it possible to express these variables in terms of state variables as follows:

$$\phi_1 = \left\{ \Lambda_o \frac{[\mu_2 v^2_{3(1)} + \mu_3 v^2_{3(2)}]^2}{w^2_{3(1)} + \mu_1 w^2_{3(2)}} \right\}^{1/6} = \left\{ \Lambda_o \frac{[\mu_2 M^2_{(3)} + \mu_3 M^2_{(4)}]^2}{M^2_{(1)} + \mu_1 M^2_{(2)}} \right\}^{1/6}$$

(2.5.5)

$$\phi_2 = \left\{ \Lambda_o \frac{[w^2_{3(1)} + \mu_1 w^2_{3(2)}]^2}{\mu_2 v^2_{3(1)} + \mu_3 v^2_{3(2)}} \right\}^{1/6} = \left\{ \Lambda_o \frac{[M^2_{(1)} + \mu_1 M^2_{(2)}]^2}{\mu_2 M^2_{(3)} + \mu_3 M^2_{(4)}} \right\}^{1/6},$$

where

$$\Lambda_o = 4\bar{c}_{(1)}/\psi_o, \quad \mu_1 = \bar{c}_{(2)}/\bar{c}_{(1)}, \quad \mu_2 = \tilde{c}_{(1)}/\bar{c}_{(1)}, \quad \mu_3 = \tilde{c}_{(2)}/\bar{c}_{(1)}.$$

If some of the constants μ_i vanish, we have a bi- or trimodal solution, whereas a unimodal solution is impossible - at least in the formulation hitherto used.

The problem is then reduced to the solution of the non-linear boundary value problem (2.5.2) and the equation (2.3.9), the values (2.5.5) or the bounding values ϕ_{ij}(i, j=1, 2) being substituted.

Details of calculations are given by B. Bochenek [33]; the optimal functions ϕ_1, ϕ_2 and the corresponding shape of column obtained for tri- modal formulation is presented in Fig. 10.

In the plane of the clamping rigidities $\bar{\psi}_{xz}$ and $\bar{\psi}_{xy}$ reduced to the interval $0 \le \bar{\psi} \le 1$ by the formulae:

$$\bar{\psi}_{xz} = \frac{\psi_{xz}}{1+\psi_{xz}}, \qquad \bar{\psi}_{xy} = \frac{\psi_{xy}}{1+\psi_{xy}}, \qquad (2.5.6)$$

the domains of bi-, tri-, and quadri-modal optimization are determined and presented in Fig. 11.

It should be noted that the classical bimodal Olhoff-Rasmussen solution for a clamped-clamped column ($\bar{\psi}_{xz} = \bar{\psi}_{xy} = 0$) is treated here as a quadri-modal case.

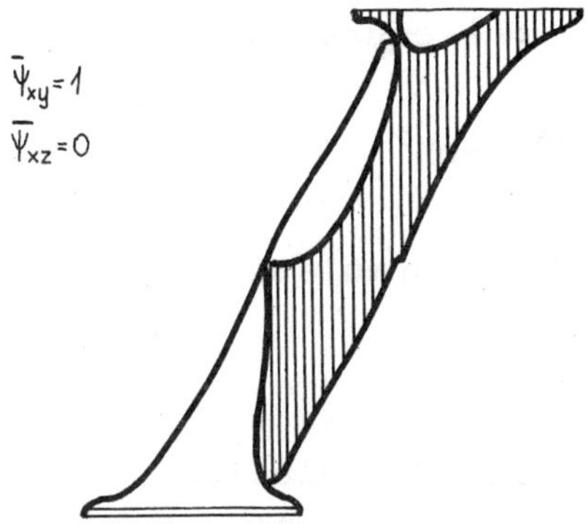

Fig. 10. Optimal column for $\bar{\psi}_{xy}=1.0$ *and* $\bar{\psi}_{xz}=0$

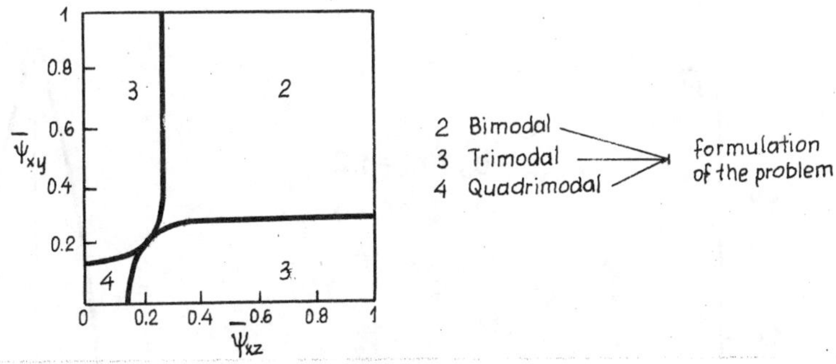

Fig. 11. The range of application of optimization modalities in terms of $\bar{\psi}_{xy}$ *and* $\bar{\psi}_{xz}$

2.5.3. The optimal design of a cantilever column with respect to one control function

If we assume that the height of the cross-section of the cantilever column is constant, $(h=h_0)$, and that only one control variable is admitted $b=b(x)$, the Hamiltonian (2.5.4) may be simplified to the case allowing for one mode in each plane only. The state equations (2.5.2) should be complemented by the following boundary conditions:

$$w_{(i)}(0)=\phi_{(i)}(0)=M_{(i)}(1)=Q_{(i)}(1)=0; \quad i=1,2, \qquad (2.5.7)$$

The optimality condition (1.2.15) confined to one control variable ϕ_1 leads to the equations:

$$\phi_1=\left\{\Lambda_0\left[M^2_{(1)}+(M^4_{(1)}+\mu M^2_{(2)})^{1/2}\right]\right\}^{1/2} \tag{2.5.8}$$

-in the case of bimodal formulation,

$$\varphi_1=(2\Lambda_0)^{1/2}\ |M_{(1)}| \tag{2.5.9}$$

-in the case of unimodal formulation.

The numerical calculations were carried out by B. Bochenek and M. Nowak [34]; the dependence of the critical forces $P_{(1)}$ and $P_{(2)}$ on the lower geometrical constraint ϕ_{11} is presented in Fig. 12 (for $b_0/h_0=1.2$).

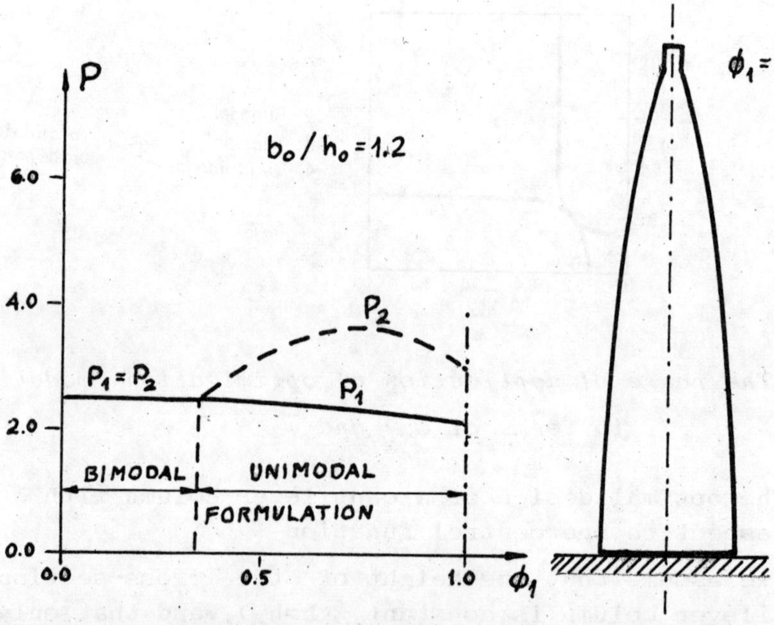

Fig. 12. The dependence of critical forces on geometrical constraint and optimal shape of a column

Some other related multimodal optimization problems are presented by S. Prager and W. Prager [35], E. J. Haug [17], N. Olhoff [36], H. L. Lam, E. J. Haug and K. K. Choi [37], K. K. Choi and E.J.Haug [38], E.J.Haug and K.K.Choi [39], W.Teschner [40], N.W.Banichuk and A.A.Barsuk [41], A.P.Seyranian [42] (the exact solution of the Olhoff-Rasmussen problem), E.F.Masur [43] (the exact solution of the Olhoff-Rasmussen problem).

3. MULTIMODAL OPTIMAL STRUCTURAL DESIGN OF ARCHES

3.1. General remarks

Problems of the optimal structural design of plane arches under stability constraints require the consideration of various forms of instability especially those due to bifurcation and snap- through. Moreover, the instability may occur in the plane of the arch and/or outside the plane. As a rule, therefore, multimodal formulation of the optimization problems should here be considered.

However, early papers on the optimal design of arches with respect to their stability discuss only unimodal formulations of the problem. J.Błachut and A.Gajewski [44] have drawn attention to the necessity for bimodal formulation for clamped- clamped funicular arches. In the general case of the precritical state, in particular if stretchability of axis is allowed for, snap-through as well as bifurcation may occur. Unimodal and bimodal optimization of extensible arches under snap-through, bifurcation and vibration constraints was carried out by J.Błachut and A.Gajewski [45] and J.Błachut [46].

Furthermore, if out-of-plane buckling is allowed for, then even bimodal formulation of optimization problems of arches may be insufficient. This problem has been discussed in detail by B.Bochenek and A.Gajewski [47] for an inextensible clamped-clamped circular arch; three- and four-modal solutions were obtained.

All the above-mentioned papers employed the magnitude of the cross-section as a design variable. B.Bochenek and A.Gajewski [48] and B.Bochenek [49] have tackled new and more complicated problems, namely the optimal design of arches for which both the depth and width of a rectangular cross-section are treated as two independent design functions. The arches are optimized with respect to plane and spatial buckling and their axes are assumed to be inextensible.

Here it should be mentioned that optimization of the shape of the axis of a prismatic arch under stability constraints was carried out by R.H.Plaut and N.Olhoff [50]. Both design variables -the magnitude of the cross-section and the shape of the axis- have as yet not been used simultaneously when designing for stability, but such a general approach was employed by N.Olhoff and R.H.Plaut [51] for vibration constraints.

3.2. The stability of non-prismatic arches

3.2.1. Introductory remarks

Stability analysis of linearly-elastic arches, based on Bernoulli's hypothesis of plane cross-sections, allowing for geometrical nonlinearities is considered in numereous papers, e.g. by G.Rakowski and R.Solecki [52], R.Schmidt [53], T.Irie, G.Yamada and I.Takahashi [54], J.Błachut [55], K.Suzuki, T.Kosawada and S.Takahashi [56]. Making use of these results, we can derive the governing equations for in-plane and out-of-plane buckling (or vibration).

3.2.2. General non-linear governing equations for in-plane motion

Let us consider a plane arch, with initial curvature $\bar{\varkappa}=\bar{\varkappa}(\bar{s})$, where $\bar{s}$ denotes the material variable measured along the axis; bars over a symbol refer to physical (dimensional) quantities. The axis is described in its undeformed state by the following parametric equations:

$$\bar{X} = \int_0^{\bar{s}} \cos \Theta(\vartheta)\, d\vartheta,$$

$$\bar{Y} = \int_0^{\bar{s}} \sin \Theta(\vartheta)\, d\vartheta, \tag{3.2.1}$$

$$\bar{Z} = 0,$$

where ϑ denotes the variable of integration, and Θ - the slope with respect to the axis $\bar{X}$, Fig.13.

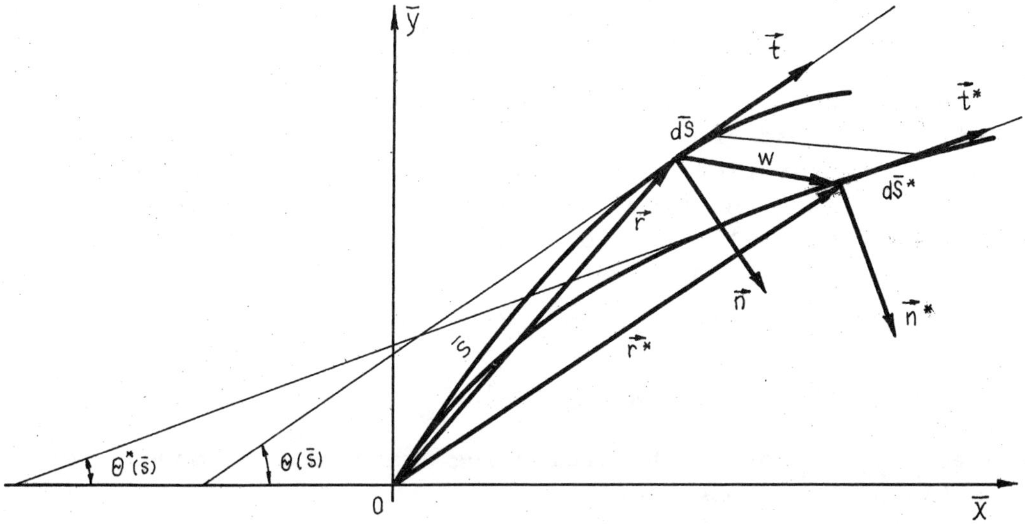

Fig. 13. Undeformed and deformed states of an arch

Allowing for stretchability of the axis we assume:

$$d\bar{s}^{*} = (1+\varepsilon_0)d\bar{s}, \tag{3.2.2}$$

where the quantities related to the deformed system are starred.

The geometrical relations between the initial and the deformed configurations, the definition of the angle φ (Fig.13) and the conditions of equilibrium of the deformed arch element, as well as Hooke's law, result in a system of six governing quasi-linear differential equations written in the following form:

$$\begin{aligned}
&u' = \varkappa v - (1-\cos\varphi) + \varepsilon_0\cos\varphi \\
&v' = -\varkappa u - (1+\varepsilon_0)\sin\varphi \\
&\varphi' = \frac{-M}{S_M} - \varepsilon_0\varkappa \\
&M' = -m - (1+\varepsilon_0)(Q\cos\varphi + N\sin\varphi) - \alpha\rho J\varphi^{\cdot\cdot} \\
&Q' = -q - \varkappa N + \rho\phi v^{\cdot\cdot} \\
&N' = -p + \varkappa Q + \rho\phi u^{\cdot\cdot}
\end{aligned} \tag{3.2.3}$$

where

$$\varepsilon_o=\alpha S_N^{-1}(N\cos\varphi-Q\sin\varphi+\varkappa M); \qquad (3.2.4)$$

p and q here denote the coordinates of the external distributed load **f** in the undeformed basis **t** and **n**.

In the equations (3.2.3) and (3.2.4) we have introduced dimensionless quantities as in Sec.2.2.1., namely those defined by (2.2.3) - (2.2.6), (2.2.8) and also:

$$\begin{aligned} &s=\bar{s}/l, \quad \varkappa=-\bar{\varkappa}l=l/R, \quad J=\tilde{I}_{2\varkappa}/I_o, \quad \alpha=I_o/A_o l^2, \\ &S_N=EA/E_oA_o, \qquad S_M=EI_{2\varkappa}/E_oI_o, \end{aligned} \qquad (3.2.5)$$

where $I_{2\kappa}$ denotes the following generalized moment of inertia of the cross-section:

$$I_{2\varkappa}=\iint\limits_A \frac{\bar{z}^2}{1+\bar{\varkappa}\bar{z}}\, dA \quad \text{and} \quad \tilde{I}_{2\varkappa}=\iint\limits_A \bar{z}^2(1+\bar{\varkappa}\bar{z})\, dA. \qquad (3.2.6)$$

Introducing the vector

$$\mathbf{X} = (u,v,\varphi,M,Q,N)^T \qquad (3.2.7)$$

the equations of state (3.2.3) may be written in the more compact notation:

$$\mathbf{X}' = \mathbf{F}(s,\mathbf{X},\mathbf{X}^{\cdot\cdot}) \qquad (3.2.8)$$

The relevant boundary conditions may be written in the form:

$$\mathbf{B}^{(0)}[\mathbf{X}(0,t)]=0, \qquad \mathbf{B}^{(1)}[\mathbf{X}(1,t)]=0. \qquad (3.2.9)$$

3.2.3. The general precritical state and the relevant conditions of in-plane loss of stability

As in Sec.2.2.2. we here use the method of small vibrations imposed on static (large) deflections. Restricting ourselves to linearly-elastic materials (creep is not allowed for), we assume the precritical state **Y** to be independent of time t:

$$\mathbf{X}(s,t)=\mathbf{Y}(s)+\varepsilon\mathbf{Z}(s)\exp\lambda t. \qquad (3.2.10)$$

Hence, the general precritical state is described by the equations (3.2.8) and (3.2.9), in which inertia forces are neglected. The relevant quantities will be denoted by the subscript "o" as in (2.2.10).

Linear vibrations (or static perturbations) of the arch in the vicinity of the precritical state of equilibrium will be denoted by the subscript "1". Carrying out the expansions, for a small in-plane motion we then obtain:

$$
\begin{aligned}
u_1' &= \varkappa v_1-(1+\varepsilon_{00})\varphi_1 \sin\varphi_0+\varepsilon_{01}\cos\varphi_0\\
v_1' &= -\varkappa u_1-(1+\varepsilon_{00})\varphi_1 \cos\varphi_0-\varepsilon_{01}\sin\varphi_0\\
\varphi_1' &= -S_M^{-1}M_1-\varkappa\varepsilon_{01} \qquad (3.2.11)\\
M_1' &= -\alpha_{1j}Z_j-(1+\varepsilon_{00})[Q_1\cos\varphi_0+N_1\sin\varphi_0+(N_0\cos\varphi_0-Q_0\sin\varphi_0)\varphi_1]-\\
&\quad -\varepsilon_{01}(Q_0\cos\varphi_0+N_0\sin\varphi_0)-\alpha\rho\lambda^2J\varphi_1\\
Q_1' &= -\alpha_{2j}Z_j-\varkappa N_1+\rho\lambda^2\phi v_1\\
N_1' &= -\alpha_{3j}Z_j+\varkappa Q_1+\rho\lambda^2\phi u_1,
\end{aligned}
$$

where the summation convention over j holds, and

$$\varepsilon_{00}= \alpha S_N^{-1}(N_0\cos\varphi_0-Q_0\sin\varphi_0+\varkappa M_0) \qquad (3.2.12)$$

$$\varepsilon_{01}= \alpha S_N^{-1}[N_1\cos\varphi_0-Q_1\sin\varphi_0-(N_0\sin\varphi_0+Q_0\cos\varphi_0)\varphi_1+\varkappa M_1].$$

The coefficients characterizing the behaviour of distributed loadings α_{ij} ($i=1,2,3$; $j=1,2...6$) depend on the precritical state and on the variable s.

Equations of the critical state (3.2.11),(3.2.12) should be complemented by the relevant linear and homogeneous boundary conditions in the form (2.2.11). For example, for a simply supported arch on immovable supports we have $u(0)=v(0)=M(0)=u(1)=v(1)=M(1)=0$, and for an arch clamped at both ends $u(0)=v(0)=\varphi(0)=u(1)=v(1)=\varphi(1)=0$.

In what follows, we confine our considerations to the static criterion of the loss of stability, $\lambda=0$ (conservative loadings and certain types of non-conservative loadings); the kinetic criterion of stability in the narrower sense has as yet not been effectively applied to the optimal structural design of arches.

3.2.4. The momentless precritical state and the relevant conditions of in-plane buckling

In the case of an inextensible axis, $\varepsilon_o=0$, and for certain particular loadings the precritical state may be characterized by vanishing moments, shear forces and deflections (so called funicular arches). The equations (3.2.3) take the form:

$$u_o=v_o=\varphi_o=M_o=Q_o=m_o=0, \quad \varkappa N_o+q_o=0, \quad N_o'+p_o=0. \qquad (3.2.13)$$

They determine the shape of the arch $\varkappa=\varkappa(s)$ for given loads in the precritical state or, if that shape is given, they yield a certain relation between the loads p_o and q_o:

$$p_o=(q_o/\varkappa)' \qquad (3.2.14)$$

For example, this condition is satisfied for a circular arch ($\varkappa$=const.) with a uniformly distributed radial load ($p_o=0$, q_o=const.).

The equations (3.2.11),(3.2.12) are now simplified and take the form:

$$\begin{aligned} &u'=\varkappa v, \quad v'=-\varphi-\varkappa u, \quad \varphi'=-S_M^{-1}M, \\ &M'=-Q-N_o\varphi-\alpha_{11}u-\alpha_{12}v-\alpha_{13}\varphi-\alpha\rho\lambda^2 J\varphi, \qquad (3.2.15) \\ &Q'=-\varkappa N-\alpha_{21}u-\alpha_{22}v-\alpha_{23}\varphi+\rho\lambda^2\phi v, \\ &N'=\varkappa Q-\alpha_{31}u-\alpha_{32}v-\alpha_{33}\varphi+\rho\lambda^2\phi u, \end{aligned}$$

where the subscript "1" has been dropped.

3.2.5. The momentless precritical state and the relevant conditions of out-of-plane buckling

The arch under consideration may also loose its stability by out-of-pane buckling. We denote the angles of rotation of the cross-section about the tangential direction **t** and about the normal direction **n** by γ and β respectively. Binormal displacement is denoted by $\bar{w}$ which is negative if its sense coincides with the binormal **b=t×n**.

The system of equations describing small out-of-plane vibrations will be written in dimensionless form as follows:

$$w'=-\beta,\qquad \beta'=S_n^{-1}M_n-\varkappa\gamma,\qquad \gamma'=C^{-1}M_t+\varkappa\beta,$$
$$M_t'=-m_t+\varkappa M_n,\qquad M_n'=-m_n+K+N_o\beta-\varkappa M_t,\tag{3.2.16}$$
$$K'=-r^{*}+\varkappa N_o\gamma-N_o'\beta+\rho\lambda^2\phi w,$$

where the material (true) shear force Q_b^{*} has been replaced by the spatial shear force K defined by:

$$K=Q^{*}-N_o\beta.\tag{3.2.17}$$

The constitutive relations have been assumed in a form which accords with those presented by E.L.Nikolai [57], S.D.Ponomarev et al.[58] and M.Ojalvo, E.Demuts and F.Tokarz [59]. M_n denotes the dimensionless bending moment, while S_n and C denote dimensionless flexural and torsional stiffnesses respectively:

$$S_n=EI_n/E_oI_o,\qquad C=GI_t/E_oI_o.\tag{3.2.18}$$

m_n denotes the distributed external couple around **n**; m_t denotes the distributed external couple around **t** (twisting); r^{*} denotes a distributed external load acting in the direction $\mathbf{b}^{*}$.

The boundary conditions may be written in the form (2.2.11). In the simplest case of an arch clamped rigidly at both ends we have: $w(0)=\beta(0)=\gamma(0)=w(1)=\beta(1)=\gamma(1)=0$.

3.3. A general statement of the optimization problem

3.3.1. Introductory remarks

The magnitude of the cross-section $\phi(s)$,its individual dimensions or the shape of the axis $\Theta(s)$ may serve as basic design variables in arch optimization problems. Critical loading will be assumed here as the basic constraint, but some other constraints, mainly related to strength or vibrations, will also be introduced. The volume V will be the design objective to be minimized.

Geometrical characteristics of cross-sections of curved bars depend on the curvature $\bar{\varkappa}$. For example, the generalized moment of inertia $I_{2\varkappa}$, (3.2.6), for a square cross-section $a^2=A$ is as follows:

$$I_{2\varkappa} = \frac{\sqrt{A}}{\bar{\varkappa}^3} \ln \frac{2+\bar{\varkappa}\sqrt{A}}{2-\bar{\varkappa}\sqrt{A}} - \frac{A}{\bar{\varkappa}^2} \qquad (3.3.1)$$

For $\bar{\varkappa}\sqrt{A} \ll 1$ we may replace (3.3.1) by the classical formula for a straight bar, namely $I_{2\varkappa}=I=A^2/12$. Within the range $\bar{\varkappa}\sqrt{A} \le 0.20$, $(a/R \le 0.20)$, the relative error of the classical formula does not exceed one per cent. Hence, in what follows, we employ the formulae:

$$S_{Mb} = \Xi\, \phi^{\nu_b}, \qquad S_{Mn} = \Xi\, \phi^{\nu_n}, \qquad (3.3.2)$$

where $\nu=1,2,3$; either $\nu_b=\nu_n=2$ (similar sections), or $\nu_b=1$, $\nu_n=3$ (variable width), or $\nu_b=3$, $\nu_n=1$ (variable depth).

When considering out-of-plane buckling of arches we must also use the torsional stiffness $\bar{C}$. For a circular cross-section $\bar{C}=G\pi R^4/2$, whereas for a rectangular cross-section of width $\bar{b}$ and depth $\bar{h}$ we may substitute:

$$\bar{C} = G\bar{h}\bar{b}^3\left[\frac{1}{3} - \frac{64}{\pi^5}\left(\frac{\bar{b}}{\bar{h}}\right)\sum_{m=1,3,5\ldots}^{\infty}\frac{1}{m^5}\,\mathrm{tgh}\left(\frac{m\pi\bar{h}}{2\bar{b}}\right)\right]. \qquad (3.3.3)$$

In most cases the first term of the series in (3.3.3) gives sufficient accuracy and, in what follows, such an approximate formula will be used.

3.3.2. The general solution

In order to use Pontryagin's maximum principle the nonlinear equations of the precritical state (i.e. the equations (3.2.8) and (3.2.9) without the inertia forces) may be written in the form (1.3.1) and (1.3.3), while the equations of linear vibration of the arch (3.2.11), (3.2.12), (3.2.15) or (3.2.16) (in the vicinity of the precritical state of equilibrium) and the relevant boundary conditions may be written in the form (1.3.2) and (1.3.4). Assuming multimodal formulation of the optimization problem and taking into consideration L modes of vibration (or buckling) we can write the Hamiltonian in general form as follows:

$$H = \sum_{l=1}^{L} [\, \chi_\alpha^{(l)} G_\alpha^{(l)} + \psi_\alpha^{(l)} A_{\alpha\beta}^{(l)} Z_\beta^{(l)}] + \psi_o \phi \,, \tag{3.3.4}$$

where the summation convention over α and β holds and ψ_o is a constant.

The functions χ_i and ψ_i satisfy the linear differential equations of the adjoint state (1.3.6) and (1.3.7) and the relevant transversality condition (1.3.8)-(1.3.11).

The optimality condition (1.2.15) here takes the general form:

$$\sum_{l=1}^{L} \left[\chi_\alpha^{(l)} \frac{\partial G_\alpha^{(l)}}{\partial \phi_j} + \psi_\alpha^{(l)} \frac{\partial A_{\alpha\beta}^{(l)}}{\partial \phi_j} Z_\beta^{(l)} \right] + \psi_o \frac{\partial \phi}{\partial \phi_j} = 0, \tag{3.3.5}$$

in which a vector of the design variables ϕ_j has been introduced.

3.4. Funicular arches

3.4.1. In-plane buckling

The momentless precritical state (3.2.13) results in the elimination of eqs.(1.3.1) and (1.3.6) and of the terms with $\chi_\alpha^{(l)}$ in the optimality condition (3.3.5). The state equations are confined to (3.2.15). Introducing new adjoint variables $\bar{u}$, $\bar{v}$, $\bar{\varphi}$, $\bar{M}$, $\bar{Q}$, $\bar{N}$ - defined for each buckling mode l by the formulae:

$$\psi_u=-k\bar{N}, \quad \psi_v=-k\bar{Q}, \quad \psi_\phi=k\bar{M}, \quad \psi_M=-k\bar{\varphi}, \quad \psi_Q=k\bar{v}, \quad \psi_N=k\bar{u}, \tag{3.4.1}$$

(where $k=k_l>0$) - we may write the adjoint equations (1.3.7) in the form:

$$\begin{aligned} &\bar{u}'=\varkappa\bar{v}, \quad \bar{v}'=-\bar{\varphi}-\varkappa\bar{u}, \quad \bar{\varphi}'=-S_M^{-1}\bar{M}, \\ &\bar{M}'=-\bar{Q}-N_o\bar{\varphi} +\alpha_{33}\bar{u}+\alpha_{23}\bar{v}-\alpha_{13}\bar{\varphi}-\alpha\rho\lambda^2 J\bar{\varphi}, \\ &\bar{Q}'=-\varkappa\bar{N}-\alpha_{32}\bar{u}-\alpha_{22}\bar{v}-\alpha_{12}\bar{\varphi}+\rho\lambda^2\phi\bar{v}, \end{aligned} \tag{3.4.2}$$

$$\bar{N}'=\varkappa\bar{Q}-\alpha_{31}\bar{u}-\alpha_{21}\bar{v}-\alpha_{11}\bar{\varphi}+\rho\lambda^2\phi\bar{u}.$$

Moreover, the optimality condition (3.3.5) is simplified to:

$$\sum_{l=1}^{L} k_l \left[\frac{\partial S_M}{\partial \phi_j} \frac{M_l \bar{M}_l}{S_M^2} + \rho\lambda^2 \frac{\partial \phi}{\partial \phi_j} (u_l \bar{u}_l + v_l \bar{v}_l) + \alpha\rho\lambda^2 \frac{\partial J}{\partial \phi_j} \varphi_l \bar{\varphi}_l \right] +$$

$$+ \psi_o \frac{\partial \phi}{\partial \phi_j} = 0. \tag{3.4.3}$$

For all conservative loads (e.g. dead loads) - i.e. satysfying the conditions:

$$\alpha_{11}=\alpha_{12}=\alpha_{23}=\alpha_{33}=0, \qquad \alpha_{21}=\alpha_{32}, \tag{3.4.4}$$

-we obtain the adjoint boundary-value problem which is identical with the original problem (3.2.15). The adjoint variables (with bars) are therefore identical with the original variables (without bars), which renders separate solving of (3.4.2) unnecessary.

Many effective unimodal solutions to optimization problems of symmetric parabolic arches under conservative loadings were given by I.Tadjbakhsh and M.Farshad [60].

3.4.2. The in-plane buckling of a circular arch under hydrostatic loading.

One control function.

For circular arches under constant hydrostatic loading P ($p_o=0, q_o=P, N_o=-P/\varkappa$), we obtain $\alpha_{33}=P$, and the equations of state (3.2.15) are no longer self-adjoint (as understood by Pontryagin's maximum principle). However, J.Błachut and A.Gajewski [44] have proved the existence of a linear transformation from the state variables (u,v,φ,M,Q,N) into new state variables, resulting in a self-adjoint system of equations. For example, in the simplest case of a circular arch, this transformation introducing new state variables χ, K, Z, is of the form:

$$\chi=-\varphi-\varkappa u, \quad K=Q+Pu, \quad Z=N+\varkappa M. \tag{3.4.5}$$

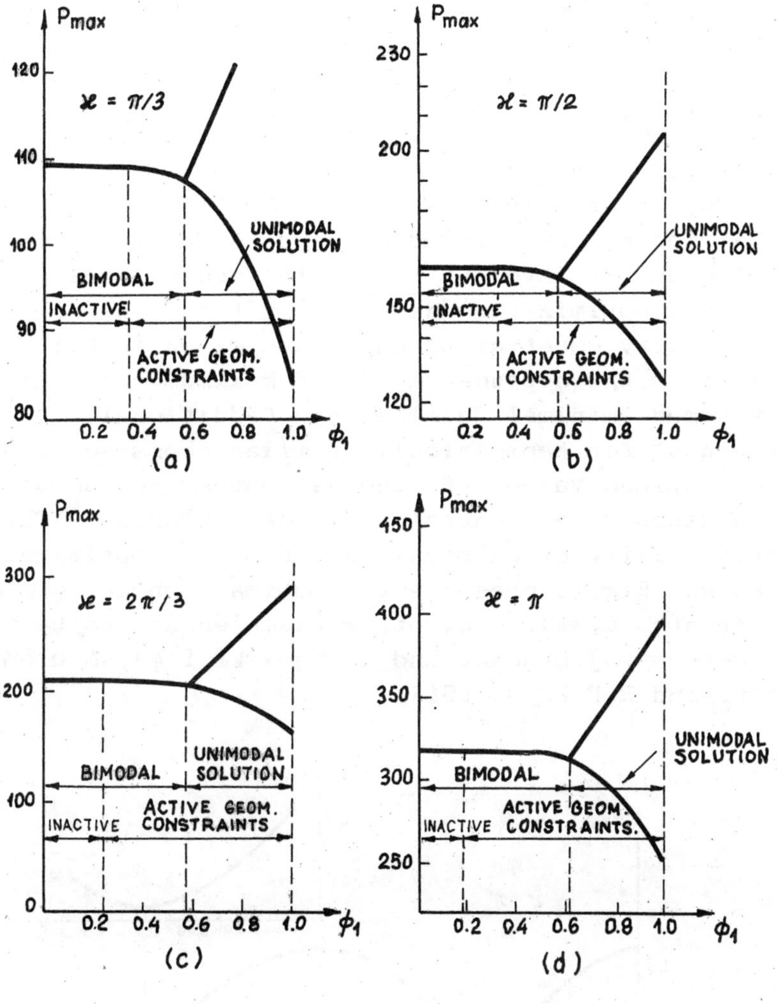

Fig. 14. Maximal (first) and second critical forces in terms of lower geometrical constraint

The optimality condition (3.4.3) is therefore here reduced to the form:

$$\sum_{l=1}^{L} k_l \left[\frac{\partial S_M}{\partial \phi_j} \frac{M_l^2}{S_M^2} + \rho\lambda^2 \frac{\partial \phi}{\partial \phi_j}(u_l^2+v_l^2)+\alpha\rho\lambda^2 \frac{\partial J}{\partial \phi_j}(\chi_l + \varkappa u_l)^2 \right] +$$

$$+ \psi_0 \frac{\partial \phi}{\partial \phi_j} = 0. \qquad (3.4.6)$$

Optimization of circular funicular arches, clamped at both ends, under stability and vibration constraints has been carried out by J. Błachut and A. Gajewski [44]. Both symmetric and antisymmetric buckling modes were considered and bimodal optimization was allowed for. The results for $\lambda=0$ (stability constraints only) are given in Fig. 14 and 15. Fig. 14 gives the dependence of the maximum critical load on the assumed minimal dimension (additional geometrical constraints) for geometrically similar cross-sections, $\nu=2$, and for various values of the steepness parameter $\varkappa$ (in fact, $\varkappa$ denotes the central angle of the arch). The ranges of applicability of unimodal and bimodal optimization are also shown. Fig. 15 presents the optimal shape of a semicircular arch ($\varkappa=\pi$). Many other examples are to be found in the papers by: J. Błachut and A. Gajewski [44], N. Olhoff [61], N. Olhoff and R. H. Plaut [51].

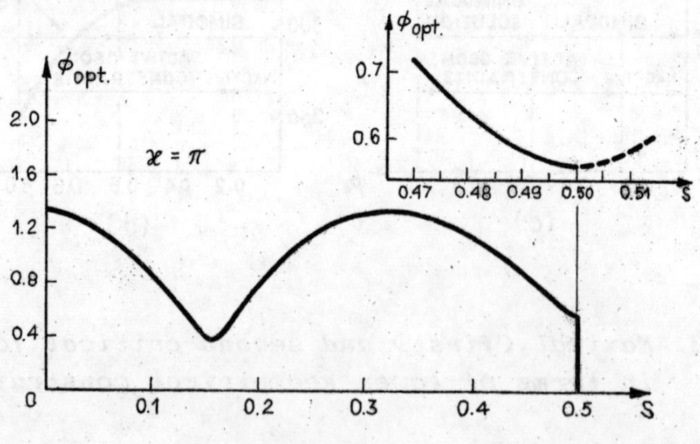

Fig. 15. Optimal shape of a semicircular arch

3.4.3. The simultaneous in-plane and out-of plane buckling of a circular arch.

One control function.

The solutions discussed above may turn out to be incorrect for some values of the parameters. For example, if the cross-section is of the form of a slender rectangle (e.g. obtained for variable depth, $\nu=3$), then the critical load corresponding to out-of-plane buckling may be lower than that assumed in optimization for in-plane buckling. Then, to obtain correct results, we have to take out-of-plane buckling into consideration and employ three- or four-modal optimization, since in both cases symmetric or anti-symmetric modes may appear.

We confine our considerations to the buckling of circular funicular arches under uniformly distributed radial loads P maintaining a constant direction in space (conservative loads which are perpendicular to the undeformed axis). Then the static criterion of stability is always sufficient, $\lambda=0$; moreover, in the equations (3.2.15) we have: $\alpha_{\iota j}=0$, $N_o=-P/\varkappa$, whereas in eqs. (3.2.16) $r^{*}=\varkappa N_o\gamma$, $m_t=m_n=0$, $N_o'=0$. Both systems of equations are self-adjoint and the optimality condition (3.3.5) may be written as follows:

$$\sum_{\iota=1}^{2} k_\iota \left(\frac{\partial S_{Mb}}{\partial \phi_j} \frac{M_{b\iota}^2}{S_{Mb}^2} \right) + \sum_{\iota=3}^{4} k_\iota \left(\frac{\partial S_{Mn}}{\partial \phi_j} \frac{M_{n\iota}^2}{S_{Mn}^2} + \frac{\partial C}{\partial \phi_j} \frac{M_{t\iota}^2}{C^2} \right) +$$

$$+ \psi_o \frac{\partial \phi}{\partial \phi_j} = 0, \tag{3.4.7}$$

where $\iota=1,2$ corresponds to the symmetric and antisymmetric in-plane buckling, and $\iota=3,4$ to the symmetric and anti-symmetric out-of-plane buckling modes, respectively. In general, eqs. (3.4.7) are transcendental with respect to design variables ϕ_j.

The results of numerical calculations for rectangular, geometrically similar cross-sections $h/b=\eta=$const. (given), as well as for h=const. and for b=const., have been obtained by B. Bochenek and A. Gajewski [47]. In the first of the above variants the design variable (e.g. h) may be expressed directly in terms of the state variables, since the series in (3.3.3) has a constant value, depending on η only:

$$\phi = \left[\sum_{l=1}^{2} (\Lambda_l M_{bl}^2) + \sum_{l=3}^{4} \Lambda_l (\varphi_1 M_{nl}^2 + \varphi_1 \varphi_2 M_{tl}^2) \right]^{\frac{1}{3}}, \quad (3.4.8)$$

where

$$\Lambda_l = \frac{-k_l}{\psi_0}, \quad \varphi_1 = \eta^{-2}, \quad \varphi_2 = \frac{1+\nu}{6f(\eta)}, \quad \eta = h/b,$$

$$f(\eta) = \frac{1}{3} - \frac{64}{\pi^5 \eta} \, \mathrm{tgh}\left(\frac{\pi\eta}{2}\right), \quad (3.4.9)$$

and ν denotes Poisson's coefficient.

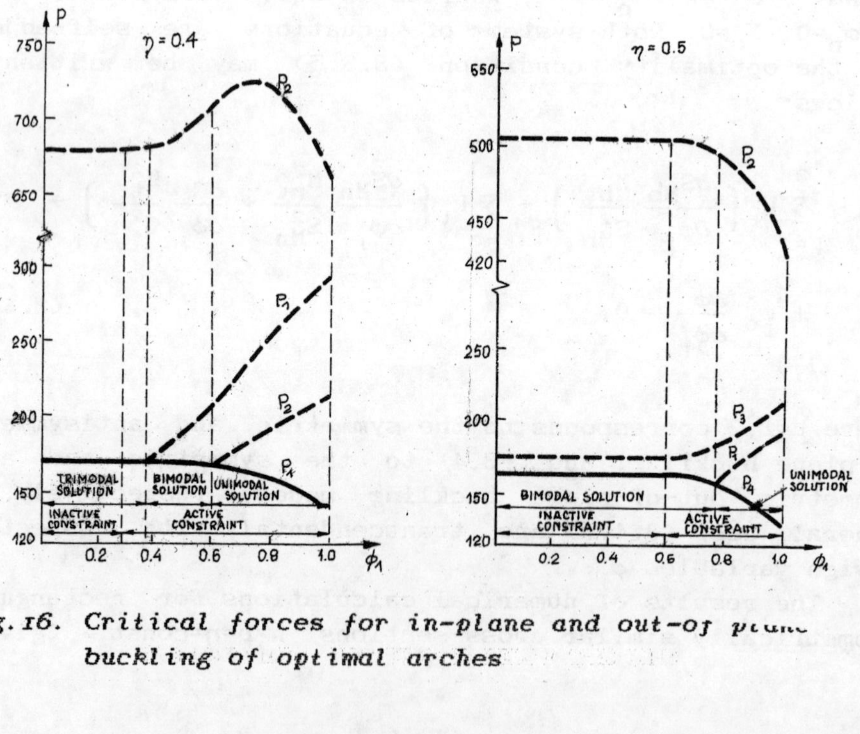

Fig. 16. Critical forces for in-plane and out-of plane buckling of optimal arches

Fig.16 shows the dependence of the critical force for optimal arches on the geometrical constraint ϕ_1 (for η=0.5 and η=0.4), while Fig.17 gives the domains of uni-, bi-, and tri-modal optimization in the system of coordinates (η, ϕ_1).

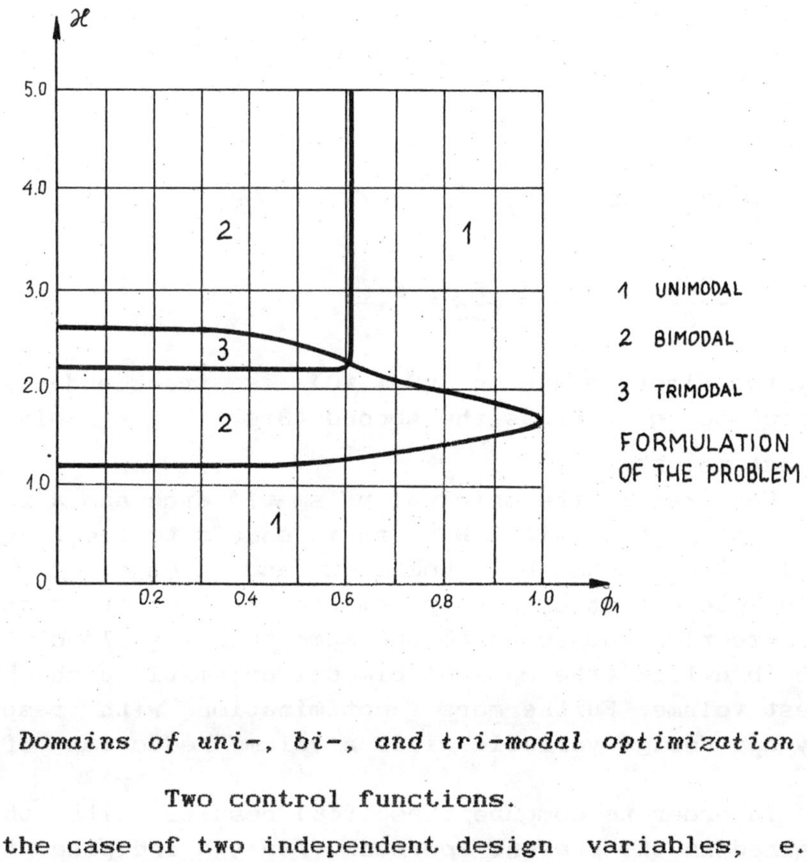

Fig. 17. Domains of uni-, bi-, and tri-modal optimization

Two control functions.

In the case of two independent design variables, e.g. h(s) and b(s) for a rectangular cross-section, the calculations are much more complicated. For each step of numerical integration of the state equations one has to solve the system of transcendental equations (3.4.7) with respect to h and b. However, the optimality conditions may here be reduced to the form:

$$\eta^2 = \frac{\mu_1 M_{b1}^2 + \mu_2 M_{b2}^2}{\mu_3 M_{n3}^2 + \mu_4 M_{n4}^2 + c\,(\mu_3 M_{t3}^2 + \mu_4 M_{t4}^2)\ \dfrac{f - \eta \frac{df}{d\eta}}{f^2}} \tag{3.4.10}$$

$$b^6 = \frac{1}{\Lambda \eta^4} \Big\{ \mu_1 M_{b1}^2 + \mu_2 M_{b2}^2 + \tag{3.4.11}$$

$$+ \eta^2 \left[3\,(\mu_3 M_{n3}^2 + \mu_4 M_{n4}^2) + c\,(\mu_3 M_{t3}^2 + \mu_4 M_{t4}^2)\ \frac{3f - \eta \frac{df}{d\eta}}{f^2} \right] \Big\} ,$$

where:

$$\mu_1 = k_1, \quad \mu_2 = k_2/k_1, \quad \mu_3 = k_3/k_1, \quad \mu_4 = k_4/k_1,$$

$$\Lambda = \frac{-\psi_o}{k_1}, \qquad c = \frac{E}{12G} = \frac{1+\nu}{6} . \tag{3.4.12}$$

Only the first equation (3.4.10) is transcendental with respect to η, whereas the second (3.4.11) is solved with respect to b^6.

For $\varepsilon = \pi/2$ the original prismatic arch has a rectangular cross-section with b/h ratio equals to 1.62. From the conclusions reached by B. Bochenek and A. Gajewski [48] it would appear that of all prismatic arches of rectangular cross-section subjected to the same buckling load the one with $b/h = 1.62$ (the optimal bimodal prismatic arch) has the lowest volume. Furthermore, optimization with respect to only one design variable gives a volume reduction of about 10% .

In order to compare these [48] results with the ones obtained in the present approach (for two independent design functions) detailed calculations for $\varepsilon = \pi/2$ have been presented by B. Bochenek and A. Gajewski [48]. The selected optimal functions $\tilde{b}(s)$ and $\tilde{h}(s)$ and the shape of arch for certain geometrical constraints are shown in Fig. 18.

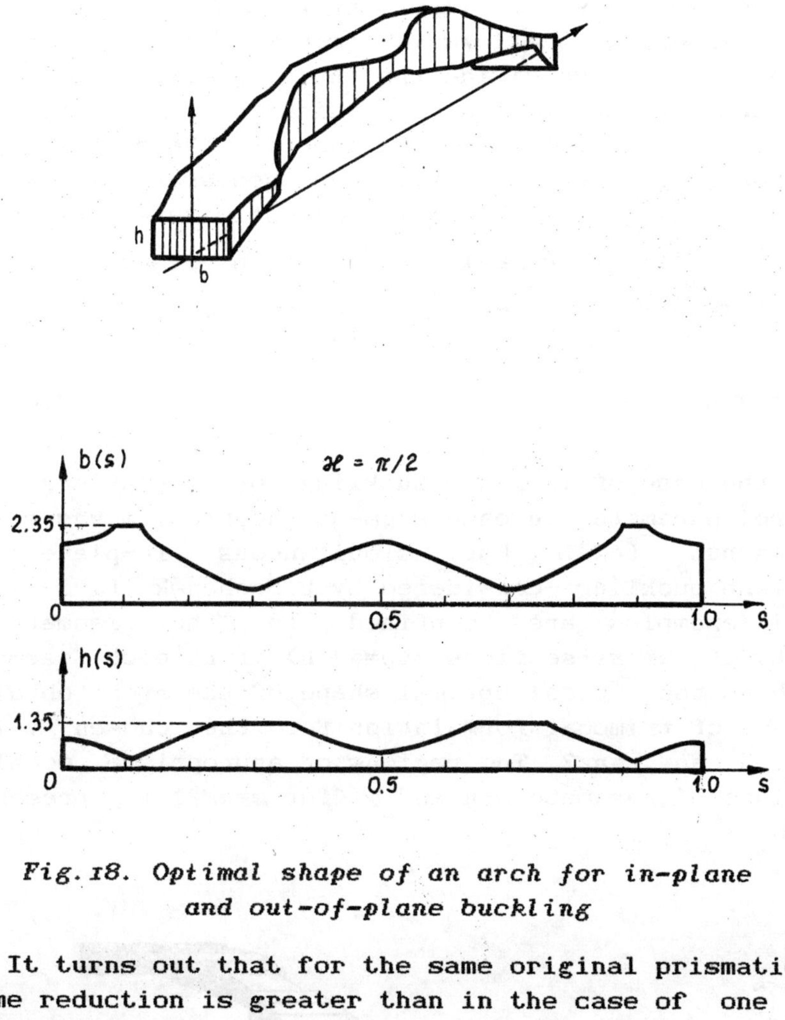

Fig. 18. Optimal shape of an arch for in-plane and out-of-plane buckling

It turns out that for the same original prismatic arch volume reduction is greater than in the case of one design variable. On the other hand optimal shapes are more complicated as far as mass distribution is concerned.

3.4.4. The multimodal parametrical optimal design of funicular arches

The optimization problems presented in sections 3.4.2., 3.4.3., and 3.4.4. deal with optimal shapes whose mass is distributed in a continuous manner. However, the problem of optimal shaping by the formation of a finite number of prismatic segments is very important from a practical point of view. This approach allows very simple production of such optimal arches. Let us divide the whole length [0,1] of the arch into n equal pieces in which ϕ_i , $i=1..n$, have constant values. According to the general theory presented by L.S.Pontryagin et al. [2] the necessary optimality condition in the case of parametrical optimization requires an extremum of $\bar{H}$ to be calculated with respect to the design parameters ϕ_i, where $\bar{H}$ is defined as:

$$\bar{H} = \int_0^1 H \, ds . \tag{3.4.13}$$

For the case of in-plane buckling (or vibration) with one control parameter in each segment the problem was solved by J.Błachut [62]. For simultaneous in-plane and out-of-plane buckling -considered by B.Bochenek [49]-, the numerical examples are confined to the geometrical similarity of cross-sections ($b=\gamma$ h) for each segment. Fig.19 shows the typical optimal shape of the arch obtained in the case of trimodal formulation for the chosen values $n=10$, $\gamma=2.5$ and $\varepsilon=\pi/2$. The regions of appropriate modality in the plane of parameters n and γ (for $\varepsilon=\pi/2$) is presented in Fig.20.

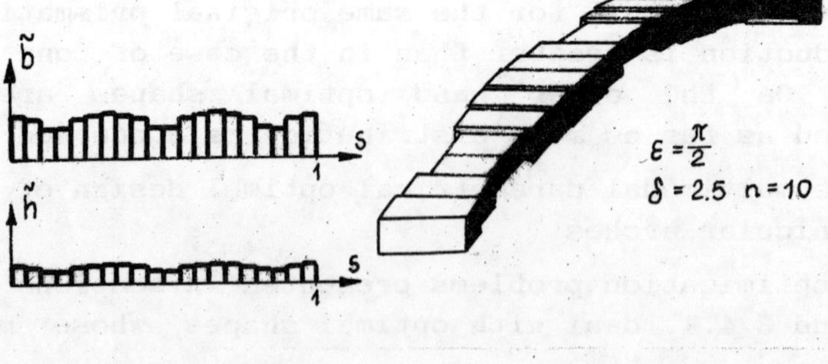

Fig.19. The optimal arch: bimodal solution

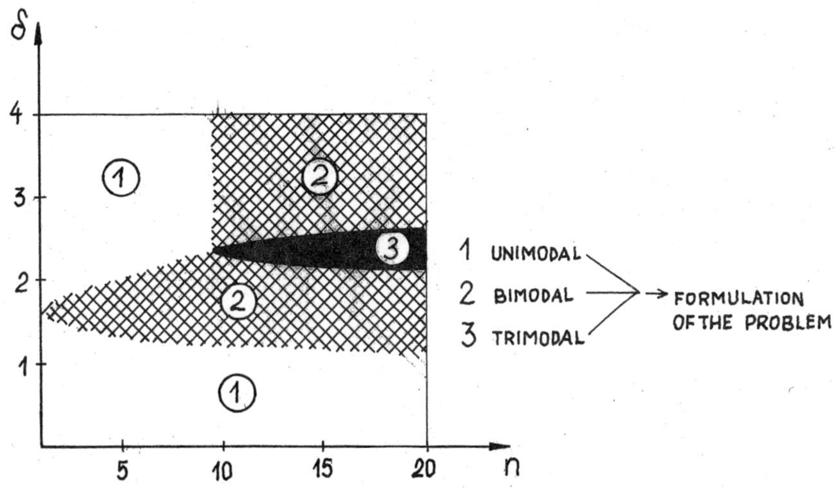

Fig. 20. Domains of uni-, bi-, and tri-modal optimization in the plane of parameters n *and* δ

3.5. Extensible arches optimized for in-plane bifurcation and snap-through

Omitting the assumption of a funicular state and introducing stretchability of axis brings interesting new problems to the optimization of arches. Buckling may thus occur either by bifurcation or by snap-through, depending on the support conditions, the initial shape of the axis and the steepness of the arch. Geometrically non-linear stability is necessary here and in most cases bimodal optimization should be employed, equating the critical forces of bifurcation and snap-through.

A full analysis allowing for small vibrations imposed on the precritical state and based on eqs. (3.2.3),(3.2.4) and (3.2.11),(3.2.12) was given by J. Błachut and A. Gajewski [45] and by J. Błachut [46]. The optimality condition (3.3.5) was simplified by dropping the terms with adjoint variables of the precritical state $\chi_{\alpha}^{(1)}$. It was shown that the error was much less than one per cent. Numerical calculations were carried out for clamped-clamped circular arches under hydrostatic pressure P (p_0=P $\sin\varphi_0$, q_0=P $\cos\varphi_0$, p_1=P $\varphi_1\cos\varphi_0$ q_1=-P φ_1 $\sin\varphi_0$, the non-self-adjoint problem), for geometrically similar cross-sections (ν=2), and for various slenderness ratios. Very laborious calculations made it possible to find optimal shapes for various vibration frequencies and various loading parameters. In the limiting case λ=0 the optimal shape for simultaneous bifurcation and snap-through was obtained. The results for h_0/l=1/20 and $\varkappa$=1/R=π/2 are shown in Fig.21.

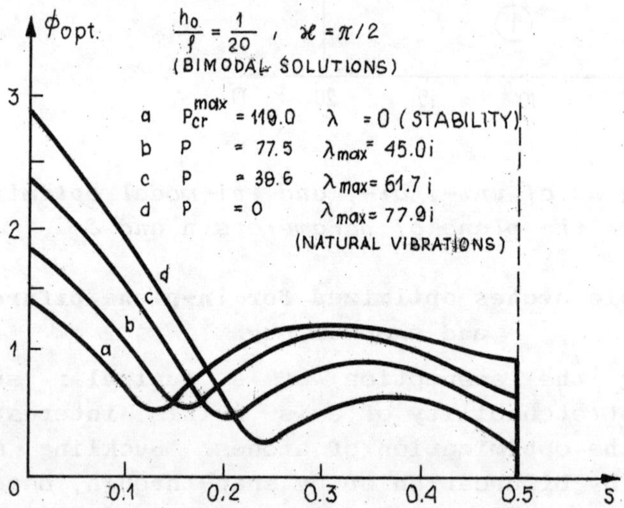

Fig.21. Optimal extensible arches for buckling and vibration

4. NON-CONSERVATIVE OPTIMIZATION PROBLEMS OF ANNULAR PLATES

4.1. Introductory remarks

The optimal structural design of plate elements compressed by non-conservative forces is usually treated within the framework of aeroelastic problems and is seen as the minimization of the plate volume for a constant critical velocity of gas flow (calculated for flutter or bifurcation). Previous solutions of the problem have been confined to simple panels or rectangular plates. They are presented and discussed e.g. by B.L.Pierson [63] and by A.P.Seyranian [64].

The optimal structural design of annular plates compressed by non-conservative forces has as yet not been investigated. The papers by J.C.Frauenthal [65], V.B.Grinev and A.P.Filippov [66], K.Rzegocińska-Pełech and Z.Waszczyszyn [67] deal with circular or annular plates compressed by conservative forces of constant direction in space during buckling.

The aim of this section is to present a new and -from a theoretical point of view- interesting optimization problem of annular plates compressed by uniformly distributed non-conservative forces. Both the precritical membrane state and the small transverse vibration are taken into account. In general, the kinetic criterion of stability has to be applied.

The vibration and stability of annular plates of variable thickness subjected to a follower distributed load have been considered -without any optimization problem-by T.Irie, G.Yamada and Y.Kaneko [68] who have not, however, taken into account the membrane precritical state.

4.2. The governing equations of state in polar coordinates

Let us consider an isotropic elastic annular plate loaded at the inner boundary r=a by $\bar{P}_a$ and at the outer

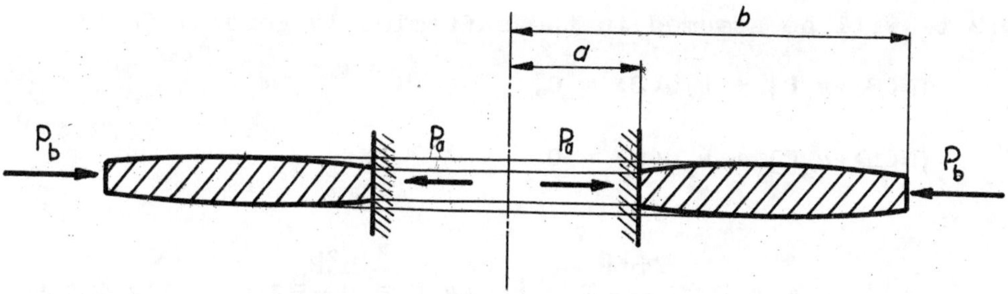

Fig.22. Non-uniform annular plate subjected to a follower force

boundary r=b by $\bar{P}_b$ (Fig.22). The in-plane loadings $\bar{P}_a$ and $\bar{P}_b$ are assumed to be uniformly distributed and to be positive if they are compressive. We also assume that the plate thickness h(r) is circularly symmetric. Then the precritical state is also circularly symmetric and is described by the membrane forces $\bar{N}_r$ and $\bar{N}_\Theta$ (positive in the case of tension) and by a radial displacement $\bar{u}(r)$-all of which are determined by the solution of the following linear boundary value problem written in the dimensionless form suggested by V.B. Grinev and A.P. Filippov [66]:

$$u' = \frac{1-\nu^2}{xS} N - \frac{\nu}{x} u, \qquad N' = \frac{\nu}{x} N + \frac{S}{x} u, \tag{4.2.1}$$

where:

$$x = \frac{r}{b}, \quad u = \frac{\bar{u}}{\alpha b}, \quad N = \frac{b^2}{D_o} x N_r, \quad S = \frac{\bar{S}}{S_o}, \quad \bar{S} = Eh, \tag{4.2.2}$$

$$S_o = Eh_o, \quad D_o = \frac{Eh_o^3}{12(1-\nu^2)}, \quad \alpha = \frac{D_o}{S_o b^2} = \frac{1}{12(1-\nu^2)} \left(\frac{h_o}{b}\right)^2,$$

and the prime denotes differentiation with respect to x.

The remaining membrane force N_Θ may be evaluated from the equilibrium condition:

$$\bar{N}_\Theta = D_o N'. \tag{4.2.3}$$

The boundary conditions relevant to the equations (4.2.1) will be assumed in the sufficiently general form:

$$\begin{aligned} &[N(\beta)+\pi_1 P] + \varkappa_1 u(\beta) = 0, \\ &[N(1)+\pi_2 P] + \varkappa_2 u(1) = 0, \end{aligned} \tag{4.2.4}$$

where:

$$\beta = \frac{a}{b}, \quad P_a = \pi_1 P = \frac{\beta b^2 \bar{P}_a}{D_o}, \quad P_b = \pi_2 P = \frac{b^2 \bar{P}_b}{D_o}, \tag{4.2.5}$$

the constant parameters $\varkappa_1$ and $\varkappa_2$ characterize elastic clampings with respect to the radial displacements of the edges of the plate, while the parameters π_1 and π_2 determine the ratio of the loads acting on the external and internal edges. It has therefore been assumed that the only load parameter is the value of P.

The well-known equation of small vibrations of plates superimposed on the precritical state may be transposed to the following set of four ordinary differential equations -after a separation of time, radial and circumferential variables- as suggested by V.B.Grinev and A.P.Filippov [66]:

$$w' = \varphi,$$

$$\varphi' = \left(\frac{\nu m^2}{x^2} + \frac{N}{2xD}\right) w - \frac{\nu}{x}\varphi - \frac{1}{xD} M,$$

$$M' = \left[(1-\nu)(3+\nu)\frac{m^2 D}{x^2} + \frac{Su}{2x}\right] w - (1-\nu)(1+\nu+2m^2)\frac{D}{x}\varphi + $$
$$+ \frac{\nu}{x} M + Q, \tag{4.2.6}$$

$$Q' = \left[(1-\nu)(2+m^2+\nu m^2)\frac{m^2 D}{x^3} - \frac{N^2}{4xD} + \frac{m^2 Su}{x^2} - \rho\lambda^2 x\phi\right] w -$$
$$- \left[(1-\nu)(3+\nu)\frac{m^2 D}{x^2} + \frac{Su}{2x}\right]\varphi + \left(\frac{\nu m^2}{x^2} + \frac{N}{2xD}\right) M,$$

where m denotes the number of circumferential waves, λ - the non-dimensional square of the frequency of vibrations, ρ - the non-dimensional density of material and $\phi=h/h_0$ - the control variable. The stiffnesses S and D are related to ϕ by the relations:

$$S = \phi, \qquad D = \phi^{\delta}, \tag{4.2.7}$$

in which $\delta=1$ for sandwich plates and $\delta=3$ for solid plates. We will add the following general, linear and homogeneous boundary conditions to the equations (4.2.6):

$$\frac{1}{\beta}\left[M(\beta)-\frac{1}{2}N(\beta)w(\beta)\right]+\varkappa_3\varphi(\beta)+\alpha_1\pi_1 Pw(\beta)+\beta_1\pi_1 P\varphi(\beta)=0,$$

$$\frac{1}{\beta}\left[Q(\beta)+\frac{1}{2}N(\beta)\varphi(\beta)\right]+\varkappa_4 w(\beta)+\alpha_2\pi_1 Pw(\beta)+\beta_2\pi_1 P\varphi(\beta)=0, \tag{4.2.8}$$

$$[M(1)-\frac{1}{2}N(1)w(1)]+\varkappa_5\varphi(1)+\alpha_3\pi_2 Pw(1)+\beta_3\pi_2 P\varphi(1)=0,$$

$$[Q(1)+\frac{1}{2}N(1)\varphi(1)]+\varkappa_6 w(1)+\alpha_4\pi_2 Pw(1)+\beta_4\pi_2 P\varphi(1)=0,$$

where the constants $\varkappa_3 \ldots \varkappa_6$ characterize elastic rigidities of edges with respect to transversal and rotational displacements, and the constants $\alpha_1 \ldots \alpha_4$ and $\beta_1 \ldots \beta_4$ characterize the behaviour of distributed loads during vibrations.

The boundary value problems (4.2.1),(4.2.4) and (4.2.6)-(4.2.8) determine the so-called characteristic curves, i.e. the relations between the loading parameter P and the frequency of vibration λ (see also section 2.2.2). The value of P corresponding to the equalization of two frequencies of vibration (not equal to zero) is the critical value of flutter, whereas the value of P corresponding to λ=0 is the critical value of bifurcation. In non-conservative cases of loading both these possibilities may occur. This depends not only on the behaviour of the load during vibration but also on the modes of support, on the variation of plate stiffness and so on.

4.3. The optimization problem

The optimization problem will here be formulated as the maximization of a critical load under the constraint of constant volume. The method of solution will be based on the sensitivity analysis which has already been described in section 1.3.2. The non-linear precritical state equations (1.3.1),(1.3.3) are here confined to the equations of the membrane state (4.2.1),(4.2.4). The boundary value problem of small vibrations (1.3.2),(1.3.4) takes the form of (4.2.6) and (4.2.8).

In order to obtain more symmetrical formulae we introduce new adjoint variables by the following substitutions:

$$\chi_1=-\bar{N},\quad \chi_2=\bar{u},\quad \psi_1=-\bar{Q},\quad \psi_2=\bar{M},\quad \psi_3=-\bar{\varphi},\quad \psi_4=\bar{w}. \qquad (4.3.1)$$

Thus, the adjoint equations of the precritical state (1.3.6) may be written in the form:

$$\bar{u}' = \frac{-\nu}{x}\,\bar{u} + \frac{1-\nu^2}{xS}\,\bar{N} - \frac{1}{2xD}(\bar{M}w+M\bar{w}-Nw\bar{w}),$$

$$\bar{N}' = \frac{\nu}{x}\,\bar{N} + \frac{S}{x}\,\bar{u} - \frac{S}{2x}(w\bar{\varphi}+\bar{w}\varphi) + \frac{m^2S}{x^2}\,w\bar{w}, \tag{4.3.2}$$

while the adjoint equations (1.3.7) are identical with the equations (4.2.6).

The following boundary conditions of the adjoint state may be obtained from the transversality condition (1.3.7)-(1.3.10):

$$\bar{N}(\beta)+\varkappa_1\,[\bar{u}(\beta)-\tfrac{1}{2}w(\beta)\bar{\varphi}(\beta)-\tfrac{1}{2}\bar{w}(\beta)\varphi(\beta)]=0,$$

$$\bar{N}(1)+\varkappa_2\,[\bar{u}(1)-\tfrac{1}{2}w(1)\bar{\varphi}(1)-\tfrac{1}{2}\bar{w}(1)\varphi(1)]=0, \tag{4.3.3}$$

and

$$\frac{1}{\beta}[\bar{M}(\beta)-\tfrac{1}{2}N(\beta)\bar{w}(\beta)]+\varkappa_3\bar{\varphi}(\beta)-\beta_2\pi_1P\bar{w}(\beta)+\beta_1\pi_1P\bar{\varphi}(\beta)=0,$$

$$\frac{1}{\beta}[\bar{Q}(\beta)+\tfrac{1}{2}N(\beta)\bar{\varphi}(\beta)]+\varkappa_4\bar{w}(\beta)+\alpha_2\pi_1P\bar{w}(\beta)-\alpha_1\pi_1P\bar{\varphi}(\beta)=0,$$

$$[\bar{M}(1)-\tfrac{1}{2}N(1)\bar{w}(1)]+\varkappa_5\bar{\varphi}(1)-\beta_4\pi_2P\bar{w}(1)+\beta_3\pi_2P\bar{\varphi}(1)=0, \tag{4.3.4}$$

$$[\bar{Q}(1)+\tfrac{1}{2}N(1)\bar{\varphi}(1)]+\varkappa_6\bar{w}(1)+\alpha_4\pi_2P\bar{w}(1)-\alpha_3\pi_2P\bar{\varphi}(1)=0,$$

which, in general, are different from (4.2.4) and (4.2.8).

However, it may be observed that the boundary conditions of the adjoint critical state (4.3.4) and the original boundary conditions (4.2.8) become identical if the following conditions are satisfied:

$$\alpha_1+\beta_2=0, \qquad \alpha_3+\beta_4=0. \tag{4.3.5}$$

In such a case the boundary value problem (4.2.6),(4.2.8) is self-adjoint.

Confining our considerations to the annular plate rigidly clamped at the inner edge, and compressed by the distributed, non-conservative forces acting at the free outer edge we have:

$$\varkappa_1 \Longrightarrow \infty, \quad \varkappa_3 \Longrightarrow \infty, \quad \varkappa_4 \Longrightarrow \infty, \quad \varkappa_2=0, \quad \varkappa_5=0,$$
$$\varkappa_6=0, \quad \pi_1=0, \quad \pi_2=1, \quad \alpha_3=0, \quad \alpha_4=0, \quad \beta_3=0, \quad \beta_4=\eta, \tag{4.3.6}$$

where the parameter η characterizes the behaviour of the load during vibration (the so-called tangency coefficient). For $\eta=0$ we have a distributed conservative load of constant direction, while for $\eta=1$ the distributed load is tangential to the slope of the external edge (the follower force).

The basic equation of the sensitivity analysis (1.3.11) - for the parameters (4.3.6) - takes the form:

$$\int_\beta^1 g(x)\ \delta\phi\ dx + C_1\delta P + C_2\delta\lambda = 0, \tag{4.3.7}$$

where:

$$g(x) = \frac{dS}{d\phi}\left[-\frac{(1-\nu^2)}{xS^2}\,N\bar{N} + \frac{1}{x}\,u\bar{u} + \frac{m^2u}{x^2}\,w\bar{w} - \frac{u}{2x}(w\bar{\varphi}+\bar{w}\varphi)\right] +$$
$$+ \frac{dD}{d\phi}\left[-\frac{1}{xD^2}(M-\tfrac{1}{2}Nw)(\bar{M}-\tfrac{1}{2}N\bar{w}) + (1-\nu)(2+m^2+\nu m^2)\frac{m^2}{x^3}w\bar{w} +\right.$$
$$\left. + (1-\nu)(1+\nu+2m^2)\,\frac{1}{x}\varphi\bar{\varphi} - (1-\nu)(3+\nu)\frac{m^2}{x^2}(w\bar{\varphi}+\bar{w}\varphi)\right] +$$
$$- \rho\lambda^2 xw\bar{w}\ , \tag{4.3.8}$$

$$C_1 = \bar{u}(1)-\tfrac{1}{2}w(1)\bar{\varphi}(1)-\tfrac{1}{2}\bar{w}(1)\varphi(1)+\eta\bar{w}(1)\varphi(1), \tag{4.3.9}$$

$$C_2 = -2\rho\lambda\int_\beta^1 x\phi w\bar{w}\ dx\ . \tag{4.3.10}$$

The equation (4.3.7) constitutes the relationship between small variations of the compressive force δP, the frequency of vibration $\delta\lambda$ and the thickness distribution $\delta\phi$. For a given thickness distribution $\phi(x)$- i.e. $\delta\phi=0$ -we obtain the flutter condition -resulting from the relation $\delta P/\delta\lambda=0$- which for the annular plate under consideration takes the form:

$$\int_{\beta}^{1} x\phi w\bar{w}\, dx = 0. \tag{4.3.11}$$

The optimal design of the plate entails an iterational improvement of the plate thickness (control function ϕ) according to the formulae:

$$\phi^{(n+1)} = \phi^{(n)}+\varepsilon(x)\,[\mu_1 g_1(x)+\mu_2 g_2(x)+.....+\Lambda x], \tag{4.3.12}$$

where $\varepsilon(x)$ is an arbitrary positive function usually assumed to be constant (the so called gradient step); where $\mu_1,\mu_2.....$ are constants to be determined from the conditions of equalization of the critical flutter and/or bifurcation loads and where Λ is a constant to be determined from the constant volume condition (1.3.14). The gradients $g_1(x)$, $g_2(x)$,... have to be calculated at critical points of flutter or bifurcation. In both cases the constant C_2 is equal to zero.

The procedure described above leads to a great increase of the critical load (over 3 times). By applying consecutive iterations, the typical effects already known from the optimal design of columns (see for example T.A.Weishaar and R.H.Plaut [19]) such as continuous and discontinuous switchings may be observed. Some new effects connected with the evaluation of characteristic curves depending on the number of circumferential waves may also be observed.

Numerical calculations have been carried out for an annular plate for which β=0.2, ν=0.3 and η=1 making use of the fourth order Runge-Kutta-Gill integration method.

The shapes of the plate and the related characteristic curves obtained in successive iterations are presented in Figs. 23 and 24. It may be seen that further optimization is possible using the multimodal formulation.

More details will be given in the paper by A.Gajewski and P.Cupiał [69].

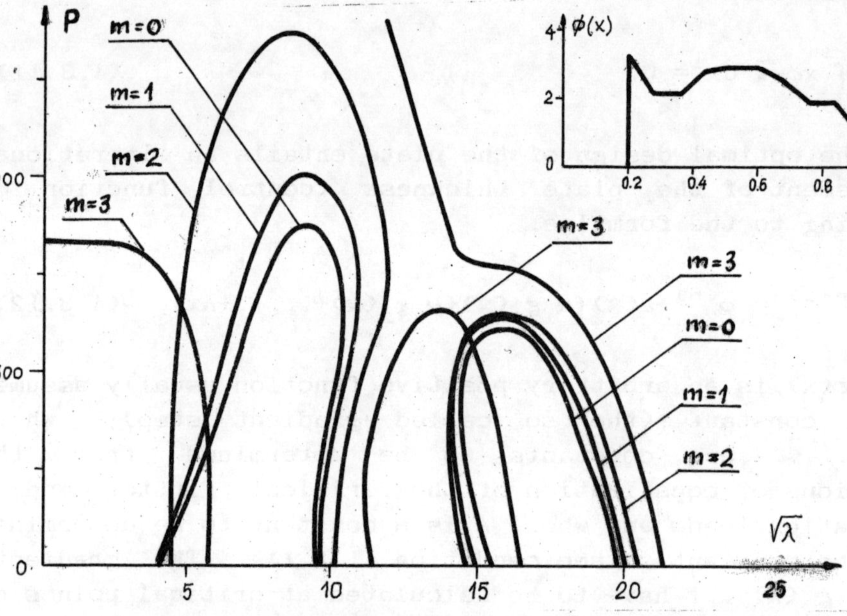

Fig. 23. Characteristic curves for a plate presented in figure

Fig. 24. Characteristic curves for a plate presented in figure

5. THE PARAMETRICAL OPTIMIZATION OF SIMPLE BAR SYSTEMS IN CERTAIN DYNAMIC PROBLEMS

5.1. The parametrical optimization of a viscoelastic column compressed by a follower force with respect to its stability

A cantilever column and a Ziegler model shown in Fig.25 are subjected to the compressive, tangential and periodic forces:

$$P(t) = P_0+P_1\cos \Theta t, \tag{5.1.1}$$

where the parameters P_0, P_1 and Θ are assumed to be known.

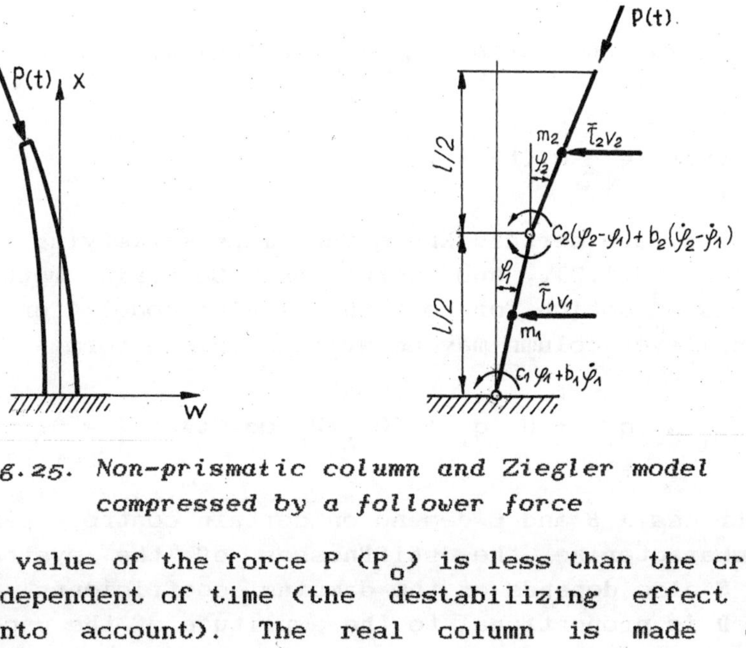

Fig.25. Non-prismatic column and Ziegler model compressed by a follower force

The mean value of the force P (P_0) is less than the critical force independent of time (the destabilizing effect being taken into account). The real column is made of a viscoelastic material of the Kelvin-Voigt type, while the Ziegler model is subjected to internal damping in its joints. The motion of both systems is also damped by external viscous forces.

The motion of the cantilever column is determined by the well-known partial differential equation of the fifth order with the appropriate boundary conditions:

$$\frac{\partial^2}{\partial x^2}\left[EI(x)\frac{\partial^2 w}{\partial x^2}+\lambda I(x)\frac{\partial^3 w}{\partial x^2 \partial t}\right]+P(t)\frac{\partial^2 w}{\partial x^2}+\gamma\frac{\partial w}{\partial t}+\rho A(x)\frac{\partial^2 w}{\partial t^2} = 0 \tag{5.1.2}$$

$$w(0,t)=\frac{\partial w}{\partial x}(0,t)=\left[EI(x)\frac{\partial^2 w}{\partial x^2} + \lambda I(x)\frac{\partial^3 w}{\partial x^2 \partial t}\right]_{(1,t)} = 0,$$

$$\left\{\frac{\partial}{\partial x}\left[EI(x)\frac{\partial^2 w}{\partial x^2} + \lambda I(x)\frac{\partial^3 w}{\partial x^2 \partial t}\right]\right\}_{(1,t)} = 0, \tag{5.1.3}$$

where $w(x)$ denotes deflection of the column, λ and γ denote damping coefficients and ρ denotes mass density per unit length.

Assuming the deflection of the column to be of the form:

$$w(x,t) = \sum_{i=1}^{I} q_i(t)\ w_i(x), \tag{5.1.4}$$

where $w_i(x)$ are certain known functions satisfying boundary conditions (5.1.3), and using the Galerkin method, the equations of motion for both the Ziegler model (for I=2) and the cantilever column may be written in the form:

$$\sum_{j=1}^{I}\left[\ A_{ij}\ddot{q}_j + B_{ij}\dot{q}_j + (C_{ij}+D_{ij}\cos\theta t)q_j\ \right] = 0, \quad i=1..I. \tag{5.1.5}$$

The matrices **A**, **B** and **C** depend on certain control parameters which characterize the stiffnesses of the systems. The matrix **B** also depends on the damping coefficients and the matrix **D** is proportional to the amplitude of the oscillating component of the force P(t).

Instability regions for the column or the Ziegler model may be obtained from the equation (5.1.5). We confine our considerations to the first instability region which occurs in the vicinity of the double value of the lowest natural frequency of undamped vibrations. To determine the boundaries of the instability region we look for solutions to the equations (5.1.5) in the following form:

$$q_i(t) = E_i \sin\frac{\Theta t}{2} + F_i \cos\frac{\Theta t}{2}, \qquad i=1,2, \tag{5.1.6}$$

where E_i and F_i are constants to be determined (V.V.Bolotin [70], A.S.Volmir [71]).

Substituting the equations (5.1.6) into (5.1.5) we obtain a system of four linear and homogeneous algebraic equations for E_i and F_i. A non-zero solution exists if the corresponding fourth-order determinant vanishes.

The optimal design of the systems is here confined to the parametrical type, with only one control parameter ϕ. We look for an optimal value of this parameter for which the value of the excitation coefficient P_1/P_0 reaches its maximum at a given value of P_0 and for a fixed volume of the system. We consider linearly tapered cantilever columns whose cross-sectional moment of inertia and volume are expressed by the formulae:

$$I(x) = I_0\left(1-\phi_c \frac{x}{l}\right)^4, \qquad V=\frac{3A_0 l}{3-3\phi_c+\phi_c^2}, \tag{5.1.7}$$

and the Ziegler model, for which we assign the following volume (as suggested by M.Życzkowski and A.Gajewski [72]):

$$V = \frac{1}{4}klc_2^{\varkappa}\left(\frac{3n+1}{n+1} + \phi_z^{\varkappa}\right), \tag{5.1.8}$$

where $\phi_z=c_1/c_2$, while the parameters $k,n,\varkappa$ depend on the manner in which the stiffness of the real column is ascribed to the rigidities of the hinges c_1 and c_2. We assume here that $\varkappa=0.5$ and $n=4$.

The boundaries of the first instability region have been calculated numerically for various values of the control parameter ϕ_c (or ϕ_z) and certain damping parameters λ,γ. Particular solutions are presented in Fig.26 where it can be seen that the optimal value of the control parameter ($\phi_c=1$) corresponds to the cone column of zero cross-section at its free end. This result unexpectedly differs from results obtained by J.L.Claudon [73], J.Błachut and A.Gajewski [16] and M.Hanaoka and K.Washizu [74] for columns compressed by a constant tangential force. The paper by A. Foryś and A. Gajewski [75] contains more details of the calculations involved.

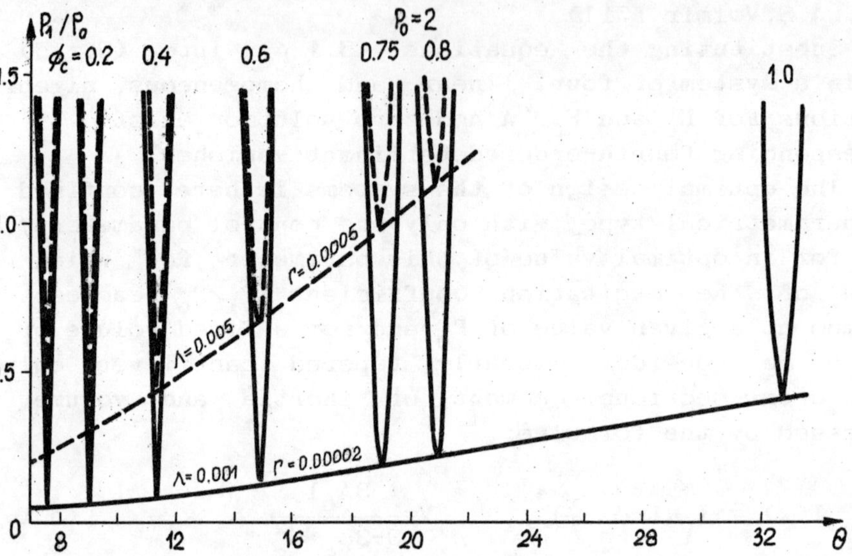

Fig. 26. Dynamic instability domains for a column

5.2. The parametrical optimization of a plane system of rods in conditions of internal resonance

A plane system of three rods connected by joints presented in Fig. 27 is subjected to a uniformly distributed periodic load acting on its beam. The transverse symmetrical vibrations, the periodic or non-periodic combination resonance and the internal resonance of a similar system -with constant stiffnesses of rods- have been analized by Anna Foryś [76,77] and Anna Foryś and J. Nizioł [78]. The couplings of elements of the system through internal longitudinal forces which are transverse forces at the ends of neighbouring rods have been taken into account. In order to avoid infinite amplitudes of vibration in the internal resonance state some geometrical non-linearities have been introduced. In the paper by Anna Foryś and J.Nizioł [78] the appropriate equations of motion for prismatic rods have been derived from Lagrange equations of the second kind. The equations obtained by Anna Foryś [76] and Anna Foryś and J.Nizioł [78] have been adopted by Anna Foryś and A.Gajewski [79] and later by Anna Foryś and A.Foryś [80] for a similar system with rods of variable cross-sections.

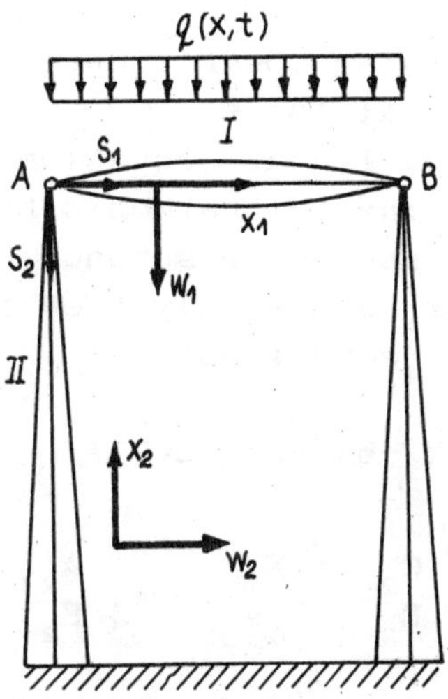

Fig.27. Dynamically loaded system of non-prismatic rods

After separating time and spatial variables and after a certain discretization, the motion of the system is described by the non-linear equations:

$$A_1\ddot{T}_1 + B_1T_1 - C_1T_1T_2 + D_1\dot{T}_1 + E_1\dot{T}_1T_2^2 = \Gamma_1\sin\omega t$$
$$(5.2.1)$$
$$A_2\ddot{T}_2 + B_2T_2 - C_2T_1T_2 + D_2\dot{T}_2 + E_2\dot{T}_2T_1^2 = \Gamma_2\sin\omega t,$$

where:

$$A_i = \int_0^{l_i} m_i(x)Y_i^2(x)\,dx, \qquad B_i = E_i\int_0^{l_i} I_i(x)\,[Y_i''(x)]^2dx,$$

$$C_1 = \frac{\partial}{\partial x}[E_2I_2Y_2'']_{x=l_2}\int_0^{l_1} Y_1'^2dx\ ,$$

$$C_2 = \frac{\partial}{\partial x}[E_1 I_1 Y_1'']_{x=0} \int_0^{l_2} Y_2'^2 dx , \qquad (5.2.2)$$

$$D_i = \eta_i \int_0^{l_i} I_i(x) Y_i''^2 dx, \qquad E_i = K_i \left[\int_0^{l_i} Y_i'^2 dx \right]^2 ,$$

$$\Gamma_i = \int_0^{l_i} \gamma_i Y_i \, dx, \qquad i=1,2.$$

where η_i and K_i are damping coefficients and γ_i denote the amplitudes of the external loads.

The system of non-linear equations (5.2.1) may be approximately solved using the harmonic balance method valid for a steady state of internal resonance. The amplitudes of vibration of the rods R_1 and R_2 may then be calculated from the following algebraic equations:

$$R_1 = 2\sqrt{X_1} , \qquad aX_1^3 + bX_1^2 + cX_1 + d = 0, \qquad (5.2.3)$$

$$R_2 = \sqrt{X_2} , \quad X_2 = \left[\left(\frac{C_2}{E_2} \right)^2 \frac{X_1}{\omega_{02}^2} - 4\left(\frac{A_2}{E_2}\right)^2 \left(\omega_{02} - \frac{\omega}{2}\right) \right]^{\frac{1}{2}} - \frac{D_2}{E_2} , \qquad (5.2.4)$$

where:

$$\omega_{02}^2 = \frac{B_2}{A_2} , \qquad a=1, \qquad b=2\frac{D_1}{E_1} , \qquad (5.2.5)$$

$$c = 4\left(\frac{A_1}{E_1}\right)^2 \left(\sqrt{B_1/A_1} - \omega\right)^2 + \left(\frac{D_1}{E_1}\right)^2 , \qquad d = \frac{-A_1 \Gamma_1^2}{4B_1 E_1^2} .$$

They are dependent on the distribution of masses and the stiffnesses of the elements of the system.

In the paper by Anna Foryś and A. Gajewski [79] the amplitude R_2 of the element subjected to parametrically forced vibrations has been assumed as the objective function. The volume of the particular element or of the whole system may be used as constraints. Some constant parameters describing the variations of the cross-sectional areas of the rods have served as control variables. To simplify calculations it has been assumed that the side of the square cross-section of the beam (element I) varies according to the square parabola while the side of the columns (element II) varies according to the linear function. By successively changing the values of the control parameters the considerable influence of the latter on the objective function R_2 and also on the amplitude R_1 may be observed. A typical situation is presented in Fig. 28 where the amplitudes R_1 and R_2 against the frequency of vibration ω obtained for various values of control variables are shown. A considerable reduction of both amplitudes can be seen. The optimization problem presented above has been extended to include a system with additional concentrated masses at the hinges by Anna Foryś and A. Foryś [80]. It should be noted that the approximate solutions obtained in this section are capable of improvement - something which should come about in the near future.

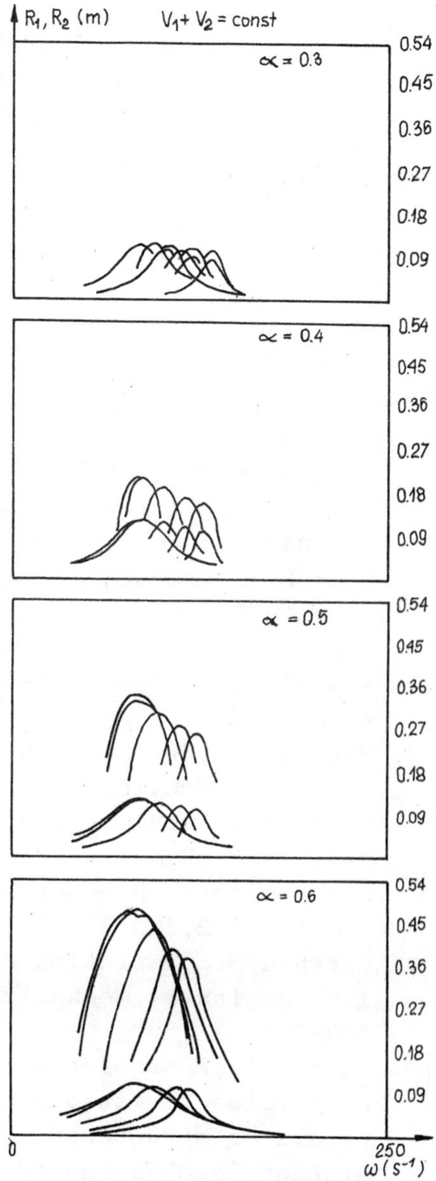

Fig. 28. The amplitudes of rods in internal resonance conditions

REFERENCES

1. Gajewski,A., Życzkowski,M.: Optimal structural design under stability constraints, Kluwer Academic Publishers, Dordrecht-Boston-London 1988.
2. Pontryagin,L.S., Boltyanskii,V.G., Gamkrelidze,R.V., Mishchenko, E.F.: The mathematical theory of optimal processes (in Russian), Fizmatgiz, Moskva 1961. English translation: Wiley, New York 1962.
3. Pedersen,P.: A unified approach to optimal design, in: Optimization methods in structural design, Euromech-Colloquium 164, Siegen 1982 (Eds. H.Eschenauer, N.Olhoff), Bibliogr. Inst., Mannheim-Wien-Zürich 1983, 182-187.
4. Adelman,H.M.,Haftka,R.T.: Sensitivity analysis of discrete structural systems, AIAA Journal,**24** (1986),5, 823-832.
5. Haug,E.J.,Arora,J.S.: Applied optimal design, Wiley, New York 1979.
6. Haug,E.J.,Choi,K.K.,Komkov,V.: Design sensitivity analysis of structural systems, Academic Press, New York 1986.
7. Wittrick,W.H.: Rates of change of eigenvalues with reference to buckling and vibration problems, J.Roy.Aero. Soc., **66** (1962), 590-591.
8. Plaut,R.H.,Huseyin,K.: Derivatives of eigenvalues and eigenvectors in non-selfadjoint systems, AIAA Journal, **11** (1973),2,250-251.
9. Farshad,M.: Variation of eigenvalues and eigen-functions in continuum mechanics, AIAA Journal, **12** (1974),4, 560-561.
10. Haug,E.J.,Rousselet,B.: Design sensitivity analysis of eigenvalue variations, in: Optimization of distributed parameter structures (Eds. E.J.Haug and J.Cea),Vol.I-II, Sijthoff and Noordhoff, Alphen aan den Rijn 1981.
11. Seyranian,A.P., Sharanyuk,A.V.: Sensitivity and optimization of critical parameters in dynamic stability problems (in Russian), Mekh.Tv.Tela,**18** (1983), 5,174-183.

12. Claudon,J.-L.,Sunakawa,M.: Sensitivity analysis for continuous mechanical systems governed by double eigenvalue problems, in: Optimization of distributed parameter structures (Eds. E.J.Haug and J.Cea), Vol.I-II, Sijthoff and Noordhoff, Alphen aan den Rijn 1981.
13. Pedersen,P.,Seyranian,A.P.: Sensitivity analysis for problems of dynamic stability, Int.J.Solids Structures **19** (1983),4,315-335.
14. Szefer,G.: Analiza wrażliwości i optymalizacja układów dynamicznych z rozłożonymi parametrami, Zesz.Nauk. AGH, Kraków, Mechanika, **1** (1982),4,5-36.
15. Demyanov,V.F.,Malozemov,V.N.: Introduction to minimax (in Russian), Nauka, Moskva 1972. English translation: Wiley 1974.
16. Błachut,J.,Gajewski,A.: A unified approach to optimal design of columns, Solids Mech.Archives, **5** (1980),4,363-413.
17. Haug,E.J.: Optimization of distributed parameter structures with repeated eigenvalues, in:New Approaches to Nonlinear Problems in Dynamics, (ed.P.J.Holmes), SIAM Publications, 1980.
18. Plaut,N.: Optimal design with respect to structural eigenvalues, Proc.15th Int.Congr.Theor. and Appl.Mech. (Toronto 1980), Prepr. Amsterdam 1980, 133-149.
19. Weishaar,T.A.,Plaut,R.H.: Structural optimization under nonconservative loading, in: Optimization of distributed parameter structures (Eds. E.J.Haug and J.Cea), Vol.I-II, Sijthoff and Noordhoff, Alphen aan den Rijn 1981.
20. Olhoff,N.,Rasmussen,S.H.: On single and bimodal optimum buckling loads of clamped columns, Int.J.Solids Structures, **13** (1977),7,605-614.
21. Tadjbakhsh,I.,Keller,J.B.: Strongest columns and isoperimetric inequalities for eigenvalues, J.Appl.Mech, **29** (1962),1,159-164.
22. Gajewski,A.,Życzkowski,M.: Optimal design of elastic columns subject to the general conservative behaviour of loading, ZAMP, **21** (1970),5,806-818.
23. Farshad,M.,Tadjbakhsh,I.: Optimum shape of columns with general conservative end loading, JOTA, **11** (1973),4,413-420.

24. Gajewski,A.: A note on unimodal and bimodal optimal design of vibrating compressed columns, Int.J.Mech.Sci. **23** (1981),1,11-16.
25. Bochenek,B.,Gajewski,A.: Jednomodalna i dwumodalna optymalizacja ściskanych prętów drgających, Mech.Teor.Stos., **22** (1984),1/2,185-195.
26. Gajewski,A.: Bimodal optimization of a column in an elastic medium with respect to buckling or vibration, Int.J.Mech.Sci., **27** (1985),1/2,45-53.
27. Kiusalaas,J.: Optimal design of structures with buckling constraints, Int.J.Solids Structures, **9**(1973),7,863-878.
28. Repin,S.I.: Shape optimization of a bar on elastic foundation for multiple solutions (in Russian), Prikl. Mat., Tula 1979, 44-50.
29. Larichev,A.D.: Problem of optimization of a clamped beam on elastic foundation (in Russian), Issled. po Stroit.Konstr., Moskva, 1982.
30. Plaut,R.H.,Johnson,L.W.,Olhoff,N.: Bimodal optimization of compressed columns on elastic foundations, J.Appl. Mech., **53** (1986),3,130-134.
31. Shin,Y.S., Plaut,R.H., Haftka,R.T.: Simultaneous analysis and design for eigenvalue maximization, Proc. of the AIAA/ASME/ASCE/AHS Structures, Structural Dynamics and Materials Conference, Monterey, California, Vol.1, April 1987, 334-342.
32. Shin,Y.S.,Haftka,R.T.,Watson,L.T.,Plaut,R.H.: Tracing structural optima as a function of available resources by a homotopy method, Int.J.Comp.Meth.Appl.Mech. and Ing., (in print).
33. Bochenek,B.: Multimodal optimal design of a compressed column with respect to buckling in two planes, Int.J. Solids Structures, **23** (1987),5,599-605.
34. Bochenek,B.,Nowak,M.: Optymalne kształtowanie słupów z uwagi na wyboczenie w dwóch płaszczyznach (submitted to print).
35. Prager,S.,Prager,W.: A note on optimal design of columns, Int.J.Mech.Sci., **21** (1979),4,249-251.
36. Olhoff,N.: Optimization of columns against buckling, in: Optimization of distributed parameter structures (Eds. E.J.Haug and J.Cea), Vol.I-II, Sijthoff and

Noordhoff, Alphen aan den Rijn 1981.

37. Lam,H.L.,Haug,E.J.,Choi,K.K.: Optimal design of structures with constraints on eigenvalues, Materials Division, The University of Iowa, Techn. Report No 79, Jan.1981, 1-71.

38. Choi,K.K.,Haug,E.J.: Repeated eigenvalues in mechanical optimization problems, (Meeting on Probl.Elastic Stab. and Vibr., Pittsburgh 1981), Providence 1981, 61-86.

39. Haug,E.J.,Choi,K.K.: Systematic occurence of repeated eigenvalues in structural optimization, JOTA, **38** (1982), 2, 251-274.

40. Teschner,W.: Minimum weight design for structural eigenvalue problems by optimal control theory, in: Optimization methods in structural design, Euromech-Colloquium 164, Siegen 1982 (Eds. H.Eschenauer, N.Olhoff), Bibliogr. Inst., Mannheim-Wien-Zürich 1983, 424-429.

41. Banichuk,N.V.,Barsuk,A.A.: On a certain method of optimization of elastic stability in the case of multiple critical loadings (in Russian), Prikl.Probl.Prochn. Plast., **24** (1983),85-89.

42. Seyranian,A.P.: On a certain solution of a problem of Lagrange (in Russian), Dokl.AN SSSR, **271** (1983),3, 337-340.

43. Masur,E.F.: Optimal structural design under multiple eigenvalue constraints, Int.J.Solids Structures, **20** (1984),3,211-231.

44. Błachut,J.,Gajewski,A.: On unimodal and bimodal optimal design of funicular arches, Int.J.Solids Structures, **17** (1981),7,653-667.

45. Błachut,J.,Gajewski,A.: Unimodal and bimodal optimal design of extensible arches with respect to buckling and vibration, Optimal Control Appl. Meth., **2** (1981),4, 383-402.

46. Błachut,J.: Unimodalna optymalizacja drgających i narażonych na utratę stateczności łuków o osi wydłużalnej, Rozpr.Inż., **30** (1982),1,37-55.

47. Bochenek,B.,Gajewski,A.: Optimal design of funicular arches with respect to in-plane and out-of-plane

buckling, J.Struct.Mech., **14** (1986),3,257-274.

48. Bochenek,B.,Gajewski,A.: Multimodal optimization of arches under stability constraints with two independent design functions, Int.J.Solids Structures, (in print).
49. Bochenek,B.: On multimodal parametrical optimization of arches against plane and spatial buckling, Eng. Optim.,**14** (1988),27-37.
50. Plaut,R.H.,Olhoff,N.: Optimal forms of shallow arches with respect to vibration and stability, J.Struct.Mech., **11** (1983),1,81-100.
51. Olhoff,N.,Plaut,R.H.: Bimodal optimization of vibrating shallow arches, Int.J.Solids Structures,**19** (1983), 6,553-570.
52. Rakowski,G.,Solecki,R.: Pręty zakrzywione: obliczenia statyczne, Arkady, Warszawa 1965.
53. Schmidt,R.: Postbuckling behaviour of uniformly compressed circular arches with clamped ends, ZAMP, **30** (1979), 553-556.
54. Irie,T.,Yamada,G.,Takahashi,I.: In plane vibration of Timoshenko arcs with variable cross-section, Ing.-Archiv, **48** (1979),5,337-346.
55. Błachut,J.: Analiza stateczności pryzmatycznych łuków o osi odkształcalnej, Mech.Teor.Stos., **20** (1982), 1/2, 141-157.
56. Suzuki,K.,Kosawada,T.,Takahashi,S.: Out-of-plane vibrations of curved bars with varying cross-section, Bull. JSME, **26** (1983),212,268-275.
57. Nikolai,E.L.: On the stability of a circular ring and of a circular arch under uniformly distributed normal loading (in Russian), Izv.Petrogradskogo Polit.Inst., **27** (1918).
58. Ponomarev,S.D.,Biderman,V.L.,Likharev,K.K.,Makushin,V.M. Malinin,N.N.,Feodosyev,V.I.: Fundamentals of contemporary methods of strength calculations in mashine design (in Russian), Mashgiz, Moskva 1952/1954, (1957/1959).
59. Ojalvo,M.,Demuts,E.,Tokarz,F.: Out-of-plane buckling of curved members, Proc.ASCE, J.Struct.Div., **96** (1969), ST10, 2305-2316.
60. Tadjbakhsh,I.,Farshad,M.: On conservatively loaded funicular arches and their optimal design, in: Optimization

in structural design (Eds. A.Sawczuk and Z.Mróz),IUTAM Symposium, Warsaw 1973, Springer, Berlin-New York 1975.

61. Olhoff,N.: Bimodality in optimizing the shape of a vibrating shallow arch, in: Optimization methods in structural design, Euromech-Colloquium 164, Siegen 1982 (Eds. H.Eschenauer and N.Olhoff), Bibliogr. Inst., Mannheim-Wien-Zürich 1983, 182-187.

62. Błachut,J.: Parametrical optimal design of funicular arches against buckling and vibration, Int.J.Mech.Sci., **26** (1984),5,305-310.

63. Pierson,B.L.: Panel flutter optimization by gradient projection, Int.J.Num.Meth.Engng.,**9** (1975),271-296.

64. Seyranian,A.P.: Optimization of structures subjected to aeroelastic instability phenomena, Arch.Mech.Stos., **34** (1982),2,133-146.

65. Frauenthal,J.C.: Constrained optimal design of circular plates against buckling, J.Struct.Mech., **1** (1972),2, 115-127.

66. Grinev,V.B.,Filippov,A.P.: Optimal design of circular plates against buckling (in Russian), Stroit.Mekh.Rasch. Sooruzh., (1972),2,16-20.

67. Rzegocińska-Pełech,K., Waszczyszyn,Z.: Numerical optimum design of elastic annular plates with respect to buckling, Computers and Structures, **18** (1984),2,369-378.

68. Irie,T.,Yamada,G.,Kaneko,Y.: Vibration and stability of a non-uniform annular plate subjected to a follower force, J.Sound and Vibration, **73** (1980),2,261-269.

69. Gajewski,A.,Cupiał P.: Optimal structural design of an annular plate compressed by non-conservative forces (submitted to print).

70. Bolotin,B.B.: Dynamic stability of elastic systems (in Russian), Gostekhizdat, Moskva 1956. English translation: Holden-Day, San Francisco 1964.

71. Volmir,A.S.: Stability of deformable systems (in Russian), Nauka, Moskva 1967.

72. Życzkowski,M.,Gajewski,A.: Optimal structural design in non-conservative problems of elastic stability, in: Instability of continuous systems (Ed.H.H.E.Leipholz), IUTAM Symposium, Herrenalb 1969, Springer, Berlin-

Heidelberg-New York 1971.
73. Claudon,J.-L.: Characteristic curves and optimum design of two structures subjected to circulatory loads, Journal de Mécanique **14** (1975),3,531-543.
74. Hanaoka,M.,Washizu,K.: Optimum design of Beck's column, Comp. and Struct., **11** (1980),6,473-480.
75. Foryś,A.,Gajewski,A.: Parametryczna optymalizacja pręta lepkosprężystego ze względu na stateczność dynamiczną, Rozpr.Inż., **35** (1987),2,297-308.
76. Foryś,Anna: Vibrations and dynamical stability of some system of rods in nonlinear approach, Nonlin.Vibr. Problems, **22** (1984),215-231.
77. Foryś,Anna: Periodic and non-periodic combination resonance in a non-linear system of rods, J.Sound and Vibration, **105** (1986),3,461-472.
78. Foryś,Anna, Nizioł,J.: Internal resonance in a plane system of rods, J.Sound and Vibration, **95** (1984),3,361-374.
79. Foryś,Anna, Gajewski,A.: Analiza i optymalizacja układu prętowego o zmiennych przekrojach w warunkach rezonansu wewnętrznego, Rozpr.Inż., **32** (1984),4,575-598.
80. Foryś,Anna, Foryś,A.: Optymalizacja parametryczna układu prętów z uwzględnieniem nieliniowości, Rozpr. Inż., **34** (1986),4,399-518.

PART III

N. Olhoff
University of Aalborg, Aalborg, Denmark

ABSTRACT
Part III deals with problems of elastic optimal design of conservative mechanical systems with respect to fundamental and higher order eigenvalues. The structural eigenvalues are used as objective functions or applied as behavioural constraints. They identify, e.g., natural frequencies of axial or torsional vibrations of rods and bars, transverse vibrations of beams, critical whirling speeds of rotating shafts, or structural buckling loads. A unified variational formulation for optimal design of one-dimensional continuum systems with respect to simple fundamental eigenvalue is presented first, and this serves as a framework for a thorough discussion of extended problems involving optimization with respect to bi- or multimodal fundamental eigenvalues, and higher order eigenvalues.

INTRODUCTION

Problems of optimal design of thin, elastic, one-dimensional continuum structures of given length are considered. The objective is to determine the distribution of structural material such that maximum values of natural frequencies of vibration, critical speeds (of rotating shafts), or structural buckling loads are obtained for a prescribed amount of material. Equivalently, the structural volume or cost is minimized for a specified vibration frequency, critical speed, or buckling load, respectively.

The practical significance of such problems is that they provide designs of minimum cost against resonance due to external excitation of given frequency or frequency range, failure due to whirling instability at service speeds, or buckling failure subject to external axial loads.

The text is subdivided into four chapters with the following titles:

1. Optimal Design of One-Dimensional, Conservative, Elastic Continuum Systems with Respect to a Fundamental Eigenvalue

2. Optimal Design of Euler Columns Against Buckling. Multimodal Formulation

3. Optimization of Transversely Vibrating Bernoulli-Euler Beams and Rotating Shafts with Respect to the Fundamental Natural Frequency or Critical Speed

4. Optimization of Bernoulli-Euler Beams and Shafts with Respect to Higher Order Eigenfrequencies

Although physically different, the problems are strongly related in that the natural vibration frequencies, critical speeds and buckling loads, with associated deflection modes, constitute eigenvalues and eigenfunctions, respectively, of mathematical eigenvalue problems. The problems considered are conservative, so that Rayleigh's minimum principle holds, and the governing equations for the problems are derived by variational analysis of the expression for the Rayleigh quotient.

Chapter 1 presents a generalized variational formulation and analysis for a broad class of structural optimization problems pertaining to fundamental eigenvalues, and this chapter constitutes the main basis for the discussion of the extended problems in the subsequent chapters. The optimization problems considered in these chapters are governed by coupled, non-linear fourth-order ordinary integro-differential eigenvalue problems, for which closed form solutions cannot be expected. The exposition and the several results presented, lend themselves on work by the author and co-workers, and the reader is referred to the original papers for details on the numerical solution procedures, which are omitted in this text for reasons of brevity. In all cases, the problems are solved via constructing a convergent scheme for numerical solution by successive iterations on the basis of a formal integration of the governing equations.

A list of references connected with the topies discussed is given by the end of the text. Comprehensive general reviews of the field of optimal structural design have, e.g. been published by Wasiutynski and Brandt [1] in 1963, Sheu and Prager [2] in 1968, Prager [3] in 1971, Niordson and Pedersen [4] in 1973, Rozvany and Mróz [5] in 1977, Venkayya [6] in 1978, Schmit [7] Lev [8] and Haug [9] in 1981, Vanderplaats [10] and Ashley [11] in 1982, Olhoff and Taylor [12] in 1983, and Levy and Lev [13] in 1987. The rapid sequence of reviews and the publication of a number of textbooks and proceadings, e.g. [14-39], witness the recent progress and increasing importance of the field. The reader is referred, e.g., to the recent textbook [37] by Gajewski and Zyczkowski for an exhaustive compilation of monographs, textbooks, proceedings, and articles on structural optimization.

1. OPTIMAL DESIGN OF ONE-DIMENSIONAL, CONSERVATIVE, ELASTIC CONTINUUM SYSTEMS WITH RESPECT TO A FUNDAMENTAL EIGENVALUE

1.1 Introduction

For one-dimensional, conservative structural systems made of linearly elastic material, problems of optimal design with respect to a fundamental eigenvalue have by now been studied quite intensively. In this chapter, we consider eigenvalues of self-adjoint and full-definite structural eigenvalue problems, as for example the first frequency of free axial, torsional or transverse vibrations of rods and beams, the first critical speed of rotating shafts (excluding gyroscopic effects), the critical Euler buckling load of columns, or the critical torsional divergence velocity of a wing. A unified variational formulation for optimal design with respect to a fundamental eigenvalue will be presented in the following, and we outline the basic concepts and characteristics of such problems.

Optimal design with respect to structural eigenvalues was already considered by Lagrange [40] in 1770, and later by Clausen [41] in 1851, but it was Keller's paper [42] from 1960 that provided inspiration for the considerable contemporary research efforts in the area. Refs. [40-42] deal with optimal design of columns against Euler buckling, i.e. optimization under static loads.

Optimal design with respect to eigenvalues of structures under dynamic loading conditions is generally more complex, because the loading changes with changes in the design. This field of optimization was opened by the significant paper [43] published by Niordson in 1965. Surveys have since been prepared by Ashley and McIntosh [44], Pierson [45], Reitman and Shapiro [46], Rao [47], and by the author [48,49].

1.2 Preliminary Considerations

We follow Ref. [49], and consider a straight, one-dimensional, single-purpose structure with a coordinate axis x embedded. The structure has given length L and variable cross-sections with common directions for principal axes of inertia. Let A(x) denote the cross-sectional area of the structural material, that is, linearly elastic material contributing to the specific structural stiffness s(x) , where the term specific refers to unit length of the structure. We then consider relationships between s(x) and A(x) in the form

$$s(x) = c\ E\ A^p(x) \qquad 0 \leq x \leq L \tag{1.1}$$

where Young's modulus E , the factor c and the power p are positive constants, that are assumed to be given.

Although (1.1) restricts the cross-sectional variation, it covers a large class of structural types and behaviour. Thus, axial deformation is covered by $c = p = 1$. In torsional divergence or vibration, thin-walled cross-sections of variable thickness but constant planform are covered by $p = 1$, and geometrically similar solid cross-sections, by $p = 2$. For the bending associated with transverse vibration of beams, whirling instability of rotating shafts, and Euler buckling of columns, Eq. (1.1) models the following types of cross-sections: Fixed width sandwich cross-sections with uniform thickness cores of zero stiffness covered by two identical thin face sheets of variable thickness are modeled for $p = 1$, and so are solid cross-sections of constant thickness and varying width. Solid cross-sections of variable size but fixed shape correspond to $p = 2$, and solid rectangular cross-sections of fixed width and variable thickness are modeled for $p = 3$.

A generalized expression for the Rayleigh quotients associated with conservative problem formulations for free axial, torsional or transverse vibrations, whirling instability, divergence or buckling instability, respectively, of our one-dimensional structure can be written in the form

$$\Lambda = \frac{\int_0^L cEA^p(x)e[y(x)]dx}{\int_0^L \left[b\rho A^r(x) + q(x)\right] f[y(x)]dx + \sum_i Q_i f[y(x_i)]} , \tag{1.2}$$

where the stiffness representation (1.1) is used. In (1.2), Λ denotes the square of the fundamental angular frequency ω for a particular vibration problem, the square of the first critical angular speed ω for a whirling problem, the Euler load P for a buckling problem, or the square of the critical free-stream velocity v multiplied by a factor for a torsional divergence problem for a wing, and $y(x)$ represents the corresponding deflection mode. According to Rayleigh's principle for self-adjoint and full-definite eigenvalue problems, Λ is stationary and equal to the fundamental eigenvalue at the fundamental mode $y(x)$ among all other kinematically admissible deflection functions.

Type of problem	Characteristics	e[y]	f[y]	Differential equation
Torsional divergence of a straight wing	$p=1$, $r=0$, $q(x)\equiv 1$, $b=0$, $Q_i=0$	y'^2	y^2	$-\{EAy'\}' = \Lambda y$
Axial vibration of rods	$p=r=b=c=1$	y'^2	y^2	$-\{EAy'\}' = \Lambda\{\rho A+q\}y$
Torsional vibration of rods	$p=r\geq 1$	y'^2	y^2	$-\{cEA^p y'\}' = \Lambda\{b\rho A^p+q\}y$
Transverse vib. of beams or whirling instability of rotating shafts	$p\geq 1$, $r=b=1$	y''^2	y^2	$\{cEA^p y''\}'' = \Lambda\{\rho A+q\}y$
Buckling of columns	$p\geq 1$, $r=0$, $q(x)\equiv 1$, $b=0$, $Q_i=0$	y''^2	y'^2	$-\{cEA^p y''\}'' = \Lambda y''$

Table 1.

The deflection $y(x)$ and its derivatives are contained in positive definite quadratic forms in the symbols $e[y]$ and $f[y]$, which must be interpreted according to the particular type of problem, cf. Table 1. The expressions given for $e[y]$ in the cases of transverse vibrations, whirling instability and buckling instability, respectively, are those consistent with Bernoulli-Euler beam theory. It is assumed by the form of (1.2) that linearly independent, homogeneous boundary conditions and conditions at possible interior supports are specified for the structure. Elastic supports are excluded for brevity.

The problem of torsional divergence instability of a straight airplane wing with elastic axis perpendicular to the airstream and constant cross-sectional profile along the span, is covered by (1.2) with $q(x) \equiv 1$, $b = 0$, $Q_i = 0$ and $p = 1$, if aerodynamic strip theory is used and the dominating contribution to the torsional stiffness comes from the skin, the (small) thickness of which is assumed to vary along the span.

For vibration problems, the term $b\rho A^r(x)$ in (1.2), with $p \geq r \geq 1$, b a given positive constant and ρ the mass density, represents the specific structural mass for rectilinear vibration types, or the specific structural polar mass moment of inertia for torsional vibration. The former types are associated with $r = b = 1$, whereas $r = p$ for the latter. Similarly, $q(x)$ represents the specific mass or specific polar mass moment of inertia of distributed dead mass loading and/or non-structural material (e.g. core filler in sandwich structures). Eq. (1.2) also covers vibrating structures carrying lumped dead mass loads at specified points $x = x_i$. The constants Q_i identify their masses or mass moments of inertia. Both Q_i and $q(x)$ are assumed to be given.

In the classical problems of buckling instability under external axial compressive forces, the own weight distribution of the structure can be disregarded, and it is then characteristic that $A(x)$ does not appear in the denominator of the Rayleigh quotient. The problem is included in (1.2) if we take $q(x) \equiv 1$, $b = 0$ and $Q_i = 0$. Allowance can be made for buckling caused by own structural weight [50] and/or lumped dead weights attached along the column span, by changing the denominator of (1.2) to a slightly more general expression, but it is omitted here for reasons of brevity.

1.3 Maximum Fundamental Eigenvalue for Prescribed Volume

For given structural length and material, cross-sectional style, given type of single-modal behaviour and given boundary conditions, the optimal design problem may be stated in the form:

> With the cross-sectional area $A(x)$ of structural material as the *design variable* and the fundamental eigenvalue Λ as the *objective function*, determine the design that *maximizes* Λ subject to the integral constraint of *given structural volume* $V = \int_0^L A(x)dx$, and subject to the *geometric constraint* that $A(x)$ may nowhere be less than a prescribed *minimum* value $\underline{A}$, i.e. $A(x) \geq \underline{A}$.

Using variational formulation, the optimal design $A(x)$, its associated mode $y(x)$ and optimal eigenvalue Λ are identified with stationarity of the following augmented form of the functional (1.2),

$$\Lambda^* = \frac{\int_0^L cEA^p(x)e[y(x)]dx}{\int_0^L \left[b\rho A^r(x)+q(x)\right]f[y(x)]dx + \sum_i Q_i f[y(x_i)]}$$

$$- \kappa\left[\int_0^L A(x)dx - V\right] - \int_0^L \beta(x)[g^2(x) - A(x) + \underline{A}]dx . \qquad (1.3)$$

Here, the quantities κ and $\beta(x)$ are Lagrangian multipliers, and the geometric minimum constraint $A(x) \geq \underline{A}$ has been converted to an equality constraint by means of the real slack variable $g(x)$ defined by $g^2(x) = A(x) - \underline{A}$. In optimal design problems, minimum cross-sectional area constraints were first considered by Taylor [51] in 1968.

Now, Rayleigh's principle and the introduction of the Lagrangian multipliers permit the variation of Λ^* with respect to variations of $y(x)$, $A(x)$ and $g(x)$ to be taken independently.

Noting that the stationarity of Λ^* (1.3) with respect to all admissible variation of $y(x)$ is equivalent to the stationarity of the Rayleigh quotient with respect to variation of $y(x)$, we first obtain the differential equation (cf. Table 1) and the natural boundary conditions for the problem, along with the jump conditions at inner supports and at the points x_i where the concentrated loads Q_i are attached to the structure.

Next, stationarity of Λ^* for arbitrary admissible variation of $g(x)$ gives

$$\beta(x)g(x) = 0 \tag{1.4}$$

Finally, stationarity of Λ^* with respect to all admissible variation of the design variable $A(x)$ yields the *optimality condition*

$$pcEA^{p-1}(x)e[y(x)] - rb\rho\Lambda A^{r-1}(x)f[y(x)] = \kappa(1 - \beta(x)) \tag{1.5}$$

after applying (1.2) and redefining the Lagrangian multiplier κ by dividing it by the denominator in expression (1.2).

In order to formulate the governing equations without explicit appearance of $\beta(x)$ and $g(x)$, we exploit that either $g(x) = 0$ or $g(x) \neq 0$. Denoting by x_c the (unions of) sub-intervals in which $g(x) = 0$ may take place, and denoting by x_u the remaining sub-intervals (where we have $g(x) \neq 0$), Eq. (1.4) gives us that $A(x) = \underline{A}$ (constrained) for $x \in x_c$ and that $A(x) > \underline{A}$ (unconstrained) for $x \in x_u$. In the latter sub-interval(s), Eq. (1.4) can only be satisfied if $\beta(x) = 0$, which clearly reduces the optimality condition (1.5) for $x \in x_u$.

A complete set of *governing equations for optimality,* which by their derivation are *necessary* conditions for a *possible* optimal solution, may now for convenience be listed as follows,

Rayleigh quotient expression for Λ , Eq.(1.2) (1.6a)

Differential equation for problem type, cf. Table 1 $0 \leq x \leq L$ (1.6b)

$$pcEa^{p-1}(x)e[y(x)] - br\rho\Lambda A^{r-1}(x)f[y(x)] = \kappa \qquad x \in x_u \tag{1.6c}$$

$$A(x) > \underline{A} \quad x \in x_u \ , \quad A(x) = A \quad x \in x_c \tag{1.6d}$$

$$\int_0^L A(x)dx = V \ . \tag{1.6e}$$

Note that the form (1.6c) of the so-called optimality condition is restricted to the unions of a priori unknown sub-intervals x_u where the design variable $A(x)$ is unconstrained. Clearly, the unions of sub-intervals x_u and x_c make up the entire interval $0 \leq x \leq L$. Specific forms of the optimality condition for particular problems are easily identified with the help of Table 1. It is noteworthy that the second term on the left-hand side of the optimality equation is not present in torsional divergence and classical buckling optimization, for which $b = 0$. The first term of the optimality condition is interpreted as the average strain energy density in the design fibres, i.e. the fibres that are affected by a change in the design. Constancy of this energy density is found to be a general principle in geometrically unconstrained optimal design under *static loads*, see Masur [52].

Along with the boundary conditions and the other conditions mentioned above, Eqs. (1.6a-e) constitute a coupled, non-linear, ordinary integro-differential eigenvalue problem, where the unknowns to be determined are the optimal eigenvalue Λ , the optimal distribution of structural material $A(x)$ (which includes determination of the subintervals x_c and x_u), the associated fundamental mode $y(x)$, and the Lagrangian multiplier κ.

It is noted that the cross-sectional constants p and r play a fundamental role in the coupling and the non-linearities of the equations. Evidently, cases of $p = r = 1$ are the easiest to deal with, because such problems are linear in the design variable $A(x)$, which even vanishes from the optimality condition (1.6c). Although (1.6c) remains non-linear in the deflection, it is often possible to obtain analytical solutions to problems with $p = r = 1$, see for example Prager and Taylor [53]. For buckling problems, where $b = 0$, so that terms involving r drop out, analytical solutions have even been obtained for $p = 2$, as for example in [42]. For vibration optimization problems associated with values of p other than unity, however, the coupling and the non-linearities of the governing equations generally only permit numerical solution.

1.4 Minimum Volume Design for Prescribed Fundamental Eigenvalue

An alternative, *dual* formulation for optimal design is the following:

> With the cross-sectional area $A(x)$ of structural material as the *design variable* and the structural volume $V = \int_0^L A(x)dx$ as the *objective function*, determine the design that *minimizes* V *subject to the behavioural constraint of specified fundamental eigenvalue* Λ and the *geometric minimum constraint* $A(x) \geq \underline{A}$ where $\underline{A}$ is given. Again, the structural length and material, the cross-sectional style, the type of single-modal behaviour, and the boundary conditions, are assumed to be given.

The set of necessary governing equations for a possible optimal solution to this formulation are easily derived by variational analysis of the functional

$$V^* = \int_0^L A(x)dx - \gamma \left[\frac{\int_0^L cEA^p(x)e[y(x)]dx}{\int_0^L \left[b\rho A^r(x) + q(x) \right] f[y(x)]dx + \sum_i Q_i f[y(x_i)]} - \Lambda \right]$$

$$- \int_0^L \mu(x) \; [g^2(x) - A(x) + \underline{A}]dx \; ,$$

where the behavioural constraint and the geometric minimum constraint have been adjoined to the functional V by means of Lagrangian multipliers γ and $\mu(x)$, respectively.

By variation of A(x) we find that the optimality condition takes the form

$$\gamma \left[pcEA^{p-1}(x)e[y(x)] - rb\rho\Lambda A^{r-1}(x)f[y(x)] \right] = 1 - \mu(x) \qquad (1.8)$$

which for $x \in x_u$ reduces to

$$\gamma \left[pcEA^{p-1}(x)e[y(x)] - rb\rho\Lambda A^{r-1}(x)f[y(x)] \right] = 1 \quad x \in x_u \qquad (1.9)$$

since the Lagrangian multiplier $\mu(x)$ vanishes in the sub-interval(s) where the design variable is geometrically unconstrained. The form of Eqs. (1.8) and (1.9) involves a redefinition of μ in the same manner as the Lagrangian multiplier κ in Eqs. (1.5) and (1.6c).

As the result of the complete variational analysis, we find that the present min.V fixed Λ formulation is also governed by Eqs. (1.6a-e), with the only exception that the Lagrangian multiplier κ in the optimality condition (1.6c) is replaced by $1/\gamma$. In the present formulation, the unknowns to be determined are the minimum structural volume V , the optimal distribution A(x) of structural material, the fundamental mode y(x) , and the Lagrangian multiplier γ .

1.5 On Equivalence of Dual Optimization Problems

The system of equations governing the max.Λ fixed V problem in Section 1.3 and the governing equations for the min.V fixed Λ problem considered in the foregoing section only differ by the Lagrangian multipliers κ and γ and corresponding slightly different appearences of the optimality conditions. It is therefore not surprising that these dual optimization problems, generally speaking, are equivalent. However, exceptions exist.

Comparing the optimality conditions, (1.6c) and (1.9), it is obvious that *a min.V fixed* Λ *solution will at the same time be a* solution *to a max.*Λ *fixed* V *problem, with* $\kappa = 1/\gamma$. Note that vanishing of the Lagrangian multiplier γ is excluded by the form of Eq. (1.9).

Furthermore, *a max.*Λ *fixed* V *solution associated with a non-zero Lagrangian multiplier* κ *is at the same time a solution to a min.*V *fixed* Λ *problem, with* $\gamma = 1/\kappa$.

However, a max.Λ fixed V solution with $\kappa = 0$ is not a min.V fixed Λ solution, because it is unable to satisfy the optimality condition (1.9) for a problem of the latter type.

Now, generalizing an approach of Brach [54], let us state precisely for which type of problems the equivalence may be lost. First, we multiply Eq. (1.5) by A(x) and integrate over the interval $0 \le x \le L$. Using Eqs. (1.6d-e) and employing that $\beta(x) \equiv 0$ for $x \in x_u$, we find

$$\int_0^L pcEA^p(x)e[y(x)]dx - \int_0^L rb\rho\Lambda A^r(x)f[y(x)f[y(x)]dx = \kappa\left[V - \underline{A}\int_{x_c}\beta(x)dx\right] . \tag{1.10}$$

Then, multiplying Eq. (1.6a), i.e. Eq. (1.2), by the product of r and the denominator on the right-hand side, and subtracting the resulting equation from Eq. (1.10), we obtain

$$\kappa\left[V - \underline{A}\int_{x_c}\beta(x)dx\right] = (p - r)\int_0^L cEA^p(x)e[y(x)]dx \tag{1.11}$$

$$+ r\Lambda\int_0^L q(x)f[y(x)]dx + \sum_i Q_i f[y(x_i)] \quad .$$

Here, the first integral on the right-hand side (representing twice the potential energy of the structure) is positive. Futhermore, we have $p \geq r$ for the types of problems considered, cf. Table 1. The subsequent terms on the right-hand side of (1.11) are non-negative.

Thus, κ can only vanish, i.e., *the equivalence of the dual optimization problems can only be lost if both* $p = r$, $q(x) \equiv 0$ *and* $Q_i = 0$. Note that, among the types of problems considered here, the equivalence can only be lost for vibration optimization problems.

Taylor [55] was the first to establish the equivalence of dualformulations for optimal design. The subject has also been considered in papers by Vavrick & Warner [56] and Seyranian [57].

1.6 Geometrically Unconstrained Optimal Design

We may drop the minimum constraint in the formulations considered above by setting $\beta(x) \equiv 0$ and $\mu(x) \equiv 0$, respectively, in Eqs. (1.3) and (1.7). The governing equations for the resulting *geometrically unconstrained* optimization problem are then obtained as a special case of Eqs. (1.6a-e) associated with $\underline{A} = 0$ and validity of the optimality condition (1.6c) in the entire interval $0 \leq x \leq L$ (which x_u becomes identical to). Examples of geometrically unconstrained optimal solutions are illustrated in dimensionless form in Fig. 1 for columns and in Fig. 7 for transversely vibrating beams.

A geometrically unconstrained optimal design is in general associated with maximum obtainable merit. However, geometrically unconstrained solutions must often be regarded as limiting solutions from the point of view of practical design, in other words, geometrically constrained optimal designs are preferable in practice.

Nevertheless, it is evident that knowledge of maximum obtainable efficiency is of both theoretical and practical importance, and that geometrically unconstrained solutions direct the designer towards maximum economy of material.

It is a general feature that geometrically unconstrained optimization of one-dimensional structures results in *statically determinate* solutions *when a single mode formulation is used*. This fact is connected with the occurrence of points in the optimal solutions with vanishing structural material. At these points, the derivative of the deflection appearing in the optimality condition may exhibit discontinuous behaviour in problems with $p = 1$, and in cases of $p = 2$ and $p = 3$, points of vanishing structural cross-section are associated with significant singular behaviour of the deflection and/or its derivatives. In optimal design of transversely vibrating beams with $p = 2$ or 3, the cross-section may vanish in two essentially different ways at singular points; either in a way that is found at a hinge, a so-called *Type I singularity*, see e.g. Fig. 7a, or in a manner found at a free structural end (Fig.7b) and at an inner separation (Fig. 7c), which is called a *Type II singularity*. In optimal columns, the governing equations only admit Type I singularities (cf. Figs. 1a,b,c) if the point is under axial compression, which by the way seems obvious from physical grounds.

Now, when *a priori* statically indeterminate structures are optimized without geometric minimum constraint on the basis of a single mode formulation, it is the automatic formation of hinges or separations in the structure that reduces this to a statically determinate one, cf. Figs. 1c and 7c. The singular behaviour at points of zero cross-section is studied quite intensively in [43,58-60] for different one-dimensional problems, and is considered in Chapters 2-4 of the present notes.

1.7 Sufficient Conditions of Optimality

The optimality equations (1.6a-e) are derived as necessary conditions for a possible optimal solution, and they do not, in general, state sufficient conditions for global optimality.

For problems associated with $p = r = 1$, sufficiency can be shown in specific cases, however. This was first demonstrated by Taylor [55], who proved the global optimality of a solution obtained by Turner [61]. Shortly after, sufficiency of possible solutions to geometrically unconstrained vibration problems associated with $p = r = 1$ was shown by Prager and Taylor [53]. A proof for corresponding constrained problems does not seem to be available, but might, for example, be established along lines indicated in [62]. For vibration optimization problems with $p > 1$, sufficiency is generally not ensured.

As to buckling optimization, sufficiency is proved by Taylor and Liu [63] for a geometrically constrained, statically determinate $p = 1$ column, and a proof for statically determinate $p = 2$ columns is available in Tadjbakhsh and Keller [64]. The latter authors claimed validity of their proof independently of the column boundary conditions, but Masur [65] and Popelar [66] have since noted that the proof breaks down for statically indeterminate cases, and [67] provides an illustration of this.

1.8 Existence of Solutions

Existence of optimal solutions cannot generally be assured a priori. Consequently, their possible existence cannot be demonstrated until the actual solution is arrived at.

Non-existence of optimal solutions on the contrary, can in some cases be shown. For example, *no min.V fixed Λ solutions exist to the types of vibration optimization problems considered, if* $p = r$ *and both geometrical constraints, non-structural material, and external dead loading are absent**). In this case, the eigenvalue problem defined by the differential equation (1.6b), cf. Table 1, and the boundary conditions considered, is linear and homogeneous in both $A^p(x)$ and $y(x)$.

---

*) Note, in view of the discussion in Section 1.5, that the dual problem would be associated with $\kappa = 0$, Eq. (1.11).

Thus, denoting by $\tilde{A}(x)$ a design associated with the prescribed value of Λ and satisfying (1.1) along with a vibration mode $y(x)$, the eigenvalue Λ is maintained by a design $C\tilde{A}(x)$, where C is an arbitrary constant. The volume of this design, however, can be made arbitrarily small by choosing a sufficiently small value of C.

Considering the dual problem of maximum Λ at fixed V and L specially for transversely vibrating $p = r = 1$ beams with $q(x) \equiv 0$, $Q_i = 0$ and no minimum constraint, Brach [68] demonstrated non-existence for the case of a cantilever, but existence of an optimal solution for a simply supported beam - results that have later been confirmed in [59,69]. It is interesting to note, as an illustration of the result cited at the end of Section 1.5, that for these cases of $p = r$, $q(x) \equiv 0$ and $Q_i = 0$, the equivalence is lost for the dual simply supported beam problems, whereas the equivalence holds for the cantilever problems in the sense that no optimal solution exists for either of the dual formulations.

2. OPTIMAL DESIGN OF EULER COLUMNS AGAINST BUCKLING. MULTIMODAL FORMULATION

2.1 Introduction

We consider the problem of determining the optimal design of a thin, elastic column such that the Euler buckling load attains a maximum possible value for given material volume, length and boundary conditions. We first assume the optimum buckling load to be a simple eigenvalue, and obtain the governing equations for the problem from the general theory of Chapter 1. The type of singular behaviour that may occur in geometrically unconstrained problems is investigated, and conditions for optimal location of inner singular points are stated.

For structures of some complexity or statical indeterminacy, optimization against buckling must be conducted with bimodal or even multimodal optimal buckling loads in perspective. This trend, which requires an extended formulation for optimal design, already manifests itself in the case of a doubly clamped column, and we shall discuss this problem in detail.

2.2 Single Mode Formulations for Optimal Design

Consider a thin, straight, elastic column which has the volume V, length L and Young's modulus E, and is subjected to an axial compressive force, the value of which is P at buckling. The cross-section of the column is permitted to vary along the column axis according to (1.1) with $s(x) = EI(x)$, the bending stiffness of structural material. In (1.1), $A(x)$ is the cross-sectional area of structural material, and c and p are given constants. The cross-sectional styles corresponding to $p = 1$, 2 and 3 are described in Section 1.2. We consider columns of Bernoulli-Euler type with given support conditions and exclude flexible supports for brevity.

In *geometrically constrained* form, our optimization problem consists in determining the cross-sectional area distribution $A(x)$ that maximizes the fundamental buckling load P for given values of V, L, $\underline{A}$, E, p and c. Assuming P to be a *simple eigenvalue*, we easily obtain the following governing equations for this problem from Table 1 and the general optimality equations (1.6a-e),

$$P = \frac{\int_0^L cEA^p y''^2 dx}{\int_0^L y'^2 dx} \tag{2.1a}$$

$$(cEA^p y'')'' = -Py'' \qquad 0 \le x \le L \tag{2.1b}$$

$$pcEA^{p-1} y''^2 = \kappa \qquad x \in x_u \tag{2.1c}$$

$$A(x) > \underline{A} \quad x \in x_u , \quad A(x) = \underline{A} \quad x \in x_c \tag{2.1d}$$

$$\int_0^L A(x) dx = V . \tag{2.1e}$$

Specification of a minimum constraint for the cross-sectional area of a column was introduced by Taylor & Liu [63], and has later been done in Refs. [53,67,70-75], for example. If V is minimized at fixed P, such a constraint is equivalent to a constraint on the maximum prebuckling stress.

Let us now, for convenience, nondimensionalize the coordinate x by division by L and introduce a dimensionless cross-sectional area $\alpha(x)$ and buckling load λ by

$$\alpha(x) = A(x)L/V \tag{2.2}$$

$$\lambda = \frac{PL^{p+2}}{EcV^{p}} \tag{2.3}$$

The *geometrically unconstrained* problem (where $\underline{\alpha} = \underline{A}L/V = 0$), is thus governed by the following dimensionless equations, obtainable from Eqs. (2.1a-e) where (2.1d) drops out and the optimality condition (2.1c) becomes valid in the entire interval $0 \le x \le 1$ for the dimensionless variable:

$$\lambda = \frac{\int_0^1 \alpha^p y''^2 dx}{\int_0^1 y'^2 dx} \tag{2.4a}$$

$$(\alpha^p y'')'' = -\lambda y'' \qquad 0 \le x \le 1 \tag{2.4b}$$

$$\alpha^{p-1} y''^2 = \kappa \qquad 0 \le x \le 1 \tag{2.4c}$$

$$\int_0^1 \alpha(x) dx = 1 \tag{2.4d}$$

These equations expose the Rayleigh quotient, the differential equation for Euler buckling, the optimality condition (where κ has been redefined) and the volume constraint, respectively, in dimensionless form. Normalizing the deflection $y(x)$ such that the denominator in (2.4a) is set equal to unity,

$$\int_0^1 y'^2 dx = 1 \ , \tag{2.5}$$

multiplying (2.4c) by $\alpha(x)$ and integrating over the interval $0 \leq x \leq 1$, taking (2.4d) and (2.5) into account, we find that the Lagrangian multiplier κ is simply given by

$$\kappa = \lambda \ . \tag{2.6}$$

Hence, for cases of $p = 2$ or $p = 3$, Eq. (2.4c) gives us the optimal cross-sectional area function $\alpha(x)$ in the form

$$p = 2\ ,\ 3\ :\quad \alpha(x) = \left[\frac{\lambda}{y''^2} \right]^{1/(p-1)} \ , \tag{2.7a}$$

while, for $p = 1$, Eq. (2.4e) states that the curvature $y''(x)$ of the deflection is constant throughout, except for possible sign shifts,

$$p = 1\ :\quad y''(x) = \pm \sqrt{\lambda} \ . \tag{2.7b}$$

The continuity conditions for a column are as follows. At all points free from kinematic constraint (column supports), the bending moment $m(x) = \alpha^p y''$ and the function $(\alpha^p y'')' + \lambda y'$ are continuous:

$$\langle \alpha^p y'' \rangle = 0 \tag{2.8}$$

$$\langle (\alpha^p y'')' + \lambda y' \rangle = 0 \tag{2.9}$$

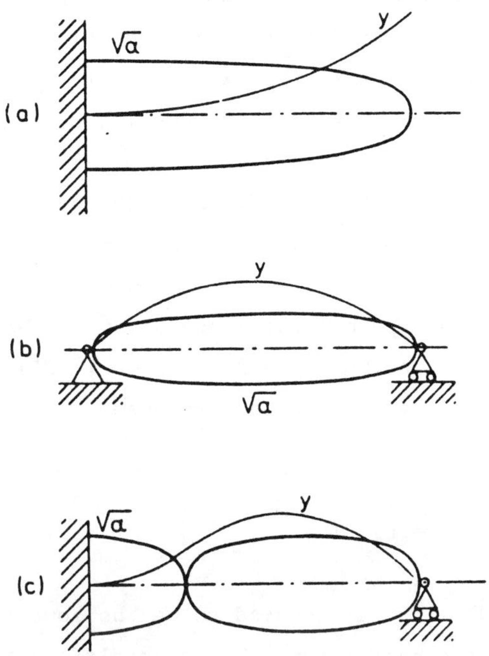

Fig. 1

The function $(\alpha^p y'')' + \lambda y'$ represents, for the buckled column, the force component of the stress resultants in the direction perpendicular to the x-axis, while the shear force $t(x) = (\alpha^p y'')'$ is the force component perpendicular to the deflected column axis.

The geometrically unconstrained problem is seen to be quite simple for $p = 1$, and solutions can be obtained analytically, see e.g. [5,53,58,76,77]. In cases of $p = 2$ and $p = 3$, where the problem becomes non-linear in $y(x)$, it is still possible to apply analytical methods of solution provided that the boundary conditions are sufficiently simple [2,3,64,65]. Else, numerical methods are available, see for example [21,50,67,71,74,79,80].

Optimal shapes $\pm\sqrt{\alpha}$ and corresponding buckling modes y are indicated in Fig. 1 for (a) cantilevered [64], (b) simply supported [42], and (c) clamped-simply supported [64] $p = 2$ columns. The buckling loads of the optimal solutions are increased by (a) 1/3, (b) 1/3 and (c) 35.1% when compared with the buckling loads of correspondingly supported *uniform* columns of the same volume, length and material [42,64].

2.3 Singularities in Geometrically Unconstrained Solutions

We now consider the behaviour of a solution to the geometrically unconstrained, single mode formulation at a point $x = x_j$ of zero bending moment $m = \alpha^p y''$. For the case of $p = 1$, zero bending moment must imply, in view of (2.7b), that α also vanishes. For $p = 2$ and $p = 3$, Eq. (2.7a) and the relation $m = \alpha^p y''$ show that the cross-section α also vanishes and that y'' tends to infinity at a point $x = x_j$ of vanishing bending moment m.

As an example, let us determine the type of the singularity at the simply supported, left-hand end $x = 0$ of the $p = 2$ column in Fig. 1b. We shall assume that the solution $y(x)$ can be expanded in the following power series

$$y(x) = a_1 x + a_2 x^2 + \ldots + bx^r + \ldots \tag{2.10}$$

in the vicinity of $x = 0$. In (2.10), r is assumed to be the lowest non-integer power of the series, and the coefficient b to be different from zero. With $\alpha = \lambda/y''^2$ and $m = \lambda^2/y''^3$ for $p = 2$, we must have $r < 2$ in order to satisfy the boundary condition $m = 0$ at the simply supported end $x = 0$. We thus have

$$y''(x) = br(r-1)x^{r-2} \ldots \tag{2.11a}$$

$$\alpha(x) = \frac{\lambda}{b^2 r^2 (r-1)^2} x^{4-2r} + \ldots \tag{2.11b}$$

$$m(x) = \frac{\lambda^2}{b^3 r^3 (r-1)^3} x^{6-3r} + \ldots \tag{2.11c}$$

in the vicinity of $x = 0$. Substitution of (2.11a-b) into the buckling differential equation (2.4b) with $p = 2$ then yields

$$(6-3r)(5-3r) \frac{\lambda^2}{b^3 r^3 (r-1)^3} x^{4-3r} + \ldots + \lambda br(r-1)x^{r-2} + \ldots = 0. \tag{2.12}$$

Here, the coefficient of the leading (lowest order) term must vanish. Assuming $r < 3/2$, we get $r = 0$ or $r = 1$, which constradicts our *a priori* assumption of r being non-integer. For $r = 3/2$, the sum of the two coefficients in (2.12) does not vanish. Finally, assuming $r > 3/2$ (but < 2 as argued above) the first term is the leading one, and we see that its coefficient vanishes for

$$r = 5/3 . \tag{2.13}$$

Hence, we have $y''(x) \propto x^{-1/3}$, $\alpha(x) \propto x^{2/3}$ and $m(x) \propto x$ in the vicinity of $x = 0$ in this example of $p = 2$.

We now return to the general discussion and consider the case where the point $x = x_j$ of $m = \alpha = 0$ is an interior point in the interval for x. Then a discontinuity of the slope y' of the deflection is possible at $x = x_j$, and in view of (2.9) such behaviour would then imply a discontinuity of the shear force $t(x) = -(\alpha^p y'')'$.

Analysing the functional behaviour predicted by Eqs. (2.4b) and (2.4c) in the right and left hand vicinities of an inner singular point $x = x_j$ by means of power series expansions, we find [58] that the leading terms in the expansions of y and α meeting continuity of y and $(\alpha_p y'')' + \lambda y'$ are given by

$$y(x) = \begin{cases} a_0 + a_1^+(x - x_j) + b^+(x - x_j)^r + \dots & x \geq x_j \\ a_0 + a_1^-(x_j - x) + b^-(x_j - x)^r + \dots & x \leq x_j \end{cases} \tag{2.13a}$$

and

$$\alpha(x) = \begin{cases} g^+(x - x_j)^s + \dots & x \geq x_j \\ g^-(x_j - x)^s + \dots & x \leq x_j \end{cases} \tag{2.13b}$$

for $p = 1$, 2 and 3, provided that $(\alpha^p y'')' + \lambda y'$ is non-vanishing at $x = x_j$. The power r in (2.13a) and the leading power s in (2.13b) are given by

$$p = 1 : \; r = 2 \quad , \; s = 1$$

$$p = 2 : \quad r = 5/3 \ , \quad s = 2/3 \qquad (2.14)$$

$$p = 3 : \quad r = 3/2 \ , \quad s = 1/2$$

respectively. For $p = 1$, higher powers than $r = 2$ are not present in (2.13a) and the coefficients b^+ and b^- are equal to $\sqrt{\lambda/2}$ or $-\sqrt{\lambda/2}$. For $p = 2$ and 3, the power r in (2.13a) is the leading non-integer power of the series, and is seen from (2.14) to cause singularity of y''. The coefficients g^+ and g^- in (2.13b) are both positive, and the slopes $\alpha'(x_j)$ and $\alpha'(x_j)$ of the cross-sectional area function are therefore of opposite signs. These slopes are finite for $p = 1$ but infinite for $p = 2$ and $p = 3$.

The functional behaviour given by Eqs. (2.13a,b) and (2.14) for $p = 1$, 2 and 3 at an arbitrary inner point $x = x_j$ of zero bending moment is associated with a possible slope jump ($a_1^+ \neq a_1^-$ in general) and jump of the shear force $t(x) = -(\alpha^p y'')'$, but the deflection $y(x)$ and the function $(\alpha^p y'')' + \lambda y'$ are continuous. Therefore, the point $x = x_j$ corresponds, physically and kinematically, to an *inner hinge* of the optimal column.

Note finally that Eqs. (2.13a,b) as special cases (with $a_0 = 0$) express the behaviour of y and α in the vicinity of a simply supported column end point $x = x_j = 0$ (or $x = x_j = 1$) for $x \geq x_j = 0$ (or $x \leq x_j = 1$).

2.4 Conditions for Optimally Placed Inner Hinges

In problems of optimizing statically determinate columns by means of a single mode, geometrically unconstrained formulation, the locations of singular points of zero bending moment are known beforehand. Thus, simply supported or free end points are predetermined to be singular. However, in *a priori* statically indeterminate problems of the type mentioned, singularities may occur at inner points. The positions of such points can be prescribed for a particular column to be optimized, while in other problems, we may consider the locations $x = x_j$, $j =$

1,...,S of the singularities (hinges) to be additional design variables. Problems of the latter type were for the first time considered by Masur [81].

For columns, the condition for optimal location $x = x_j$ of an inner hinge is [58]

$$\textit{either} \qquad y'(x_j^+) = y'(x_j^-) \tag{2.15a}$$

$$\textit{or} \qquad y'(x_j^+) = -y'(x_j^-) - 2\,\frac{y(x_j) - y(x_{j+1})}{x_{j+1} - x_j}\,, \tag{2.15b}$$

where, in (2.15b), x_{j+1} denotes the position of an adjacent inner hinge or hinged end point of the column, and it is presumed that no beam support or additional hinge are placed between the points x_j and x_{j+1}.

The reader is referred to Ref. [58] concerning the derivation of the above condition.

2.5 Discussion

The buckling load of the clamped-simply supported optimal column shown in Fig. 1c and originally determined in [64], is actually bimodal. For this type of column, we obtain the same optimal design independently of whether we use (2.15a) or (2.15b) to govern the location of the inner hinge of zero bending moment. However, as will be illustrated next, it is necessary to pay full attention to both conditions in other geometrically unconstrained problems.

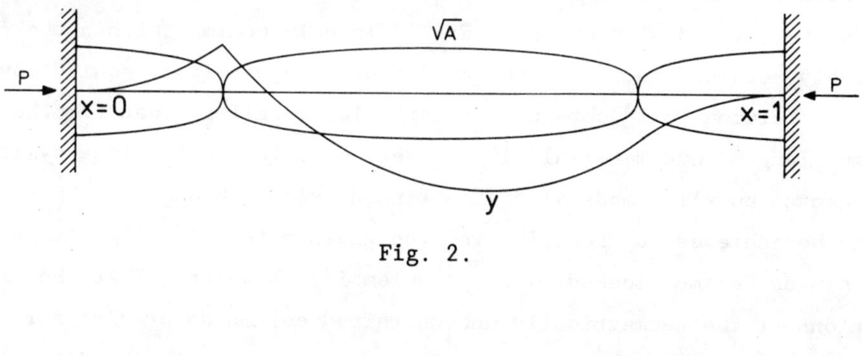

Fig. 2.

Fig. 2 shows the geometrically unconstrained optimal design of a doubly clamped column (p = 2) with two inner hinges, Refs. [67,58]. This design is also bimodal, but in this case, the optimal position of the left hand hinge is governed by condition (2.15b), while the position of the right hand hinge is governed by condition (2.15a). If, for example, (2.15a) were used for both inner hinges, their positions would change, and a slightly different design be obtained. This design would maximize the *second* buckling eigenvalue (with a symmetric mode), see Ref. [67], but it would have a much lower fundamental Euler buckling eigenvalue than the design shown in Fig. 2, and hence not be optimal in the sense of maximizing the buckling load. In fact, the doubly clamped column solution published in [64] is subject to this mistake.

2.6 Multimodal Formulation for Optimal Design

In this section, we consider a problem in which a bimodal rather than a single mode formulation is necessary in order to arrive at the correct optimal design. For reasons of generality, we present a multimodal formulation of the problem.

The example problem consists in maximizing the Euler buckling eigenvalue λ, Eq. (2.3), of a doubly clamped, solid elastic column (p = 2) of given volume and length. This problem was first considered in [64], but as is shown in [67], an erroneous solution was arrived at. The design shown in Fig. 2 constitutes the correct solution within the premises of a single mode formulation of the problem, and it replaces the design from [64], which was also obtained on the basis of a single mode formulation.

However, the design shown in Fig. 2 is only optimal within the class of doubly clamped columns with *two inner hinges*. It is quite obvious that the column would obtain a larger Euler buckling load for the same volume, length and material, if it were made to buckle in a symmetric fundamental buckling mode with a continuous slope throughout. This could easily be achieved by restributing the given material slightly so that the hinges became locked. The problem is, however, that the field equations of the geometrically unconstrained column *do predict* zero cross-section (singular behaviour) at points of vanishing bending moment, and two such points are necessarily present in a clamped-clamped column whenever a single mode formulation is used.

This clearly indicates that the single mode formulation is inadequate for the problem under consideration, and motivated a reformulation of the problem in [67], leading to a new optimality condition, that does not necessarily lead to vanishing cross-section at points of zero bending moment.

Let us now reformulate and expand the initial formulation of our optimization problem in Section 2.2 by following [82]. The eigenvalues λ_i , $i = 1, \ldots, \infty$, of our elastic column are expressed in terms of the modes $y_i(x)$ by

$$\lambda_i = \int_0^1 \alpha^p y_i''^2 dx \qquad i = 1, \ldots, \infty, \tag{2.16}$$

provided that the modes are normalized by

$$\int_0^1 y_i'^2 dx = 1 \qquad i = 1, \ldots, \infty. \tag{2.17}$$

Also the eigenfunctions for $\lambda_i \neq \lambda_j$ are orthogonal, i.e.,

$$\int_0^1 y_i' y_j' dx = 0 \qquad \text{for } \lambda \neq \lambda_j \tag{2.18}$$

but this need not be the case if $\lambda_i = \lambda_j$, $i \neq j$. However, let us take the entire set of modes $[y_k(x)]$ to be orthonormalized according to

$$\int_0^1 y_i' y_j' dx = \delta_{ij} \qquad i,j = 1, \ldots, \infty, \tag{2.19}$$

where δ_{ij} is Kronecker's delta.

The condition of given volume for the column is expressed by

$$\int_0^1 \alpha dx = 1 \ , \tag{2.20}$$

and to formulate the problem in some generality, we will consider a geometric minimum constraint for the design variable $\alpha(x)$, namely that

$$\alpha(x) \geq \underline{\alpha} \tag{2.21}$$

throughout, assuming the minimum allowable value $\underline{\alpha}$ ($0 \leq \underline{\alpha} \leq 1$) to be given.

The design problem may now be stated as

$$\max_{\alpha(x)} \ \min_{i}(\lambda_i) \tag{2.22}$$

This max-min problem is non-differentiable, however. In order to circumvent this difficulty, we use a bound formulation [82,83], which consists in introducing an extra parameter β which ensures that we have a standard differentiable problem even if multimodal eigenvalues occur. Hence, we transform the problem (2.22) into the problem of maximizing a bound β subject to the constraints $\lambda_i \geq \beta$, $i = 1, \ldots, \infty$. In this way the parameter β replaces a non-differentiable functional and is to be maximized over a constraint set in an enlarged space. The points of non-differentiability correspond to "corners" in the constraint set of the enlarged space and arise from intersections of differentiable constraints.

The bound formulation for our problem has the form:

$$\max_{\alpha(x)} \beta$$

$$\text{subject to} \quad \beta \leq \lambda_i \iff \beta - \lambda_i + h_i^2 = 0 \qquad i = 1, \ldots, \infty$$

$$\lambda_i = \int_0^1 \alpha^p y_i''^2 dx \qquad i = 1, \ldots, \infty$$

$$\int_0^1 y_i' y_j' dx = \delta_{ij} \qquad i = 1, \ldots, \infty, \quad j = 1, \ldots, i \tag{2.23}$$

$$\int_0^1 \alpha dx = 1$$

$$\alpha(x) \geq \underline{\alpha} \Leftrightarrow \underline{\alpha} - \alpha(x) + g^2(x) = 0.$$

Here the symbols h_i and $g(x)$ designate real slack variables that convert the inequality constraints to equality constraints.

To solve the problem (2.23), we construct an augmented Lagrangian

$$L = \beta - \sum_{i=1}^{\infty} \eta_i \left[\beta - \int_0^1 \alpha^p y_i''^2 dx + h_i^2\right]$$

$$- \sum_{i=1}^{\infty} \sum_{j=1}^{i} \rho_{ij} \left[\int_0^1 y_i' y_j' dx - \delta_{ij}\right] \tag{2.24}$$

$$- \Gamma\left[\int_0^1 \alpha dx - 1\right] - p\int_0^1 \sigma(x) \left[\underline{\alpha} - \alpha(x) + g^2(x)\right] dx$$

where η_i, ρ_{ij}, Γ and $\sigma(x)$ are Lagrangian multipliers. Note that the multipliers ρ_{ij} are only defined for $i \geq j$.

The condition of stationarity of L with respect to variation of β and h_i gives

$$\sum_{i=1}^{\infty} \eta_i = 1 \tag{2.25}$$

where

$$\eta_i = 0 \text{ if } \lambda_i > \beta, \quad \eta_i \geq 0 \text{ if } \lambda_i = \beta, \quad i = 1, \ldots, \infty. \tag{2.26}$$

Variation of $\alpha(x)$ and $g(x)$ yields the so-called optimality condition

$$\alpha^{p-1} \sum_{i=1}^{\infty} \eta_i y_i''^2 = \kappa - \sigma(x) \tag{2.27}$$

where $\kappa = \Gamma/p$, and

$$\sigma(x) = 0 \ \text{ if } \ \alpha(x) > \underline{\alpha} \ , \quad \sigma(x) \geq 0 \ \text{ if } \ \alpha(x) = \underline{\alpha} \ . \tag{2.28}$$

Finally , stationarity of L with respect to variation of the i-th mode y_i is expressed by

$$2\eta_i(\alpha^p y_i'')'' + 2\rho_{ii} y_i'' + \sum_{j=1}^{i-1} \rho_{ij} y_j'' + \sum_{j=i+1}^{\infty} \rho_{ji} y_j'' = 0 \tag{2.29}$$

after integration by parts, using the boundary conditions. In (2.29) and in the following, summation is only to be carried out over repeated indices when explicitly stated.

Let us now assume that a total number N of Lagrangian multipliers η_i are greater than zero, which is the same as assuming that the fundamental eigenvalue is (at least) N-fold, c.f. Eq.(2.26). Moreover, let us renumber our variables, i.e., use the first N values of an index n , i.e., n = 1,...,N , to identify the Lagrangian multipliers η_n and modes $y_n(x)$ that are associated with the N-fold eigenvalue $\beta = \lambda_n$, n = 1,...,N. Then Eqs. (2.25) - (2.29) become

$$\sum_{n=1}^{N} \eta_n = 1 \tag{2.30}$$

$$\left.\begin{array}{lll} \eta_n > 0 \ , \ \lambda_n = \beta & & n = 1,\ldots,N \\ \eta_n = 0 \ , \ \lambda_n \geq \beta & & n = N+1,\ldots,\infty \end{array}\right] \tag{2.31}$$

$$\left.\begin{array}{lll} \eta^{p-1} \sum_{n=1}^{N} \eta_n y_n''^2 = \kappa & (\text{if } \alpha > \underline{\alpha}) & x \in x_u \\ \alpha(x) = \underline{\alpha} & & x \in x_c \end{array}\right] \tag{2.32}$$

$$2\eta_n(\alpha^p y_n'')'' + 2\rho_{nn} y_n'' + \sum_{j=1}^{n-1} \rho_{nj} y_j'' + \sum_{j=n+1}^{\infty} \rho_{jn} y_j'' = 0, \quad n = 1,\ldots,\infty \tag{2.33}$$

Eqs. (2.32) are readily obtained from Eqs. (2.27) and (2.28). The symbols x_u and x_c denote the unions of sub-intervals in which we have $\alpha(x) > \underline{\alpha}$ (*unconstrained* cross-sectional area) and $\alpha(x) = \underline{\alpha}$. (*constrained* area), respectively.

In order to determine the Lagrangian multipliers ρ_{nm} , we first multiply (2.33) by y_n , integrate by parts over the interval in applying of the boundary conditions, and use (2.16) and (2.19) to obtain $\rho_{nn}=\lambda_n\eta_n$, $n = 1,\ldots,\infty$. In view of (2.31) we thus have

$$\rho_{nm} = \left[\begin{array}{ll} \beta\eta_n , & n = 1,\ldots,N , \\ 0 , & n = N+1,\ldots,\infty . \end{array}\right. \tag{2.34}$$

To determine the remaining components of ρ_{nm} (i.e., those associated with $n > m$), we first write Eq.(2.33) with index n replaced by m . Then, we multiply this equation by y_n and Eq.(2.33) by y_m , integrate both equations by parts using the boundary conditions, assume $n > m$, and apply (2.19). Subtracting and adding the two resulting equations, we finally obtain respectively.

$$\rho_{nm} = (\eta_n + \eta_m) \int_0^1 \alpha^p y_n'' y_m'' dx , \quad n > m , \quad m = 1,\ldots,\infty , \tag{2.35}$$

$$(\eta_n - \eta_m) \int_0^1 \alpha^p y_n'' y_m'' dx = 0 , \quad n > m , \quad m = 1,\ldots,\infty , \tag{2.36}$$

For values of $n > N$ we have $\eta_n = 0$ by (2.31) and it is then easily seen from (2.35) and (2.36) that

$$\rho_{nm} = 0 \ , \quad n > m \ , \ n = N + 1,\ldots,\infty \ . \tag{2.37}$$

This equation implies together with (2.31) and (2.34) that only the modes associated with the N-fold fundamental eigenvalue enter Eqs.(2.33), and that we only need to consider Eqs.(2.33) for $n = 1,\ldots,N$.

Consider now any of the equations (2.33) associated with a given $n \leq N$, and write its solution y_n as $y_n = w_n + z_n$ where the function w_n designates the solution to the eigenvalue problem consisting of the given set of boundary conditions and the differential equation $(\alpha^p w_n'')'' + \beta w_n = 0$, which is constructed from (2.33) by setting the two first terms on the left hand side equal to zero and using (2.34). The function z_n is then due to the terms under the summation signs in (2.33) The Rayleigh quotient associated with the aforementioned eigenvalue problem is defined by

$$R[u] = \int_0^1 \alpha^p u''^2 dx \Big/ \int_0^1 u'^2 dx \ ,$$ where u is an admissible function, and

Rayleigh's minimum principle implies that $\beta = R[w_n] \leq R[w_n + z_n] = R[y_n] = \lambda_n$, where the last relationship follows from Eqs.(2.16), (2.17) and the definition of R. Now, Eqs.(2.31) require strict equality of β and λ_n, i.e., $\beta = \lambda_n$ for $n = 1,\ldots,N$. Hence, we must have $y_n(x) \equiv w_n(x)$, i.e., $z_n(x) \equiv 0$, $n = 1,\ldots,N$, and as is shown in the Appendix (Section 2.9) this requires vanishing of the Lagrangian multipliers,

$$\rho_{nm} = 0 \ , \ n > m \ , \ n = 2,\ldots,N \ , \tag{2.38}$$

in Eq.(2.33).

By means of (2.31), (2.34), (2.37) and (2.38), we may now write Eqs. (2.33) as the familar differential equations for buckling

$$(\alpha^p y_n'')'' + \lambda_n y_n'' = 0 \ , \ n = 1,\ldots,N \ , \tag{2.39}$$

where $\lambda_n = \beta$, $n = 1,\ldots,N$. It is also worth noting that Eqs. (2.35), (2.31) and (2.38) imply vanishing of the integrals

$$\int_0^1 \alpha^p y_n'' y_m'' dx = 0 \ , \ n > m \ , \ n = 2,\ldots,N \ , \tag{2.40}$$

that represent the mutual bending energy of the modes y_n and y_m .

2.7 The Bimodal Case

For the type of problem treated above, the multiplicity N of the fundamental eigenvalue is not known beforehand, but must be determined as the highest possible number of positive Lagrangian multipliers η_n , which, together with their associated linearly independent modes y_n , is admitted by the optimality condition (2.32). Up to now the highest multiplicity found for an optimal Euler column buckling eigenvalue is $N = 2$, cf. [67] for the case of a doubly clamped column. $N = 1$ is usual for columns with other boundary conditions.

If $N = 1$, Eq. (2.30) gives $\eta_1 = 1$ and it is readily seen that Eq. (2.32) reduces to the traditional single mode optimality condition, Eq. (2.4c).

For the case of $N = 2$, Eqs. (2.30) and (2.32) may be combined into the following condition where we write γ in place of η_1 :

$$\alpha^{p-1} [\gamma y_1''^2 + (1 - \gamma) y_2''^2] = \kappa \qquad (\text{if } \alpha > \underline{\alpha}) \qquad x \in X_u \tag{2.41}$$

This condition is identical with the optimality condition derived by Olhoff and Rasmussen [67]. It follows directly from (2.30) and (2.31) that in (2.41) $\eta_1 = \gamma$ must lie in the interval

$$0 \le \gamma \le 1. \tag{2.42}$$

This condition has earlier been established as a sufficient

condition for local optimality by Masur and Mróz [84,85]. Other derivations of necessary and sufficient conditions for optimality of bimodal eigenvalues are available in Bratus and Seyranian [86,87].

Together with the condition $\alpha(x) = \underline{\alpha}$ for $x \in x_c$, Eq. (2.41) may be solved for $\alpha(x)$ to give

$$p = 2,3 : \quad \alpha(x) = \begin{cases} \left[\dfrac{\kappa}{(1-\gamma){y_1''}^2 + \gamma {y_2''}^2} \right]^{1/(p-1)} & (\text{if } \alpha > \underline{\alpha}) \quad x \in x_u \\ \underline{\alpha} & x \in x_c \end{cases} \tag{2.41a}$$

for $p = 2$ or $p = 3$. For the case of $p = 1$, we obtain

$$p = 1 : \quad \begin{aligned} (1-\gamma){y_1''}^2 + \gamma {y_2''}^2 &= \kappa \quad (\text{if } \alpha > \underline{\alpha}) \quad x \in x_u \\ \alpha(x) &= \underline{\alpha} \qquad\qquad\qquad\qquad x \in x_c \end{aligned} \tag{2.41b}$$

As is the case for Eq. (2.7b), we note that for $p = 1$, the optimality condition for the geometrically unconstrained sub-interval(s) does not contain the design variable α, which must therefore be determined from the buckling differential equation. A method of solution for $p = 1$ is presented in [84] and the case of $p = 1$ will not be considered further here.

For exemplification, let us follow [67] and derive convenient expressions for the Lagrangian multipliers κ and γ and the optimum buckling eigenvalue λ for the more complex cases of $p = 3$ and $p = 2$. First, we substitute (2.41a) into the volume constraint (2.20), thereby obtaining an explicit expression for κ,

$$p = 2,3 : \quad \kappa = \left[\frac{1 - \underline{\alpha} \displaystyle\int_{x_c} dx}{\displaystyle\int_{x_u} \frac{dx}{[(1-\gamma){y_1''}^2 + \gamma {y_2''}^2]^{1/(p-1)}}} \right]^{p-1} \tag{2.43}$$

Then, subtracting the two equations comprised in (2.16) for $i = 1$

and $i = 2$, substituting $\alpha(x)$ from (2.41a) and using (2.43), we find the following implicit equation for γ ,

$$p = 2,3 : \quad \int_{x_u} \frac{y_1''^2 - y_2''^2}{[(1-\gamma)y_1''^2 + \gamma y_2''^2]^{p/(p-1)}}\,dx \; + \tag{2.44}$$

$$\underline{\alpha}^p \left[\frac{\displaystyle\int_{x_u} \frac{dx}{[(1-\gamma)y_1''^2 + \gamma y_2''^2]^{1/(p-1)}}}{1 - \underline{\alpha}\displaystyle\int_{x_c} dx} \right]^p \int_{x_c} (y_1''^2 - y_2''^2)\,dx = 0$$

Finally, substitution of (2.41a) and (2.43) into first of Eqs. (2.16) gives an explicit expression for λ ,

$$p = 2,3 : \quad \lambda = \underline{\alpha}^p \int_{x_c} y_1''^2 dx + \left[\frac{1 - \underline{\alpha}\displaystyle\int_{x_c} dx}{\displaystyle\int_{x_u} \frac{dx}{[(1-\gamma)y_1''^2 + \gamma y_2''^2]^{1/(p-1)}}} \right]^p x \tag{2.45}$$

$$\int_{x_u} \frac{y_1''^2}{[(1-\gamma)y_1''^2 + \gamma y_2''^2]^{p/(p-1)}}\,dx \; .$$

Equations $(2.16)_{i=1,2}$, $(2.17)_{i=1,2}$, $(2.39)_{n=1,2}$, (2.41a), (2.43)-(2.45) comprise the complete set of necessary equations governing the bimodal optimal design problem for $p = 3$ and $p = 2$, and they constitute a strongly coupled, non-linear integro-differential eigenvalue problem. The unknowns to be determined are the optimal buckling eigenvalue λ , the optimal column cross-sectional area function $\alpha(x)$ (and thereby the sub-intervals x_u and x_c), eigenfunctions y_1 and y_2 , and the Lagrangian multipliers κ and γ , respectively. The solutions depend in general on the minimum constraint $\underline{\alpha}$, which is the only specified quantity in the non-dimensional formulation.

A method of numerical solution based on successive iterations is presented in [67]. In that paper, the modes y_1 and y_2 were not taken to be mutually orthogonal. The results are exposed in the subsequent section. Other examples where bi- or multimodality of optimal eigenvalues occur, may be found in Refs. [84-101].

The bimodal formulation for optimal design described above contains geometrically unconstrained optimization and/or single mode optimization as special cases. The principal advantage of the new formulation is that while the optimality condition (2.4c) of the single mode formulation predicts formation of hinges at points of zero bending moment in a geometrically unconstrained formulation of optimal design, the bimodal optimality condition (2.41) does not necessarily lead to zero cross-section and formation of hinges at points of zero bending moment.

2.8 Example: Bimodal Optimization of a Doubly Clamped p = 2 Column [67]

Fig. 3 illustrates optimal designs $\pm\sqrt{\alpha}$ and associated fundamental single or double modes corresponding to selected values of a geometric minimum constraint $\underline{\alpha}$ on the cross-sectional area α of a doubly In Fig. 3b, $\underline{\alpha} = 0.4$ and $\lambda = 51.775$ is simple. In Fig. 3c, $\underline{\alpha} = 0.25$ and the optimal buckling load $\lambda = 52.349$ is bimodal. Fig. 3d shows the optimal solution corresponding to any value of $\underline{\alpha}$ belonging to the interval $0 \leq \underline{\alpha} \leq 0.226$, where the constraint is no longer active in the design. The corresponding optimal buckling load $\lambda = 52.3563$ is bimodal.

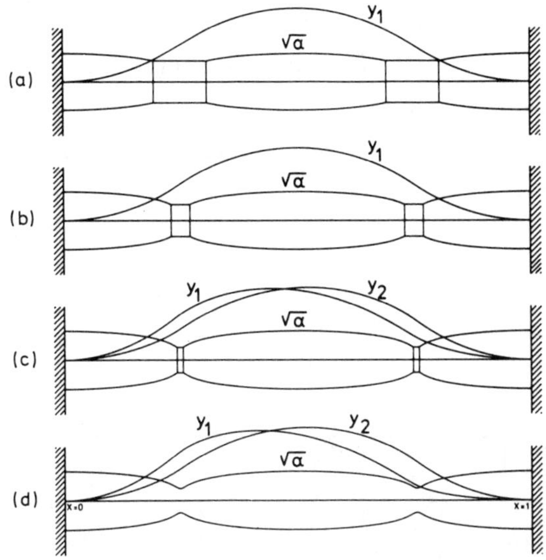

Fig. 3

In Fig. 4, curve ABCD is based on a number of solutions and shows λ as a function of the geometric minimum constraint $\underline{\alpha}$. For $0.280 < \underline{\alpha} \leq 1$, the optimal designs are associated with a simple fundamental eigenvalue λ , given by curve CD . curve CE shows the second order eigenvalues λ_2 of the simple optimal eigenvalue designs behind curve CD . At point C , the two curves are seen to coalesce at the value 0.280 for $\underline{\alpha}$, and for $0 \leq \alpha \leq 0.280$, the optimal designs are associated with a *bimodal fundamental eigenvalue*, cf. modes y_1 and y_2 in Figs. 3c and d. All the designs obtained are symmetrical (this was not assumed in the solution procedure), and purely symmetrical and antisymmetrical linear combinations of double modes y_1 and y_2 can be constructed.

As shown by curve DCB of Fig. 4, the optimal buckling eigenvalue λ increases with decreasing constraint for $0.226 \leq \underline{\alpha} \leq 1$, and for these lues of $\underline{\alpha}$ the constraint is *active* in the optimal designs, cf. Figs. 3a, b and c. However, for values of clamped $p = 2$ column. In Fig.

3a, $\underline{\alpha} = 0.7$ and $\lambda = 48.690$ is simple. α belonging to the interval $0 \leq \underline{\alpha} \leq 0.226$ the minimum constraint is *inactive* in the optimal design, and the associated bimodal fundamental buckling eigenvalue is *constant*, cf. AB in Fig. 4. For these values of $\underline{\alpha}$, $0 \leq \underline{\alpha} \leq 0.226$, the optimal design, see Fig. 3d, is the *same*, and it has *finite* variable cross-section throughout, with a minimum magnitude of $\alpha = 0.226$.

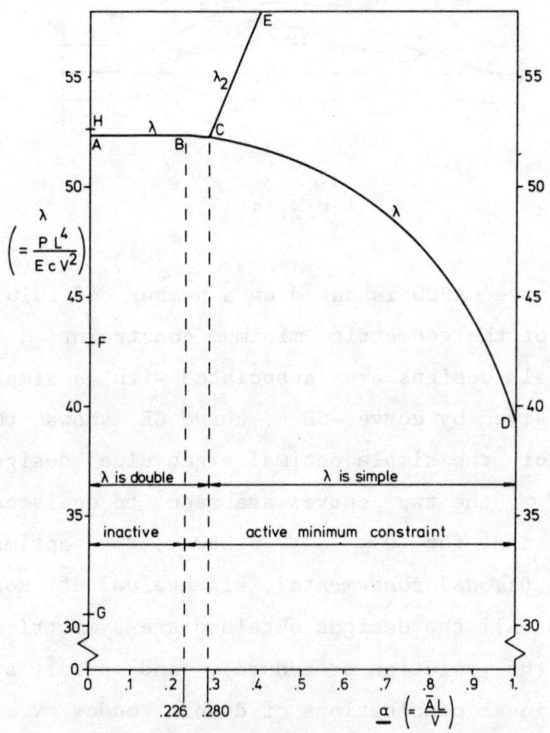

Fig. 4

This $\underline{\alpha}$ independent bimodal optimal design in Fig. 3d is the *solution to the geometrically unconstrained optimization problem for a doubly clamped column* of $p = 2$. Its fundamental, double buckling eigenvalue λ is 32.62% higher than the fundamental eigenvalue of a corresponding uniform column of the same volume, length and material. The bimodal optimal design replaces not only the solution arrived at in [64], but also the geometrically unconstrained, candidate design in Fig. 2.

The numerical result for the geometrically unconstrained bimodal column obtained in Ref. [67] has later been found to be of considerable accuracy by the exact analytical solution obtained independently by Masur [85] and Seyranian [88,89]. The result provides a noteworthy example of a *statically indeterminate* solution to a geometrically unconstrained, one-dimensional, single purpose, structural optimization problem: it constitutes an obvious exception to the "general rule" that solutions to the broad class of all such problems will always be statically determinate.

2.9 Appendix to Chapter 2

We shall consider the problem of Section 2.6 and use the same notation. See also Ref. [82].

Assume that the non-vanishing Lagrangian multipliers are not all different, so that for example

$$\begin{aligned} &\eta_1 = \eta_2 = \dots = \eta_q = \mu \\ &\eta_{q+i} \neq \eta_{q+j} \;, \quad i \neq j \;, \quad i,j = 0,\dots,N-q \;. \end{aligned} \tag{A1}$$

Eqs. (2.35) and (2.36) together with (2.37) in this case give that

$$\rho_{nm} = 0 \;, \quad n > m \;, \quad n = q+1,\dots,\infty \;. \tag{A2}$$

Eqs. (2.33) for $n = 1,\dots,q$ then reduce to

$$2\mu(\alpha^p y_n'')'' + 2\mu\beta y_n'' + \sum_{j=1}^{q} A_{nj} y_j'' = 0 \tag{A3}$$

as $\rho_{nn} = \eta_n \beta$. Here the symmetric matrix $[A_{ij}]$ is given by

$$A_{ij} = \begin{cases} \rho_{ij} & i > j \\ 0 & i = j \\ \rho_{ji} & i < j \end{cases} \tag{A4}$$

Note that $[A_{ij}]$ has trace equal to zero, so the sum of the eigenvalues of $[A_{ij}]$ is zero. Now let $\underline{a}_i = (a_{i1},\ldots,a_{iq})$ denote the eigenvectors corresponding to the eigenvalues γ_i and set

$$Y_n = \sum_{i=1}^{q} a_{ni} y_i \quad , \quad n = 1,\ldots,q \ . \tag{A5}$$

By multiplying the i'th equation of (A3) by a_{ni} and adding the resulting equations we get for Y_n

$$2\mu(\alpha^p Y_n'')'' + 2\mu\beta Y_n'' + \gamma_n Y_n'' = 0 \ , \quad n = 1,\ldots,q \tag{A6}$$

Thus we have eigenvalues

$$\beta + \gamma_n/2\mu \ , \quad n = 1,\ldots,q \ . \tag{A7}$$

It follows from the fact that $[A_{ij}]$ has trace zero that any non-zero eigenvalue γ_n is followed by a non-zero eigenvalue of opposite sign. As β is the smallest eigenvalue (cf. the discussion preceding Eq. (2.38)), we conclude that all the eigenvalues γ_n must be zero, which in turn implies that

$$\rho_{nm} = 0 \ , \ n > m \ , \ n = 2,\ldots,q \ . \tag{A8}$$

This combines with (A2) to

$$\rho_{nm} = 0 \ , \ n > m \ , \ n = 2,\ldots,\infty \ . \tag{A9}$$

3. OPTIMIZATION OF TRANSVERSELY VIBRATING BERNOULLI-EULER BEAMS AND ROTATING SHAFTS WITH RESPECT TO THE FUNDAMENTAL NATURAL FREQUENCY OR CRITICAL SPEED

3.1 Introduction

This chapter deals with problems of determining the distribution of structural material in transversely vibrating beams or rotating circular shafts, such that maximum values of natural frequencies or critical whirling speeds are obtained for a prescribed amount of material, length and boundary conditions for the beam or shaft. Equivalently, we minimize the volume of structural material for a given vibration frequency or critical speed.

The practical significance of such problems is that they provide designs of minimum weight (or cost) of material against beam vibrational resonance due to external excitation of given frequency, and against failure due to whirling instability at service speeds, within a large range from zero and up to the particular fundamental frequency or first critical speed.

3.2 Geometrically Constrained Optimal Design

An elastic Bernoulli-Euler beam of length L and structural volume V vibrates at its fundamental angular frequency ω of free transverse vibrations. The beam is made of a material with Young's modulus E and the mass density ρ, and it has variable but similarly oriented cross-sections with the relationship $I = cA^p$ between the area moment of inertia I and the area A, cf. Eq. (1.1). We restrict ourselves to the cases of $p = 2$ (geometrically similar, solid cross-sections) and $p = 3$ (solid cross-sections of fixed width and variable height), because the case of $p = 1$ (sandwich cross-sections) is often degenerate for the types of problems to be considered, cf. the discussion in Sections 1.5, 1.8, and Refs. [48,54,59,60]. The constant c for the cross-sectional shape and the value of p (p = 2 or p = 3) are assumed to be given.

For the *particular case of* p = 2 , we may conceive the structure to be a *shaft* of circular cross-sections, that rotates at its fundamental critical angular whirling speed ω, if we neglect gyroscopic effects.

We shall assume that our vibrating beam (or rotating shaft) carries no *distributed* nonstructural mass, but that a number K of given nonstructural masses (or circular disks) Q_i , $i = 1,\ldots,K$, are attached to the beam/shaft at specified points $X = X_i$, where X is the beam/shaft coordinate (which is denoted by x in Chapter 1).

Identifying Λ of Chapter 1 as ω^2 , and introducing non-dimensional quantities,

$$x = X/L \qquad 0 \le x \le 1 \tag{3.1a}$$

$$\alpha(x) = A(x)L/V \quad , \quad \underline{\alpha} = \underline{A}L/V \tag{3,1b}$$

$$q_i = \frac{Q_i}{\rho V} \qquad x = x_i \ , \ i = 1,\ldots,K \tag{3.1c}$$

$$\lambda = \omega^2 \frac{\rho\, L^{3+p}}{E\, c\, V^{p-1}} \ , \tag{3.1d}$$

i.e. coordinate x , cross-sectional area function $\alpha(x)$, non-structural masses (or circular disks) q_i and fundamental eigenvalue λ, respectively, the non-dimensional beam or shaft will have unit volume and unit length.

Our *dimensionless, geometrically constrained optimization problem* then consists in determining the function $\alpha(x) \ge \underline{\alpha}$ that maximizes λ for given minimum allowable beam/shaft cross-sectional area $\underline{\alpha}$, given positions x_i and magnitudes q_i of non-structural masses/disks, and given, homogeneous boundary conditions.

Using Eqs. (3.1a-d), the governing equations for the optimization problem are easily obtained as a special case of the general set of optimality equations (1.6a-e) with the help of Table 1. Normalizing the vibration/whirling mode according to

$$\int_0^1 \alpha y^2 dx + \sum_i q_i y^2(x_i) = 1 \ , \tag{3.2}$$

such that the denominator of the dimensionless Rayleigh quotient equals unity, the set of governing equations takes the following form

$$\lambda = \int_0^1 \alpha^p y''^2 dx \tag{3.3a}$$

$$(\alpha^p y'')'' = \lambda \alpha y \qquad 0 \le x \le 1 \tag{3.3b}$$

$$p\alpha^{p-1} y''^2 - \lambda y^2 = \kappa \qquad x \in X_u \tag{3.3c}$$

$$\alpha(x) > \underline{\alpha} \quad x \in X_u \ , \quad \alpha(x) = \underline{\alpha} \qquad x \in X_c \tag{3.3d}$$

$$\int_0^1 \alpha(x) dx = 1 \ , \tag{3.3e}$$

where we have redefined the Lagrangian multiplier κ in the optimality condition (3.3c).

In addition to Eqs. (3.2) and (3.3a-e), our optimization problem must satisfy (i) the boundary conditions, (ii) the condition of continuity of the bending moment $m = \alpha^p y''$ except at possible points of prescribed y', and (iii) the conditions of continuity of the shear force $t(x) = -(\alpha^p y'')'$ except at possible points of prescribed y and at the points $x = x_i$, $i = 1,\ldots,K$ with attached nonstructural masses q_i . At the latter points, the jumps of t are identified as $\langle t \rangle_{x_i} = -\lambda q_i y(x_i)$.

Due to the nonlinearity and coupling of the governing equations, closed form solutions cannot be expected for $p = 2$ and $p = 3$, and numerical solution procedures must therefore be applied. Such procedures are available in Refs. [102-104]. Refs. [102] and [104] present results for geometrically constrained $p = 2$ cantilevers with and without nonstructural masses, respectively, and Ref. [103] offers results for p

= 2 and p = 3 beams with various other boundary conditions.

Fig. 5 shows results obtained in [102], namely cantilever beams of geometrically similar cross-sections, p = 2 , (or rotating circular

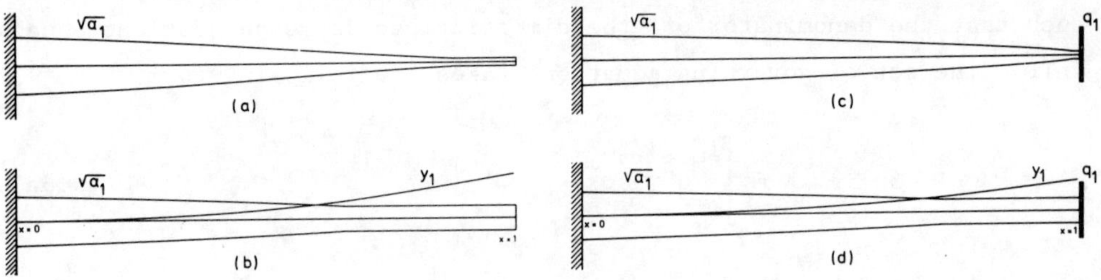

Fig. 5.

cantilever shafts) optimized with respect to the fundamental natural transverse vibration frequency (or first critical speed) $\omega = \omega_1$. The beams are illustrated by optimal *shapes* $\pm\sqrt{\alpha_1}$, where $\alpha_1(x) = \alpha(x) = A(x)L/V$, and the solutions in (a), (c) and (b), (d), respectively, correspond to minimum constraint values $\underline{\alpha} = 0.05$ and 0.5 . The dimensionless nonstructural tip mass in (c) and (d) is $q_1 = Q_1/\rho V = 0.1$. The fundamental frequencies ω_1 of the optimal designs are increased by (a) 279%, (b) 88%, (c) 81% and (d) 57%, respectively, in comparison with those corresponding to uniform designs of the same volume, length, material, and, for (c) and (d), tip mass.

Fig. 6 shows the square root $\sqrt{\lambda_1}$ of the fundamental eigenvalue $\lambda = \lambda_1$ and the square root $\sqrt{\lambda_2}$ of the next eigenvalue λ_2 as functions of the dimensionless cross-sectional area constraint $\sqrt{\underline{\alpha}} = \sqrt{\underline{A}L/V}$ for optimal cantilevers of the type in Fig. 5. Note that $\underline{\alpha} = 0$ and $\underline{\alpha} = 1$ correspond to geometrically unconstrained and fully constrained (uniform) designs, respectively, and that the square root eigenvalues are proportional to the first and second vibration frequencies (critical speeds) ω_1 and ω_2 , respectively, for given beam volume, length and material. The solid curves in Fig.6 represent optimal $\lambda = \lambda_1$

designs without nonstructural mass, see for example Fig. 5a,b, while other curves are for optimal $\lambda = \lambda_1$ designs with a dimensionless tip mass $q_1 = Q_1/\rho V$, see e.g. Fig. 5c,d.

Fig. 6 clearly illustrates that geometrically unconstrained designs are associated with maximum obtainable merits in comparison with corresponding geometrically constrained designs, cf. the discussion in Section 1.6.

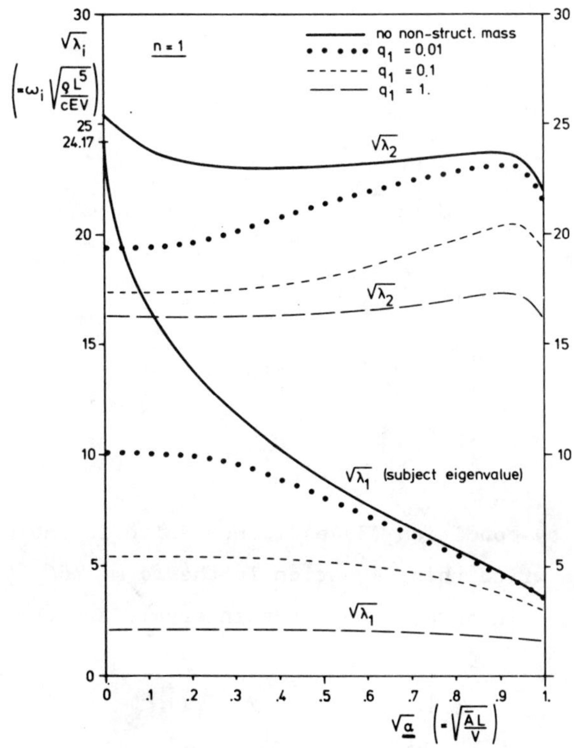

Fig. 6

3.3 Geometrically Unconstrained Optimization

We now consider the case where no geometric constraint is specified for the cross-sectional area function $\alpha(x)$ in the process of optimization. This constitutes a special case of the formulation considered in Section 3.2, and corresponds to setting $\underline{\alpha} = 0$ and the interval x_u equal to the entire interval $0 \leq x \leq 1$. Doing this, the system of Eqs. (3.2), (3.3a-e) reduces to the following system for geometrically unconstrained optimization,

$$\lambda = \int_0^1 \alpha^p y''^2 dx \tag{3.4a}$$

$$\int_0^1 \alpha y^2 dx + \sum_i q_i y^2(x_i) = 1 \tag{3.4b}$$

$$\int_0^1 \alpha dx = 1 \tag{3.4c}$$

$$(\alpha^p y'')'' = \lambda \alpha y \qquad 0 \leq x \leq 1 \tag{3.4d}$$

$$\alpha(x) = \left[\frac{\kappa + \lambda y^2}{p y''^2}\right]^{1/(p-1)} \qquad 0 \leq x \leq 1 \tag{3.4e}$$

Here the optimality condition (3.4e) is now valid in the entire interval $0 \leq x \leq 1$. If we write this equation in the form $p\alpha^{p-1} y''^2 - \lambda y^2 = \kappa$, multiply it by α, integrate over the interval, and use Eqs. (3.4a-c), we find

$$\kappa = \lambda \left[p - 1 + \sum_i q_i y^2(x_i)\right] , \tag{3.5}$$

i.e., Lagrangian multiplier κ is always positive for problems with $p > 1$.

Geometrically unconstrained solutions obtained numerically by successive iterations are available in Refs. [43,59,60,69,105]. Ref. [59] presents optimal $p = 2$ and $p = 3$ cantilevers with and without tip mass, and $p = 2$ solutions without nonstructural mass are available in Refs. [43,60] for other boundary conditions.

Fig. 7 shows examples of geometrically unconstrained optimal design of $p = 2$ beams with respect to the fundamental natural vibration frequency $\omega = \omega_1$, when no non-structural masses are considered. The design in Fig. 7a is the solution for a simply supported beam [30], Fig. 7b shows the optimal cantilever design [59], and Fig. 7c illustrates the optimal design of a doubly clamped beam [60]. The design in Fig. 7b may be compared with the constrained designs in Figs. 5a-b, and it should be noted that its optimal characteristics (maximum ω_1 for given V and L, or minimum V for given ω_1 and L) are represented by the value indicated for the solid $\sqrt{\lambda_1}$ curve at $\sqrt{\alpha} = 0$ in Fig. 6. It is also worth mentioning that the design in Fig. 7b is at the same time the optimal design of a clamped-simply supported beam [60].

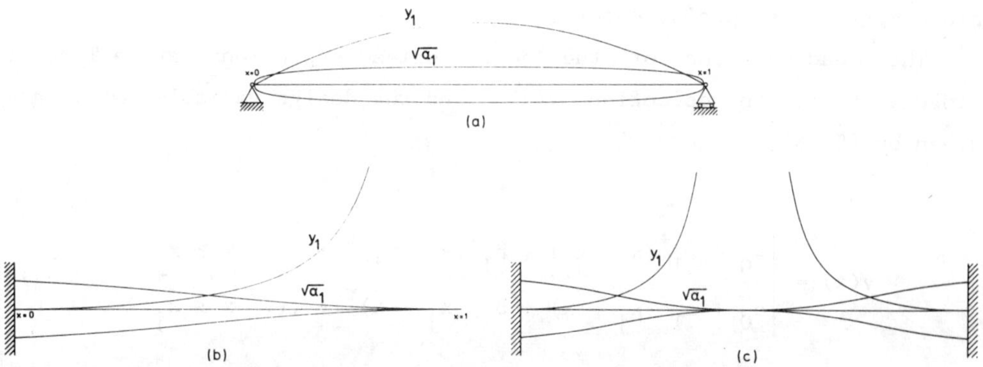

Fig. 7

The fundamental frequencies ω_1 of the optimal solutions in Fig. 7 are increased by (a) 6.6%, (b) 588% and (c) 332%, respectively, when compared with the frequencies of corresponding uniform beams of the same volume, length and material. Comparing (b) with a uniform clamped-simply supported beam, its fundamental frequency ω_1 is increased by

57%.

3.4 Types of Singular Behaviour

In problems of optimizing transversely vibrating Bernoulli-Euler beams or rotating shafts without geometric constraint, there may occur two different types of singular behaviour, both of which are associated with zero bending moment $m(x) = \alpha^p y''$ and cross-section α, but in one type (I) the shear force is finite, while in the other type (II) the shear force vanishes at the singularity. The types of the singularities are independent of whether the beam (or shaft) is optimized with respect to the fundamental frequency (first critical speed) or a higher order eigenfrequency (higher order critical speed).

The singular behaviour can be determined analytically by expanding a solution to Eqs. (3.4d) and (3.4e) in a power series (see e.g. [43,60,58]) near the singular point $x = x_j$, which may either be an inner point or an end point of the beam. In the sequel let us conceive $x = x_j$ as an arbitrary inner singular point (singular boundary points are comprised as special cases).

The leading terms of the power series expansions at a Type I singularity for the delection $y(x)$ and the design variable $\alpha(x)$ are given by [58,60]

$$y(x) = \begin{cases} a_0 + a_1^+(x - x_j) + b_1^+(x - x_j)^r + \ldots & x \geq x_j \\ a_0 + a_1^-(x_j - x) + b_1^-(x_j - x)^r + \ldots & x \leq x_j \end{cases} \tag{3.6a}$$

(which reflects continuity of y), and

$$\alpha(x) = \begin{cases} g_1^+(x - x_j)^s + \ldots & x \geq x_j \\ g_1^-(x_j - x)^s + \ldots & x \leq x_j \end{cases} \tag{3.6b}$$

respectively, where a_0, a_1^+, a_1^-, b_1^+, b_1^-, g_1^+ (>0) and g_1^-

(>0) are coefficients. Depending on the value of p, the powers r and s in (3.6a) and (3.6b) attain the following values,

$$p = 2 : \quad r = 5/3 \ , \ s = 2/3$$
$$p = 3 : \quad r = 3/2 \ , \ s = 1/2 \ . \tag{3.7}$$

In (3.6a), r is the leading non-integer power of the series, which implies that the second and higher order derivatives of y are unbounded in the neighbourhood of the singularity.

The finite shear force $t(x) = -(\alpha^p y'')'$ at a Type I singularity ensures non-zero values of the coefficients b_1^+ and b_1^- in (3.6a) as well as of the coefficients g_1^+ and g_1^- in (3.6b). With exception of problems in which an external transverse force acts at the point $x = x_j$ (and in buckling problems in general), the shear force $t(x) = -(\alpha^p y'')'$ will be continuous at $x = x_j$, and therefore we have $b_1^+ = -b_1^-$ in (3.6a) and $g_1^+ = g_1^-$ in (3.6b). Clearly the behaviour of the beam is characterized kinematically and physically by a *hinge* at the location of a Type I singularity.

Let us now discuss singularities of the second type (II) in which the behaviour is associated with zero shear force $t(x) = -(\alpha^p y'')'$ in addition to zero bending moment $m(x) = \alpha^p y''$. At such an interior singularity the deflection y as well as the slope y' may exhibit discontinuity and the point therefore corresponds to an *inner separation* of the beam.

Now, investigating Eqs. (3.4d) and (3.4e) for optimal, geometrically unconstrained transversely vibrating beams by means of power series expansions, the leading terms of the deflection $y(x)$ and of the design variable $\alpha(x)$ are found to be [37,39]

$$y(x) = \begin{bmatrix} d^+(x - x_j)^n + \dots & x > x_j \\ d^-(x_j - x)^n + \dots & x < x_j \end{bmatrix} \tag{3.8a}$$

and

$$\alpha(x) = g(x - x_j)^s + \ldots \quad x \geq x_j \ , \ x \leq x_j \qquad (3.8b)$$

respectively, in the vicinity of an arbitrary inner point $x = x_j$ with a Type II singularity. In Eqs. (3.8a) and (3.8b), d^+, d^- and g are non-vanishing coefficients (g is positive) and the powers n and s of the leading terms attain the following values,

$$\begin{aligned} p = 2 : \ n = -2 \ , \ s = 4 \\ p = 3 : \ n = -1 \ , \ s = 2 \ . \end{aligned} \qquad (3.9)$$

The type of singular behaviour is seen to be quite significant, in that the deflection $y(x)$ and all its derivatives tend to infinity at the point $x = x_j$ where the bending moment, shear force and beam cross-section vanish.

3.5 Conditions for Optimally Placed Inner Type I and II Singularities

Below, we state without derivation the necessary conditions which for $p = 2$ and $p = 3$ govern optimum location of inner Type I singularities (hinges) and inner Type II singularities (separations in optimal, transversely vibrating beams or rotating shafts. A derivation of the conditions is available in [58].

For a Type I singularity, we have the condition

$$y'(x_j^+) + y'(x_j^-) \ , \qquad (3.10)$$

that is, *the slope* y' *is continuous at an optimally placed inner Type I singularity*. This result implies that $a_1^+ = -a_1^-$ in (3.6a) if $x = x_j$ is optimally placed, and was first derived by Masur, Ref. [106].

The condition for an optimally placed Type II singularity is

$$d^+ = \pm d^- \qquad (3.11)$$

i.e., *the coefficients of the leading singular terms in the expansion*

(3.8a) of the vibration mode $y(x)$ *are either equal or equal with opposite signs at an optimally placed inner Type II singularity*. By considering the case of $p = 2$ and assuming d^+ and d^- to have equal signs, the condition $d^+ = d^-$ was first derived in Ref. [60].

In view of Eqs. (3.8a) and (3.8b) where $n = -2$ for $p = 2$ and $n = -1$ for $p = 3$, conditions (3.11) imply that, in the vicinity of the singular point, the vibration mode is either symmetrical with respect to the *line* $x = x_j$ or symmetrical with respect to the *point* $x = x_j$, if x_j is the optimal location of an inner Type II singularity. Note that an optimally placed inner Type II singularity is found in Fig. 7c.

4. OPTIMIZATION OF BERNOULLI-EULER BEAMS AND SHAFTS WITH RESPECT TO HIGHER ORDER EIGENFREQUENCIES

4.1 Introduction

Here, we shall consider an extension of the types of optimization problems studied in Chapter 3. Thus, instead of maximizing the fundamental natural frequency (or critical speed), we will deal with the problem of maximizing a particular *higher order* natural frequency (or critical speed) of given order n $(n > 1)$ for a transversely vibrating Bernoulli-Euler beam (or rotating shaft) of prescribed structural volume, length, material, and boundary conditions.

This problem is governed by a set of dimensionless equations consisting of Eqs. (3.2) and (3.3a-e) for geometrically constrained optimal design or Eqs. (3.4a-e) for geometrically unconstrained design, with λ, $\alpha(x)$ and $y(x)$ subscribed as λ_n, $\alpha_n(x)$ and $y_n(x)$ (indicating reference to the given order n of the subject eigenfrequency), and the *additional equations*

$$\int_0^1 \alpha_n y_n y_j dx + \sum_i q_i y_n(x_i) y_j(x_i) = 0 \ , \ j = 1\ldots,n-1 \tag{4.1a}$$

$$(\alpha_n y_j'')'' = \lambda_j \alpha_n y_j \;, \; j = 1,\ldots,n-1 \qquad (4.1b)$$

see Refs. [60,102]. Eqs. (4.1a) are conditions of orthogonality of $y_n(x)$ against the lower modes $y_j(x)$, $j = 1,\ldots,n-1$, and Eqs. (4.1b) constitute together with the boundary conditions n-1 eigenvalue problems for the lower modes $y_j(x)$ of the optimal design α_n .

As will be discussed in the following, a geometrically unconstrained solution α_n to the problem coincidently constitutes the optimal design to the problem of maximizing the *difference* between two adjacent natural frequencies (or critical speeds) ω_n and ω_{n-1} for given volume and length of the beam (or rotating shaft), see [60]. It is not surprising, therefore, that geometrically *constrained* solutions also exhibit large gaps between two adjacent frequencies [102], and that these gaps are very close to the maximum obtainable gaps. The point is that the geometrically constrained problem of maximum, single, higher order natural frequency is simpler to solve than the problem formulated in terms of maximum difference between two adjacent natural frequencies. The latter type of problem has been solved in Refs. [107,108] (see also Ref. [109]).

Thus, the type of problem to be considered is directly related to pratical design against resonance of beams due to external exitation and whirling instability of rotating shafts: if the structure is designed such that the external excitation frequency or the service speed is isolated in a broad gap between two consecutive higher order natural frequencies or higher order critical speeds, considerable weight savings are possible compared with designs where the excitation frequency or service speed is placed between zero and a high value of the fundamental natural frequency or first critical speed.

Another direct and very important advantage of considering optimization (geometrically constrained as well as unconstrained) with respect to a single, higher order natural frequency (or critical speed) of given order n, is that the resulting optimal design is, at the same time, the optimal design to the problem of optimizing with respect to the *fundamental* natural frequency (or first critical speed), assuming the positions of n-1 available interior supports to be design variables in addition to the cross-sectional area distribution [60,102]. According to Mróz and Rozvany [110] and Rozvany [111], zero support reaction is a necessary condition for optimum location of an interior simple support. This implies [60,102] that the optimal positions for available interior supports in a problem of optimal design with respect to the fundamental frequency, are simply identified with the n-1 nodal points of the vibration mode $y_n(x)$ of the higher order frequency optimal design.

Fig. 8 illustrates as an example the geometrically unconstrained design with optimal positions of four available inner supports that maximize the fundamental frequency of a transversely vibrating $p = 2$

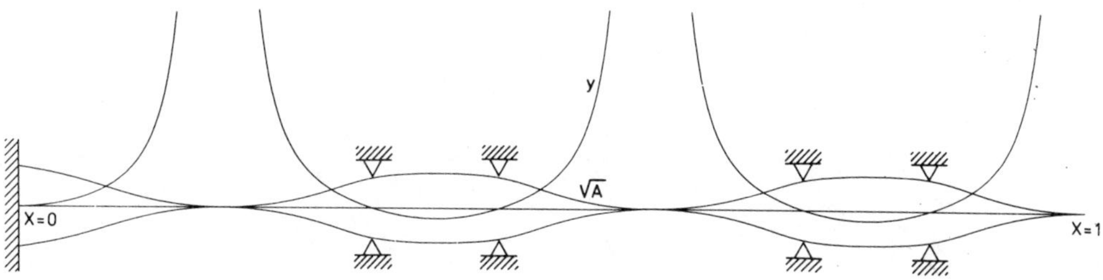

Fig. 8.

beam with clamped and free end points. The design is determined by optimizing the beam without inner supports with respect to its fifth eigenfrequency, and the inner supports are subsequently placed optimally at the four nodal points of the corresponding mode.

4.2 Results of Geometrically Constrained Problems

Let us now consider some examples of geometrically constrained solutions for transversely vibrating Bernoulli-Euler cantilever beams ($p = 2$) and rotating, cantilevered circular shafts from Ref. [102]. Results for Timoshenko beams are available in [112-114].

Fig. 9 shows cantilever beams optimized with respect to the third natural frequency, ω_3. The optimal designs are associated with $p = 2$ and are geometrically constrained with $\underline{\alpha} = \underline{A}L/V = 0.05$ for (a) and (c), and $\underline{\alpha} = 0.5$ for (b) and (d), respectively. Designs (c) and (d) are equipped with a dimensionless tip mass, $q_1 = Q_1/\rho V = 0.1$. The first four vibration modes of the optimal designs (a) and (c) are also shown in the figure. The natural frequencies ω_3 of the optimal beams are increased by (a) 129%, (b) 39%, (c) 82%, and (d) 28%, respectively, when compared with the same frequency of uniform beams of the same volume, length, material, and, for (c) and (d), tip mass. The frequency differences $\omega_3 - \omega_2$ of the optimal beams are (a) 228%, (b) 53%, (c) 156% and (d) 41% higher than the corresponding frequency differences of the uniform beams.

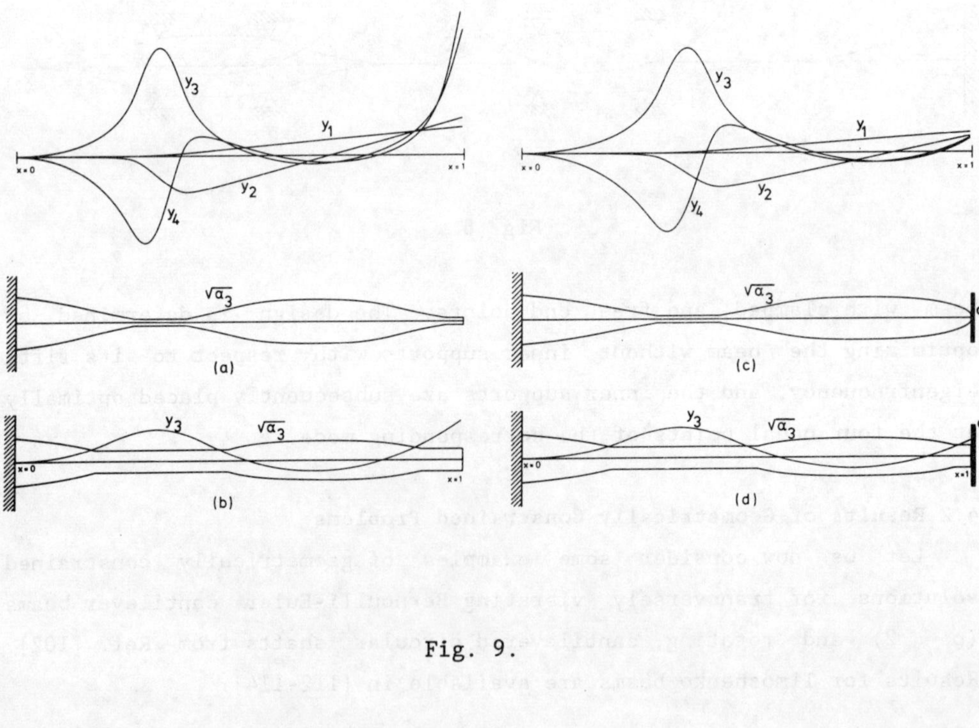

Fig. 9.

Fig. 10 summarizes, for n = 3 , the results for a number of optimal solutions of the types shown in Fig. 9. Adopting the concept of given volume V and length L , the square roots of the eigenvalues in Fig. 10 directly represent the natural *frequencies* of the optimal designs associated with n = 3 . The results in Fig. 10 may be compared with corresponding results for n = 1 in Fig. 6, noting especially the different ordinate scales and the different type of behaviour of the subject eigenfrequencies of the mass- carrying beams, substantiating advantages of optimizing with respect to higher order eigenfrequencies. Similarly, the optimal n = 3 designs in Fig. 9 may be compared with the n = 1 designs in Fig. 5.

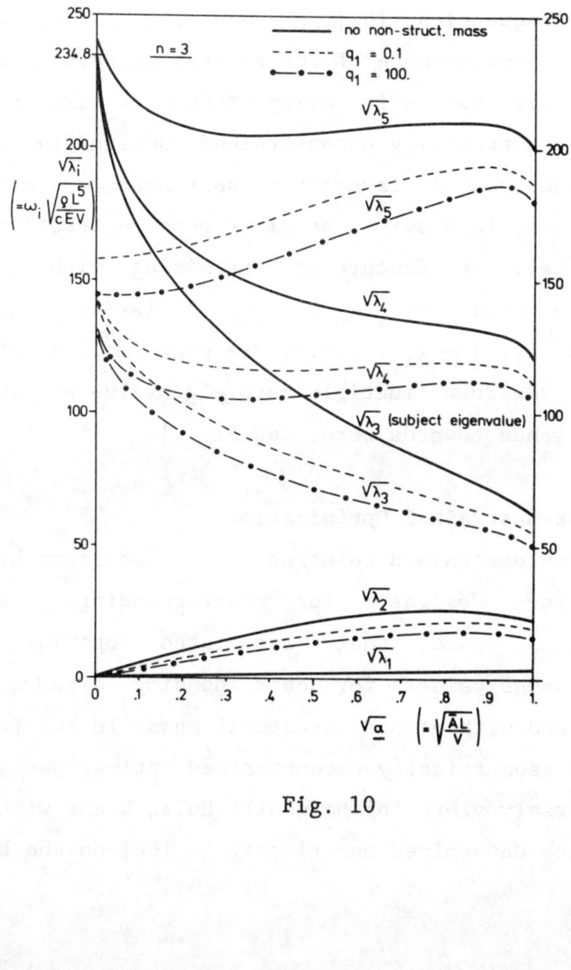

Fig. 10

As is illustrated for $n = 3$ in Fig. 10, we find for any value of $n > 1$ that both the optimal natural frequency ω_n and the distance between the consecutive natural frequencies ω_n and ω_{n-1} increase with decreasing geometric constraint. In fact, the absolute values of the differences between ω_n and ω_{n-1} increase with increasing n, and this, irrespective of whether the beams are equipped with nonstructural mass or not.

Optimizing cantilevers for $n > 3$, not only the subject frequency ω_n but also the closest subsequent natural frequencies ω_{n+1}, ω_{n+2}, etc. are pushed upwards, and the sub-spectrum consisting of these natural frequencies becomes significantly condensed for optimal solutions associated with small geometric constraint. In the limiting case of geometrically unconstrained optimization, the subject frequency may be increased to the extent that it coalesces with one or more of the subsequent natural frequencies [102], see e.g. Fig. 10. For the problems considered in [102], coalescence of the subject eigenvalue with one or more higher order eigenvalues is always found to take place in the limiting case of geometrically unconstrained design ($\underline{\alpha} = 0$). This implies the advantage that *it is not necessary here, to apply a bi- or multimodal formulation*, in order to obtain the correct optimal design.

Another characteristic feature of optimizing with respect to a higher order natural frequency ω_n, $n \geq 2$ is that all the lower natural frequencies ω_j, $j = 1,\ldots,n-1$, are kept small by this process, and that they tend toward a multiple zero eigenvalue as the geometric minimum constraint tends towards zero, see Fig. 10.

4.3 Geometrically Unconstrained Optimization

Geometrically unconstrained solutions are important because they constitute *limiting designs* for corresponding geometrically unconstrained designs, and their associated optimal eigenvalues constitute *upper bound values* for corresponding eigenvalues of all similar beams with and without non-structural mass. In the folllowing we shall consider some geometrically unconstrained optimal designs obtained in [60] for transversely vibrating Bernoulli-Euler beams with $p = 2$.

The solutions are determined numerically in [60] on the basis of a formal integration of the geometrically unconstrained formulation for optimal design, with possible types of singular behaviour appropriately allowed for in the numerical solution procedure. The positions as well as the types (I and II) of the singularities are additional design variables in the optimization process, and the optimal solutions are, in fact, determined via a path through a class of geometrically unconstrained sub-optimal solutions.

Fig. 11 provides an illustration of this in the case of optimizing a cantilever for $n = 3$. The solutions in Fig. 11a are both sub-optimal

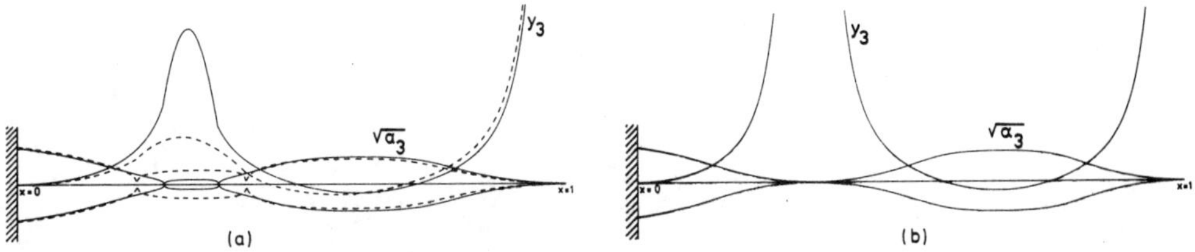

Fig. 11.

solutions. The dashed solution has inner Type I singularities at $x = 0.218$ and $x = 0.418$, and the eigenvalue λ_3 is $\lambda_3 = 1.77\cdot10^4$. For the solid solution in Fig. 11a, the type I singularities are placed at $x = 0.268$ and $x = 0.368$, and the eigenvalue is increased, $\lambda_3 = 2.87\cdot10^4$. Fig. 11b shows the resulting optimal solution with $\lambda_3 = 5.511\cdot10^4$. This solution has an inner Type II singularity, which is optimally placed at $x = 0.321$, and the optimal frequency ω_3 of the design is increased by 280% in comparison with the corresponding frequency of a uniform cantilever of the same volume, length and material. It should be noted that it is the design in Fig. 11b which lies behind the result for $\underline{\alpha} = 0$ in Fig. 10, and that it may be compared to corresponding constrained optimal designs in Figs. 9a,b.

Now, each Type I and Type II singularity introduces, respectively, one and two degrees of kinematic freedom to a geometrically unconstrained beam, and when optimizing with respect to the n'th natural

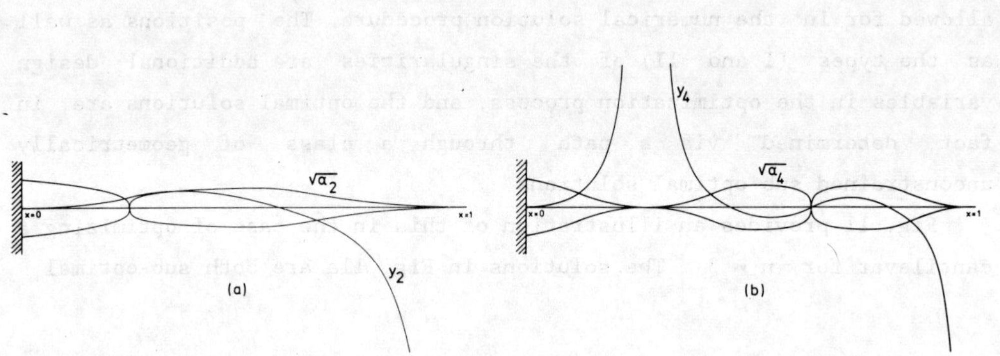

Fig. 12

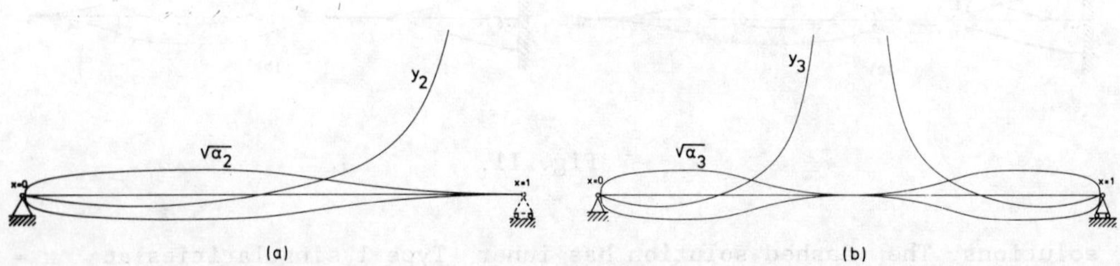

Fig. 13

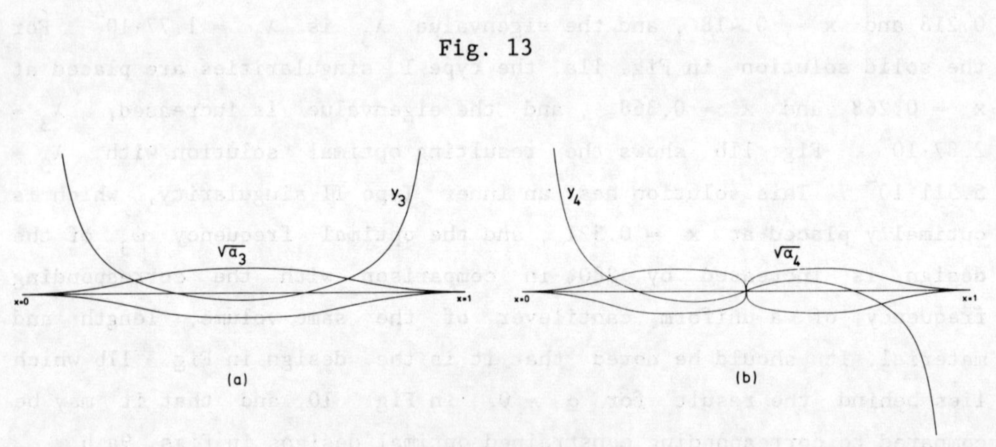

Fig. 14

frequency ω_n $(n > 1)$, it turns out [60] that the optimal design exactly possesses n-1 degrees of kinematic freedom to perform *rigid body motions*. Thus, all natural frequencies lower than the subject natural frequency of a geometrically unconstrained optimal design correspond to rigid body motions*) and attain zero value, cf. Fig. 10 (with $\underline{\alpha} = 0$).

This clearly implies that a geometrically unconstrained solution to the problem of optimizing the n'th natural frequency ω_n $(n > 1)$ is coincidently the solution to the problem of optimizing the *difference* between the two adjacent frequencies ω_n and ω_{n-1}.

The conditions of Section 3.5 for optimal locations of the two possible types of inner singularities also hold good for optimization with respect to higher order natural frequencies. These conditions are illustrated by optimally placed inner Type I singularities in Figs. 12a, 12b and 14b, and by optimally placed inner Type II singularities in Figs. 11b, 12b and 13b. Figs 12a,b show optimal cantilevers for $n = 2$ and $n = 4$, respectively. Figs. 13a,b show optimal simply supported beams for $n = 2$ and $n = 3$, respectively, and Figs. 14a,b illustrate free beams optimized for $n = 3$ and $n = 4$, respectively. (The first two natural frequencies of free beams correspond to rigid body motions, and are always zero).

---

*) Note, in Fig. 9 above, the tendency of the lower vibration modes y_1 and y_2 to become rigid body motions in the limiting case of zero constraint, and that these modes indicate the two degrees of kinematic freedom associated with an inner Type II singularity, namely jumps in both deflection and slope.

The study in Ref. [60] reveals the notable feature that *two* Type I singularities may *coalesce*, thereby forming *one* Type II singularity. Fig. 11 provides an illustration of such coalescence at an inner point. Note also that the inner beam separation formed at the resulting Type II singularity of the optimal $n = 3$ cantilever in Fig. 11b in fact divides this beam into a *scaled* optimal cantilever corresponding to $n = 1$, see Fig. 7b, and a scaled optimal free beam corresponding to $n = 3$, as shown in Fig. 14a. The coalescence may also take place at a beam end point; for example, the Type II singularity of the simply supported, optimal $n = 2$ beam in Fig. 13a results from an original inner Type I singularity coalescing with an a priori singularity of Type I at the beam end.

The formation of Type II singularities is found to constribute considerably to high subject eigenvalues, and singularities of this type consequently play a predominant role in geometrically unconstrained optimal designs. Thus, no more than *one* inner Type I singularity is found in any of the designs [60].

In fact when optimizing beams of a given type of end conditions, it is only necessary to apply the numerical solution procedure for values of n up to a particular value, N, where the first Type II singularity occurs in the corresponding optimal design. As discussed and outlined in [60], the inner Type II singularities open up the possibility of determining the optimal designs associated with higher values of n simply by assembling optimal beam elements obtained numerically for small values of n.

A so-called "Method of scaled beam elements" is developed in [60] by means of which optimal designs subject to any higher value of n are easily determined for beams with any combination of clamped, simply supported, and free end conditions.

Some geometrically unconstrained $p = 2$ designs corresponding to small values of n, determined numerically in [60] for the combinations of end conditions mentioned, are shown in Figs. 7, 11b, 12, 13 and 14 of the present notes, and their optimal eigenvalues λ_n are all given in [60]. For higher values of n, the compositions of optimal $p = 2$ beams with the sets of end conditions considered are indicated in Table

Beam type	Free beam	S. supp.-free beam	S. supp. beam	Cantilever	Clamped-s. supp. beam	Doubly clamped beam
N	5	4	3	3	2	1
λ_N	$1.875 \cdot 10^5$	$8.611 \cdot 10^4$	$3.274 \cdot 10^4$	$5.511 \cdot 10^4$	$1.837 \cdot 10^4$	$9.350 \cdot 10^3$
λ_{N+1}	$3.239 \cdot 10^5$	$1.875 \cdot 10^5$	$8.611 \cdot 10^4$	$1.138 \cdot 10^5$	$5.511 \cdot 10^4$	$\lambda_{N+3} = 2.939 \cdot 10^5$
Optimal beam composition for $n=N+2j$, $j=0,1,\ldots$	$\underbrace{\text{cc..cc}}_{j}$	$\underbrace{\text{bc..cc}}_{j}$	$\underbrace{\text{bc..cb}}_{j}$	$\underbrace{\text{ac..cc}}_{j}$	$\underbrace{\text{ac..cb}}_{j}$	$\underbrace{\text{ac..ca}}_{j}$
Optimal beam composition for $n=N+2j+1$, $j=0,1,\ldots$	$\underbrace{\text{cc..cC}}_{j}$	$\underbrace{\text{Bb..cc}}_{j}$	$\underbrace{\text{bc..cB}}_{j}$	$\underbrace{\text{ac..cC}}_{j}$	$\underbrace{\text{ac..cB}}_{j}$	$j=0$: Aa $j \geq 1$: $\text{aC}\underbrace{\text{c..ca}}_{j-1}$

Optimal beam elements

a	A	b	B	c	C
$\lambda^a = 5.844 \cdot 10^2$	$\lambda^A = 3.540 \cdot 10^3$	$\lambda^b = 2.046 \cdot 10^3$	$\lambda^B = 1.172 \cdot 10^4$	$\lambda^c = 1.172 \cdot 10^4$	$\lambda^C = 3.274 \cdot 10^4$

Table 2. Compositions and data for optimal p = 2 beams corresponding to $n \geq N$. (From [60]).

2 (from [60]. Their associated optimal eigenvalues λ_n , $n = N + 2j$ and $n = N + 2j + 1$, with $j = 0,1,\ldots$, are given by

$$\lambda_{N+2j} = \left[\lambda_N^{1/4} + j \cdot 10.4048\right]^4 \qquad j = 0,1,\ldots \qquad (4.2a)$$

$$\lambda_{N+2j+1} = \left[\lambda_{N+1}^{1/4} + j \cdot 10.4048\right]^4 \qquad j = 0,1,\ldots , \qquad (4.2b)$$

and appropriate values of N , λ_N and λ_{N+1} are available in Table 2. Eqs. (4.2a-b) hold for the beam types included in Table 2 with the single exception that for the doubly clamped beam, $\lambda_2 = 2.545\cdot10^4$, and its optimal eigenvalues λ_{N+2j+1} , $j = 1,2,\ldots$, are given by

$$\lambda_{N+2j+1} = \left[\lambda_{N+3}^{1/4} + (j-1) \cdot 10.4048\right]^4 \qquad j = 1,2,\ldots \qquad (4.2b')$$

instead of (4.2b).

The individual volumes v^i and lengths ℓ^i of the longitudinally and transversely scaled optimal elements a , A , b , B , c and C (see Table 2) meeting overall structural volume V and length L , are determined by [60]

$$\frac{v^i}{V} = \frac{\ell^i}{L} = \left[\frac{\lambda^i}{\lambda_n}\right]^{1/4} , \qquad (4.3)$$

where superscript i symbolizes lower or upper case letter of a particular beam element, and where the values of λ^i for individual elements are given in Table 2.

6. REFERENCES

1. Wasiutynski, Z. & Brandt, A.: Appl. Mech. Rev. 16, 344-350 (1963).
2. Sheu, C.Y. & Prager, W.: Appl. Mech. Rev. 21, 985-992 (1968).
3. Prager, W.: AGARD-Rept. No. 589 (1971).
4. Niordson, F.I. & Pedersen, P.: Proc. 13th Int. Cong. Th. Appl. Mech. Moscow, 264-278, Springer-Verlag (1973).
5. Rozvany, G.I.N. & Mróz, Z.: Appl. Mech. Rev. 30, 1461-1470 (1977).
6. Venkayya, V.B.: Int. J. Num. Meth. in Engrg. 13, 203-228 (1978).
7. Schmit, L.A.: AIAA J. 19, 1249-1263 (1981).
8. Lev, O.E.: American Society of Civil Engineers, New York (1981).
9. Haug, E.J.: In Ref. [23], 3-74 (1981).
10. Vanderplaats, G.N.: AIAA J. 20, 992-1000 (1982).
11. Ashley, H.: J. Aircraft 19, 5-28 (1982).
12. Olhoff, N. & Taylor, J.E.: J. Appl. Mech. 50, 1139-1151 (1983)
13. Levy, R. & Lev. O.E.: Proc. ASCE, J. Struct. Engrg. 113, 1939-62 (1987).
14. Hemp, W.S.: Optimum Structures, Clarendon, Oxford (1973).
15. Gallagher, R.H. & Zienkiewicz, O.C.: Optimum Structural Design, John Wiley, New York (1973).
16. Prager, W.: Introduction to Structural Optimization, Springer-Verlag (1974).
17. Distefano, N.: Non-Linear Processes in Engineering, Academic Press, New York (1974).
18. Sawczuk, A. & Mroz, Z. (Eds.): Optimization in Structural Design, Springer-Verlag, Berlin (1975).
19. Rozvany, G.I.N.: Optimal Design of Flexural Systems, Pergamon, Oxford (1976). Russian version: Strojizdat, Moscow (1980).
20. Reitman, M.I. & Shapiro, G.B.: Methods of Optimal Design of Deformable Bodies (in Russian), Nauka, Moscow (1976).
21. Haug, E.J. & Arora, J.S.: Applied Optimal Design, John Wiley, New York (1979).
22. Grinev, W.B. & Filippov, A.P.: Optimization of Beams Governed by Eigenvalue Problems (in Russian), Naukova Dumka, Kiev (1979).
23. Haug, E.J. & Cea, J. (Eds.): Optimization of Distributed Parameter Structures, Sijthoff & Nordhoff, Netherlands (1981).

Structures, Sijthoff & Nordhoff, Netherlands (1981).

24. Kirsch, U.: Optimum Structural Design, McGraw-Hill, New York (1981).

25. Carmichael, D.G.: Structural Modelling and Optimization, Ellis Horwood Ltd., Chichester (1981).

26. Troitskij, W.A. & Petuchov, L.W.: Optimal Design of Elastic Bodies (in Russian), Nauka, Moscow (1982).

27. Lepik, Ü.: Optimal Design of Inelastic Structures Under Dynamic Loading (in Russian; Extended summary in English), Walgus, Tallinn (1982).

28. Morris, A.J. (Ed.): Foundations of Structural Optimization: A Unified Approach, Wiley, New York (1982).

29. Eschenauer, H. & Olhoff, N. (Eds.): Optimization Methods in Structural Design, Bibliographisches Institut, Mannheim, FRG (1983).

30. Haug, E.J., Choi, K.K. & Komkov, V.: Design Sensitivity Analysis of Structural Systems, Academic Press, New York (1983).

31. Banichuk, N.V.: Problems and Methods of Optimal Structural Design, Plenum Press, New York (1983).

32. Brandt, A.M.: Criteria and Methods of Structural Optimization, Nijhoff Publishers, The Hague (1984).

33. Atrek, E., Gallagher, R.H., Ragsdele, K.M. & Zienkiewicz, O.C. (Eds.): New Directions in Optimum Structural Design, Wiley, New York (1984).

34. Vanderplaats, G.A.: Numerical Optimization Techniques for Engineering Design, McGraw-Hill, New York (1984).

35. Haftka, R.T. & Kamat, M.P.: Elements of Structural Optimization, Nijhoff Publishers, The Hague (1985).

36. Soares, C.A.M. (Ed.): Computer Aided Optimal Design: Structural and Mechanical Systems, Springer-Verlag, Berlin (1987).

37. Gajewski, A. & Zyczkowski, M.: Optimal Structural Design under Stability Constraints, Kluwer Academic Publishers, Dordrecht, Netherlands (1988).

38. Rozvany, G.I.N.: Structural Design via Optimality Criteria, Kluwer

Academic Publishers, Dordrecht, Netherlands (1989).

39. Lurie, K.A.: Applied Optimal Control of Distributed Systems, Plenum Press, New York (to appear).

40. Lagrange, J.L.: Miscellanea Taurinensia V, 123 (1770-1773).

41. Clausen, T.: Mélanges Mathématiques et Astronomiques I, 279-294 (1849-1853).

42. Keller, J.B.: Arch. Rat. Mech. Anal. 5, 275-285 (1960).

43. Niordson, F.I.: Quart. Appl. Math. 23, 47-53 (1965).

44. Ashley, H. & McIntosh, S.C. Jr.: Proc. 12th Int. Cong. Th. Appl. Mech. (Ed. M. Hetényi & W.G. Vincenti), 100-113, Stanford, Springer-Verlag (1969).

45. Pierson, B.L.: Int. J. Num. Meth. Engrg. 4, 491-499 (1972).

46. Reitman, M.I. & Shapiro, G.S.: All-Union Symp. "On the Problems of Optimization in Mech. of Solid Deformable Bodies", Vilnius, USSR, June (1974).

47. Rao, S.S.: Shock Vib. Dig. 7, 61-70 (1975).

48. Olhoff, N.: Shock Vib. Dig. 8, No. 8, 3-10 (Part I: Theory), 8, No. 9, 3-10 (Part II: Applications (1976).

49. Olhoff, N.: Proc. 15th Int. Cong. Th. Appl. Mech., Toronto, Canada, 133-149, North-Holland (1980).

50. Keller, J.B. & Niordson, F.I.: J.Math. Mech. 16, 433-446 (1966).

51. Taylor, J.E.: AIAA J. 6, 1379-1381 (1968).

52. Masur, E.F.: J. Engrg. Mech. Div., ASCE, 96, 621-640 (1970).

53. Prager, W & Taylor, J.E.: J. Appl. Mech. 35, 102-106 (1968).

54. Brach, R.M.: J. Opt. Th. & Appl. 11, 662-667 (1973).

55. Taylor, J.E.: AIAA J. 5, 1911-1913 (1967).

56. Vavrick, D.J. & Warner, W.H.: J. Struct. Mech. 6, 233-246 (1978).

57. Seyranian, A.P.: Int. J. Solids Struct. 15, 749-759 (1979).

58. Olhoff, N. & Niordson, F.I.: ZAMM 59, T16-T26 (1979).

59. Karihaloo, B.L. & Niordson, F.I.: J. Opt. Th. & Appl. 11, 638-654 (1973).

60. Olhoff, N.: J. Struct. Mech. 4, 87-122 (1976).

61. Turner, M.J.: AIAA J. 5, 406-412 (1967).

62. Olhoff, N. & Taylor, J.E.: J. Opt. Th. & Appl. 27, 571-582 (1979).

63. Taylor, J.E. & Liu, C.Y.: AIAA J. 6, 1497-1502 (1968).
64. Tadjbakhsh, I. & Keller, J.B.: J. Appl. Mech. 29, 159-164 (1962).
65. Masur, E.F.: J. Opt. Th. & Appl. 15, 69-84 (1975).
66. Popelar, C.H.: J. Struct. Mech. 5, 45-66 (1977).
67. Olhoff, N. & Rasmussen, S.H.: Int. J. Solids Struct. 13, 605-614 (1977).
68. Brach, R.M.: Int. J. Solids Struct. 4, 667-674 (1968).
69. Karihaloo, B.L. & Niordson, F.I.: Arch. of Mech. 24, 1029-1037 (1972).
70. Huang, N.C. & Sheu, C.Y.: Appl. Mech 35, 285-288 (1968).
71. Frauenthal, J.C.: J. Struct. Mech. 1, 79-89 (1972).
72. Adali, S.: Int. J. Solids Struct. 15, 935-949 (1979).
73. Simitses, G.J., Kamat, M.P. & Smith, C.V. Jr.: AIAA J. 11, 1231-1232 (1973).
74. Rasmussen, S.H.: J. Struct. Mech. 4, 307-320 (1976).
75. Olhoff, N. & Taylor, J.E.: J. Struct. Mech. 6, 367-382 (1978).
76. Taylor, J.E.: J. Appl. Mech. 34, 486-487 (1967).
77. Banichuk, N.V. & Karihaloo, B.L.: Int. J. Solids Struct. 13, 725-733 (1977).
78. Banichuk, N.V.: MTT (Mechanics of Solids), 9, 150-154 (1974).
79. Mróz, Z. & Rozvany, G.I.N.: J. Struct. Mech. 5, 279-290 (1977).
80. Szelag, D. & Mróz, Z.: ZAMM 58, 501-510 (1978).
81. Masur, E.F.: In Ref. [18], 441-453 (1975)
82. Bendsøe, M.P., Olhoff, N. & Taylor, J.E.: J. Struct. Mech. 11, 523-544 (1983-84).
83. Taylor, J.E. & Bendsøe, M.P.: Int. J. Solids Struct. 20, 301-314 (1984).
84. Masur, E.F. & Mróz, Z.: Proc. IUTAM Symp. on Variational Methods in the Mech. of Solids (Ed. S. Nemat-Nasser), Northwestern Univ., USA, Pergamon Press (1980).
85. Masur, E.F.: Int. J. Solids Struct. 20, 211-231 (1984).
86. Bratus, A.S. & Seyranian, A.P.: Prikl. Mat. Mekh. 47, 451-457 (1983).
87. Bratus, A.S. & Seyranian, A.P.: Prikl. Mat. Mekh. 48, 466-474 (1984).

88. Seyranian, A.P.: Doklady Acad. Nauk USSR 271, 337-40 (1983). English translation: Sov. Phys. Dokl. 28, 550-51 (1983).
89. Seyranian, A.P.: MTT (Mechanics of Solids) 19, 101-111 (1984).
90. Prager, S. & Prager, W.: Int. J. Mech. Sci. 21, 249-251 (1979).
91. Gajewski, A.: Int. J. Mech. Sci. 23, 11-16 (1981).
92. Blachut, J. & Gajewski, A.: Int. J. Solids Struct. 17, 653-667 (1981).
93. Blachut, J. & Gajewski, A.: Optimal Control Appl. Methods 2, 383-402 (1981).
94. Olhoff, N. & Plaut, R.H.: Int. J. Solids Struct. 19, 553-570 (1983).
95. Rezaie-Keyvan, N. & Masur, E.F.: J. Struct. Mech 13, 181-200 (1985).
96. Blachut, J.: Int. J. Mech. Sci. 26, 305-310 (1984).
97. Bochenek, B. & Gajewski, A.: J. Struct. Mech. 14, 257-274 (1986).
98. Gajewski, A.: Int. J. Mech. Sci. 27, 45-53 (1985).
99. Bochenek, B.: Int. J. Solids Struct. 23, 599-605 (1987).
100. Plaut, R.H., Johnson, L.W. & Olhoff, N.: J. Appl. Mech. 53, 130-134 (1986).
101. Seyranian, A.P. & Sharanyuk, A.V.: MTT (Mechanics of Solids) 22, 34-38 (1987).
102. Olhoff, N.: J. Struct. Mech. 5, 107-134 (1977).
103. Kamat, M.P. & Simitses, G.J.: Int. J. Solids Struct. 9, 415-429 (1973).
104. Vepa, K.: Quart. Appl. Mech. 31, 329-341 (1973/74).
105. Seyranian, A.P.: MTT (Mechanics of Solids), 11, 147-152 (1976).
106. Masur, E.F.: In Ref. [18], 99-102 (1975)
107. Olhoff, N. & Parbery, R.: Int. J. Solids Struct. 20, 63-75 (1984).
108. Bendsøe, M.P. & Olhoff, N.: Optimal Control Appl. & Meth. 6, 191-200 (1985).
109. Troitskii, V.A.: MTT (Mechanics of Solids), 11, 145-152 (1976).
110. Mróz, Z. & Rozvany, G.I.N.: J. Opt. Th. & Appl. 15, 85-101 (1975).
111. Rozvany, G.I.N.: J. Struct. Mech. 3, 359-402 (1974/75).
112. Pierson, B.L.: J. Struct. Mech. 5, 147-178 (1977).

113. Pedersen, P.: J. Struct. Mech. 10, 243-271 (1981/83).

114. Kibsgaard, S.: Structural Optimization 1 (1989) (to appear).

PART IV

J. Rondal
University of Liège, Liège, Belgium

ABSTRACT

Part IV is devoted to optimal design of thin-walled bars under stability constraints. The aim of this part is to review all the single and compound buckling modes which must be taken into account in the optimal design of thin-walled bars. The optimal design of hollow and open thin-walled profiles is studied by means of efficiency charts. At last, some examples of optimization using mathematical programming are shown.

1. INTRODUCTION

Three main families of structural members are used in steel construction [1] :

- hot-rolled shapes, which are of common use and were born at the same time as the steel construction itself, belong to the first family ;
- the second family contains built-up members, i.e. plate and box girders composed by plates which are connected by bolts, rivets or welding ;
- the third family, of growing importance, is composed of sections cold-formed from steel sheets by roll-forming machines or by press-braking.

The manufacturing process of hot-rolled sections does not allow to produce the light profiles which are now extensively used in the manufacture of automobiles, road and rail transports, aircrafts, ships, storage racks,... and in standardized, or not standardized, buildings and industrialized housings.

On the contrary, built-up members and cold-formed sections can easily be produced with walls of moderate or small thickness.
This leads to an important weight saving, more especially for relatively low loadings and/or short spans, and to simplification of handling and erection.

In addition, cold-formed structural members provide interesting advantages such as [2] :

- many secondary operations can take place which are associated directly to the forming (piercing, notching, marking, painting,...) ;
- nestable sections can be obtained easily, allowing for compact packaging and economical transportation ;
- and, of first importance for our purpose, it is easy to produce sections the shapes of which are unusual but are the most appropriate to the function of the profile.

This last point gives us a wonderful field of application for optimization techniques. Typical profile shapes obtained by cold-forming are shown in Fig. 1.1.

Cold-formed profiles can be produced :

- either, by a discontinuous process, for small series of sections, with a leaf-press brake (Fig. 1.2.a) or a coin-press brake (Fig. 1.2.b) ;
- or, by a continuous forming, for more important series, by passing through successive pairs of rolls (Fig. 1.2.c).

The manufacturing procedure has an influence on the geometrical and mechanical characteristics of the profiles. This point is discussed in details in chapter 2 because these characteristics have an influence on the buckling behaviour of these profiles.

Thin-walled sections leads to major problems for what regards joints (by bolts, fasteners or welds or by adhesives for very thin sheets), on the one hand, and stability, on the other hand.

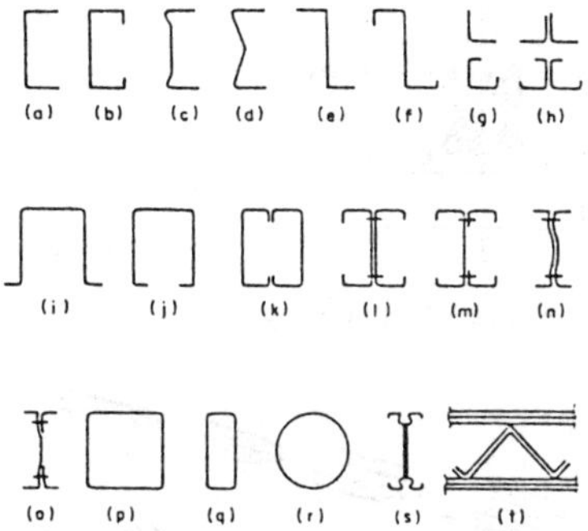

sections.

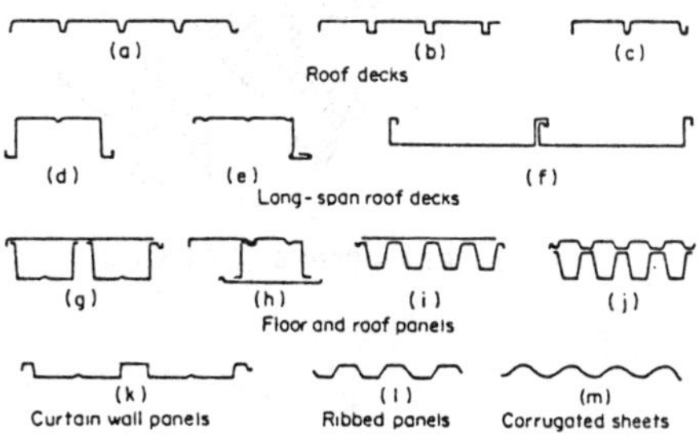

decks, panels and corrugated sheets

Figure 1.1. - Cold-formed steel profiles.

The optimization of thin-walled profiles must not only take account of the single modes of buckling (local plate buckling, flexural column buckling, torsional buckling, lateral buckling) but, also, of the interactive buckling between single modes. As first shown by Skaloud [3] and Koiter [4] in 1962, interactive buckling results from unavoidable imperfections of industrially manufactured profiles and yields a decrease of the ultimate load with respect to the corresponding load obtained by considering individually each of these instability phenomena.

In the next chapters, it will be referred extensively to cold-formed profiles because of their greatest economical importance. However, optimization of some built-up sections will be studied too.

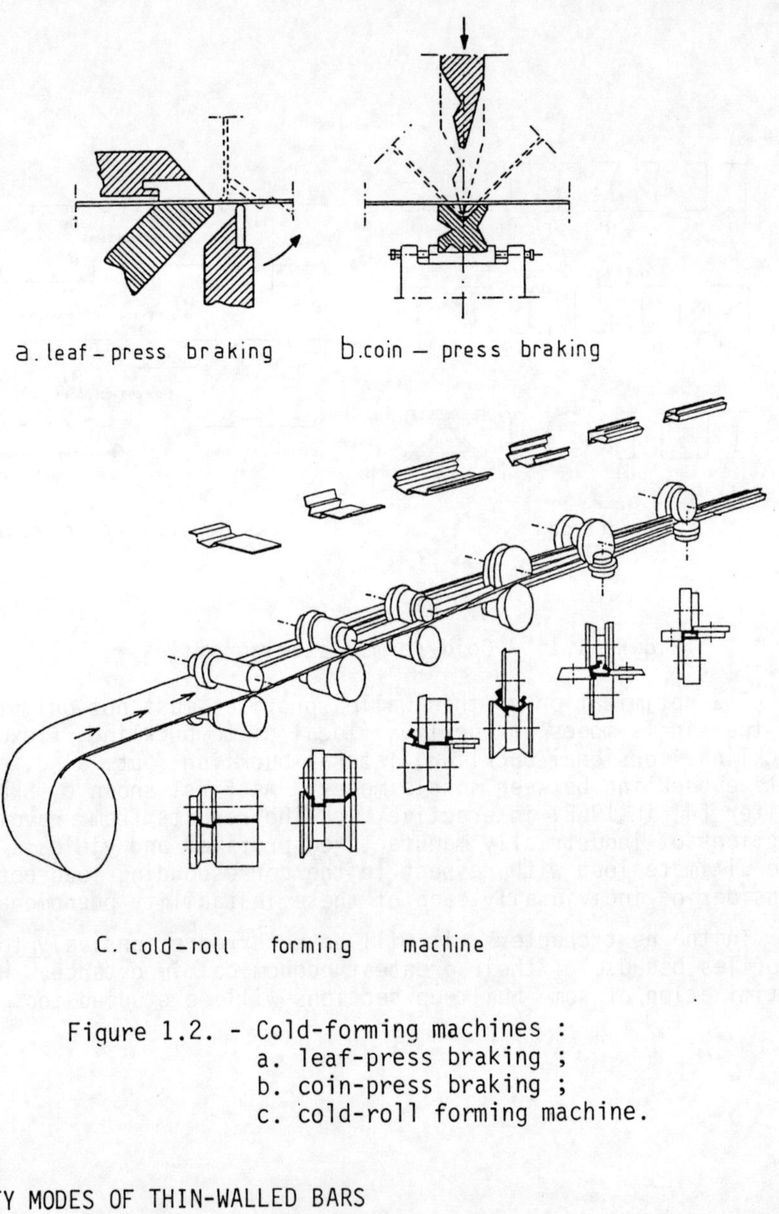

Figure 1.2. - Cold-forming machines :
a. leaf-press braking ;
b. coin-press braking ;
c. cold-roll forming machine.

2. STABILITY MODES OF THIN-WALLED BARS

2.1. Effect of cold-forming on the buckling behaviour of the profiles.

2.1.1. Geometrical imperfections.

Industrial structures and components show some deviations from the ideal geometrical form, termed as geometrical imperfections. Generally, variations in cross-sectional dimensions do not play an important role on the stability behaviour of steel structures. On the contrary, the out-of-straightness of plates and columns and the unavoidable eccentricities of the loads can sometimes yield an important erosion of the critical load.

Table 2.1. gives the characteristic value (mean + two standard deviations) of measurements of the out-of-straightness on hot-rolled and cold-formed profiles [5]. These measurements show that the way the profile is produced play an important role.

Forming method	Hot - rolling	Cold - forming	
		Cold-rolling	Press-braking
(f_o/l)	$\frac{1}{800}$ to $\frac{1}{1000}$	$\frac{1}{600}$ to $\frac{1}{1400}$	$\frac{1}{2100}$ to $\frac{1}{2700}$

Table 2.1. - Characteristic values of measured initial out-of-straightnesses.

The measurements on profiles have indicated a great variety of initial curvatures as shown in Fig. 2.1. Besides double curvature forms, some profiles have their maximum out-of-straightness in the quarter points rather than in the center. Nevertheless, in practice, one frequently refers to an equivalent sine curve. This point will be justified later.

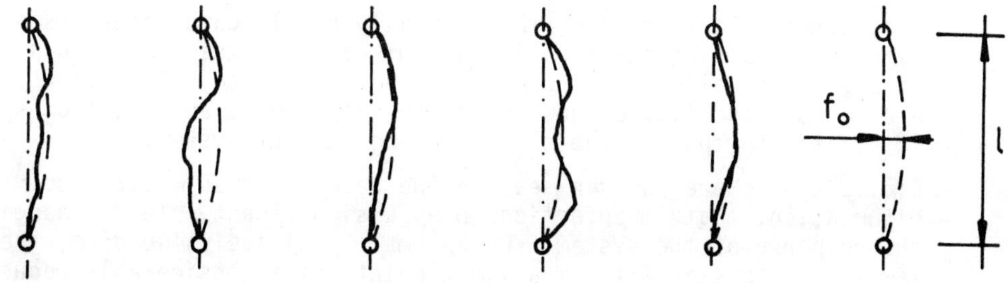

Figure 2.1. - Measured and assumed crookednesses.

Initial twist of profiles can affect lateral and torsional buckling but Table 2.2. shows that, in practice, this geometrical imperfection is rather small.

Type of profile	Hot - rolled I [6]	Welded I [6]	Cold - rolled angles [5]
(θ_o/l)	1/500000	1/200000	1/100000

Table 2.2. - Characteristic values of measured initial twists (rad/mm).

As mentioned above, the imperfections can sometimes drastically, decrease the critical load of a structure. It is therefore of first importance to know if a system is imperfection sensitive.

In accordance with Euler's concept, it was long believed that all that needed to be done in point of structural stability was to determine the critical bifurcation load. In 1945, Koiter proved that besides determining the critical bifurcation load, the postcitical behaviour of the structure has to be studied because the sensitivity to imperfections is closely connected to the type of postcritical behaviour [7].

On base of the general theory of elastic stability and using a perturbation analysis [8], one can defined four different situations (Fig. 2.2.) where N is the applied load, δ a displacement and ξ the amplitude of the imperfection. The system corresponding to $\xi = 0$ is described as the perfect system and systems corresponding to non-zero values of ξ are described as imperfect systems. Heavy solid lines in Fig. 2.2. represent the equilibrium paths of the perfect system while light solid lines represent the equilibrium paths of the imperfect systems, continuous lines representing stable equilibrium paths and dotted lines representing unstable equilibrium paths :

- Fig. 2.2.a : In this case, the initially stable path loses its stability when reaching the locally maximum value of the load. This behaviour, often called snaptrough buckling, is typical of shallow arches and domes. The figure shows that the response of an imperfect system is similar to that of the corresponding perfect system.
- Fig. 2.2.b : Here, we can see, in the case of an asymmetric point of bifurcation, that imperfections play a significant role in changing the response of the system. For a small positive value of ξ, the system loses its stability at a limit point, at a considerably reduced value of the load. On the contrary, for a small negative imperfection, the system exhibits no instability in the vicinity of the critical point and follows a rising stable path. There is here an important sensitivity to initial positive imperfections. This behaviour can be observed, for example, in the well-known Koiter frame [9] and in bridge and roof trusses [10].

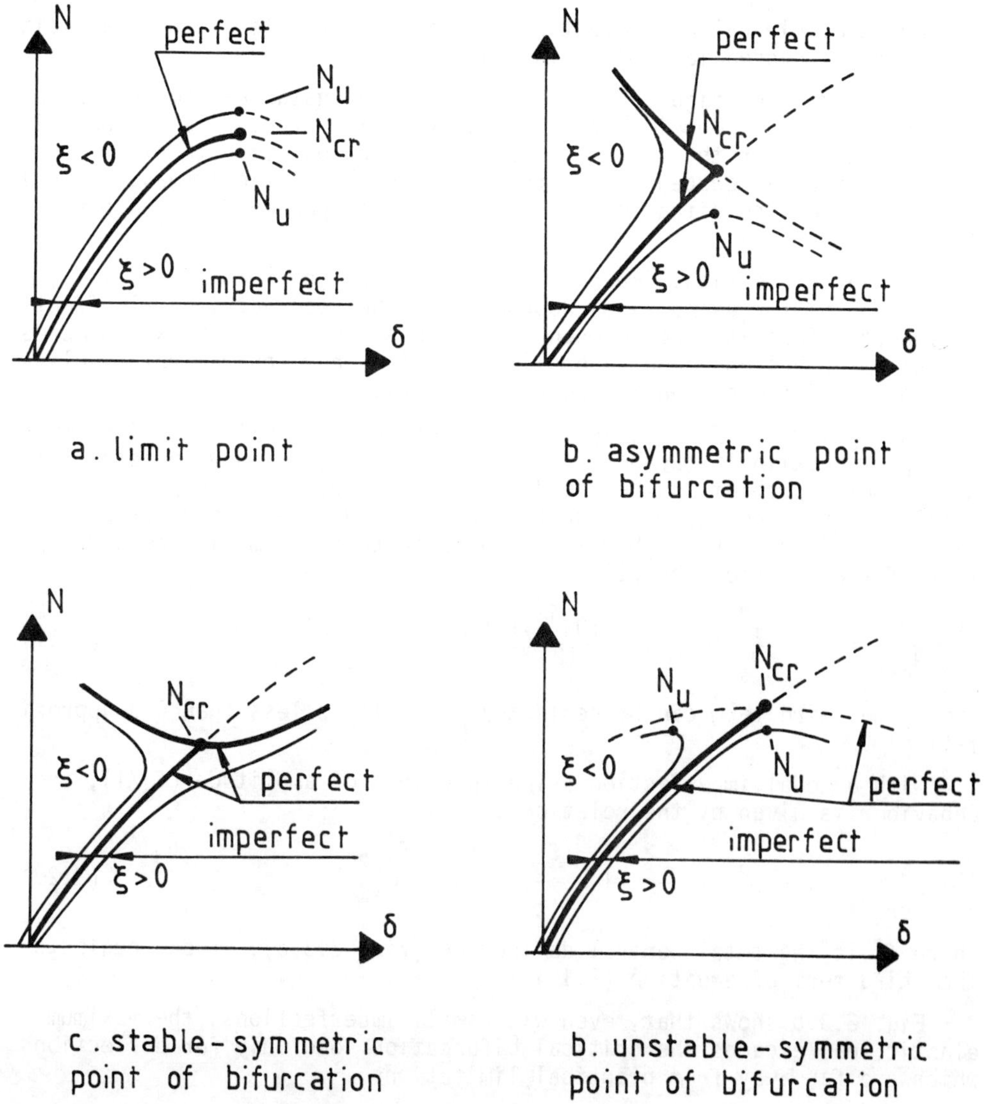

Figure 2.2. - Behaviour of perfect and imperfect systems.

- Fig. 2.2.c : This is the case of the stable-symmetric point of bifurcation. Small positive and negative imperfections have similar effects and yield a stable and rising equilibrium path. The buckling is characterized by a rapid growth of the deflections when the critical load of the perfect system is approached. This is the case for compressed columns in flexural or torsional buckling, for beams in lateral buckling and for compressed plates in plate buckling. This case is thus of great importance in practice but one must remind that these conclusions are only valid in elastic stability. The role of the plasticity will be discussed later in this chapter.

- Fig. 2.2.d : The figure shows the case of an unstable-symmetric point of bifurcation. The imperfections play in important role in modifying the behaviour of the system. Small imperfections of both signs induce a failure at a reduced load with regard to the critical load. This is, for example, the situation in some systems composed of hinged bars [11].

Initial geometrical imperfections can be of any form (Fig. 2.1.), but as shown, for example, by Roorda [12], the most detrimental effect of imperfections is reached when these are affine to the buckling modes (modal imperfections) and others imperfections have only a relatively unimportant influence on the buckling behaviour.

As mentioned before, elastic imperfect bars exhibit a stable-symmetric post-buckling behaviour. It is interesting to study in details the most popular problem in the field which is, of course, the pin-ended axially compressed column. For this elementary case, the post-critical behaviour of the perfect bar is given, with three terms in the development of the solution, by [11] :

$$\frac{N}{N_{cr}} = 1 + \frac{\pi^2}{8}\left(\frac{f}{l}\right)^2 + \frac{19\pi^4}{512}\left(\frac{f}{l}\right)^4 \tag{2.1.}$$

where the third term can be neglected if (f/l) is less than 0.3 approximatively.

With a modal imperfection (sine function) of amplitude (f_o/l), the behaviour is given by the relation :

$$\frac{N}{N_{cr}} = 1 - \frac{f_o}{f_t} + \frac{\pi^2}{8}\left(1 - \frac{f_o}{f_t}\right)\left(\frac{f_t}{l} - \frac{f_o}{l}\right)^2 \tag{2.2.}$$

where f_t is the total central deflection (fig. 2.3.a), if one neglects the third term of equation (2.1.).

Fig. 2.3.b shows that, even with small imperfections, the maximum elastic load exceeds the critical bifurcation load only for deflections which are far away from practical limitations.

Steel cannot be considered as an undefinite elastic material and failure in a compressed steel column is generally assumed to occur when the material yield stress is first reached at any point in the column. In fact, the failure occurs at a somewhat larger load because of a remaining reserve of strength as plasticity spreads through the cross-section. However, at least for thin-walled members, this strength reserve is very small and can generally be neglected. The Ayrton-Perry approach provides a simple way to take account the effect of the yield stress on the post-critical behaviour of the column [13].

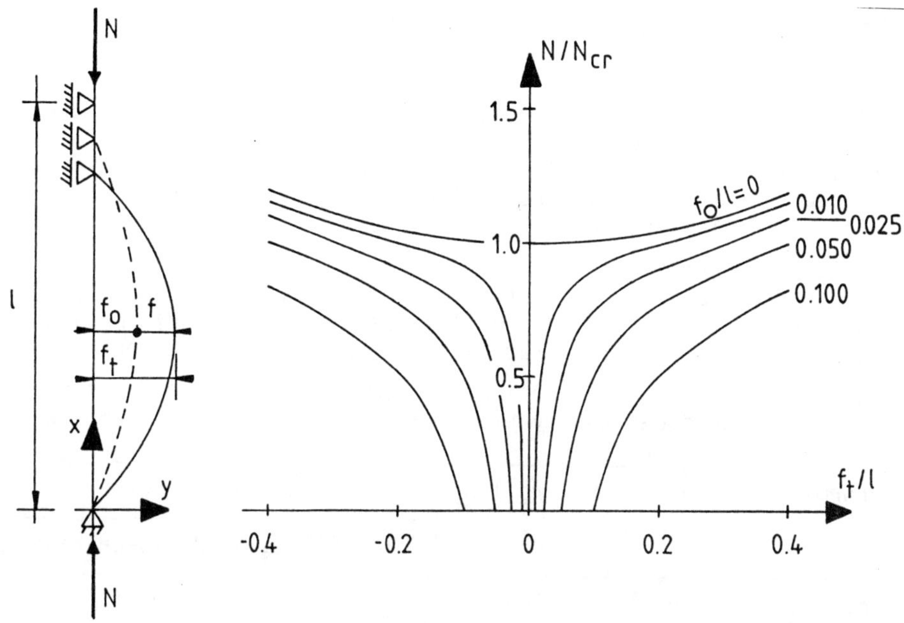

Figure 2.3. - Post-critical behaviour of an elastic strut.

The maximum bending moment (Fig. 2.3.a) in the bar occurs at x = 1/2 and is given by $M = N\,(f_o + f) = N.\,f_t$ (2.3.)

By superposing the compressive axial and maximum bending stress components and limiting it to the maximum elastic stress f_y yields :

$$\sigma \pm \frac{\sigma .\, f_t .A.v}{I} = f_y \tag{2.4.}$$

where A is the cross-section area, I is the moment of inertia and f_y is the yield stress. The plus or minus signs relate to positive or negative deflections.

After some rearrangements, equation (2.4.) can be written as :

$$\frac{N}{N_{cr}} = \frac{f_y}{\pi^2 E\,[(\frac{i}{l})^2 \pm (\frac{f_t}{l})(\frac{v}{l})]} \tag{2.5.}$$

where i is the radius of gyration and v the distance from the centroid, to the extreme fibre.

This equation may be designated as the elasticity boundary in the load-deflection space of the column. Fig. 2.4. gives, for a hollow section with a steel grade of 355 N/mm^2, the solution which appears for the three cases

$$\sigma_{cr} \gtreqless f_y .$$

This figure leads to the conclusion that the yield stress cuts drastically the effect of the post-buckling, on the first hand, and that, even for small imperfections (f_o/l = 1/1000 as considered in the European Recommendations for Steel Construction [14]), the maximum load is less than the critical load, on the second hand.

2.1.3. Residual stresses.

In contrast to hot-rolled steel members which are subject to residual cooling stresses, cold-formed profiles are affected by deformational residual stresses in both longitudinal and transverse directions [15].

Whereas hot-rolling induces mainly membranar residual stresses, cold-forming introduces mostly flexural residual stresses (Table 2.3.).

Forming method	Hot-rolling	Cold-forming	
		Cold-rolling	Press-braking
Membranar residual stresses (σ_{rm})	high	low	medium
Flexural residual stresses (σ_{rf})	low	high	medium

Table 2.3. - Longitudinal residual stresses.

These residual stresses play an important role when stability is governing. Therefore, they must be accounted for in the design.

As shown by Costa Ferreira and Rondal [16], flexural residual stresses have a less detrimental effect on the stability than membrane ones.

2.1.4. Strain-hardening.

The manufacturing process leads to a modification of the stress-strain curve of the steel. With respect to the virgin material, cold-rolling provides an increase of the yield stress and, sometimes, of the ultimate stress that is important in the corners and still appreciable in the flat parts of a profile, while press-braking let these characteristics nearly unchanged in the flat parts [17]. Obviously, these modifications have no sense for hot-rolled shapes (Table 2.4.).

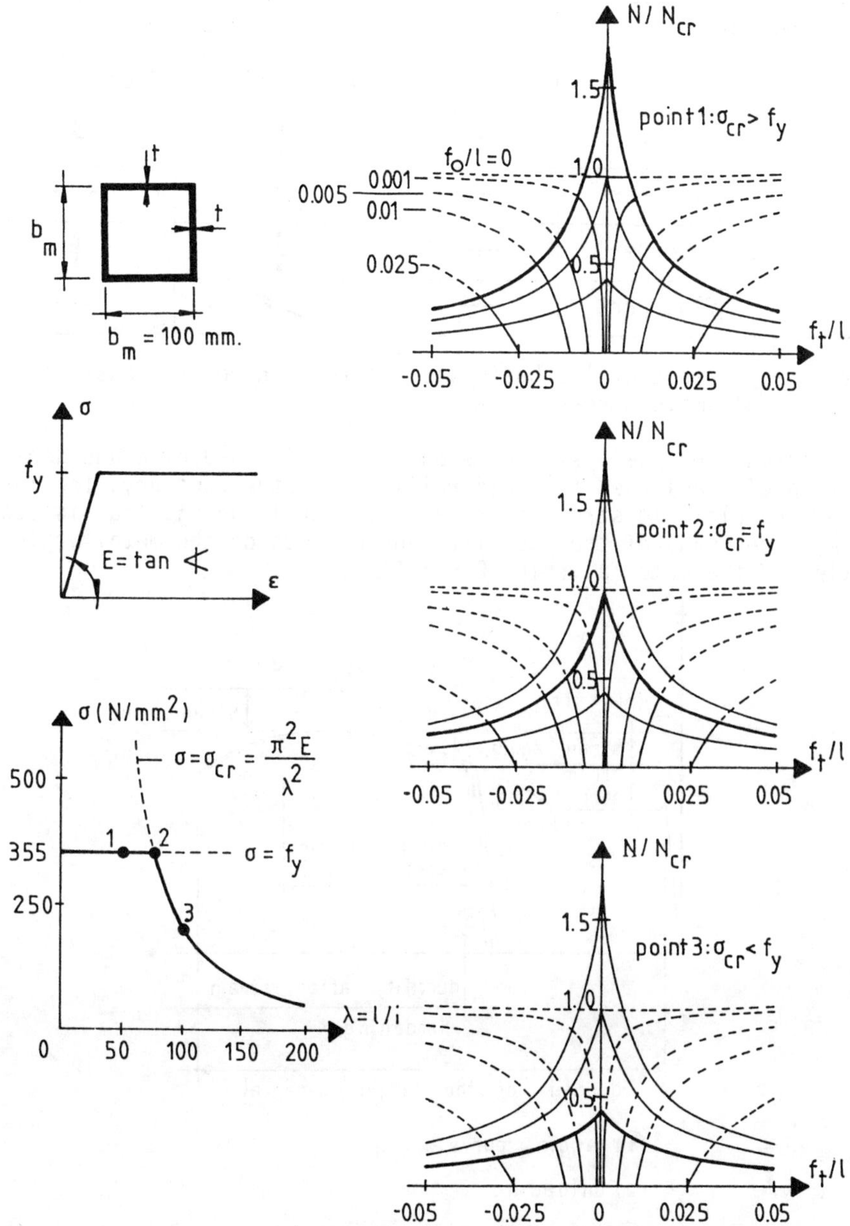

Figure 2.4. - Effect of the yield stress on the post-critical behaviour of a column with a square hollow cross-section.

Forming method		Hot-rolling	Cold-forming	
			Cold-rolling	Press-braking
Yield stress (f_y)	corner	—	↑	↑
	flat part	—	↗	—
Ultimate stress (f_u)	corner	—	↑	↑
	flat part	—	↗	—

Table 2.4. - Increase of the basic yield stress and of the basic ultimate stress.

The increase of yield stress is due to strain hardening and depends on the type of steel used for cold-rolling. On the contrary, the increase of the ultimate stress is related to strain-aging, that is accompanied by a decrease of the ductility and depends on the metallurgical properties of the material (Fig. 2.5.) [18].

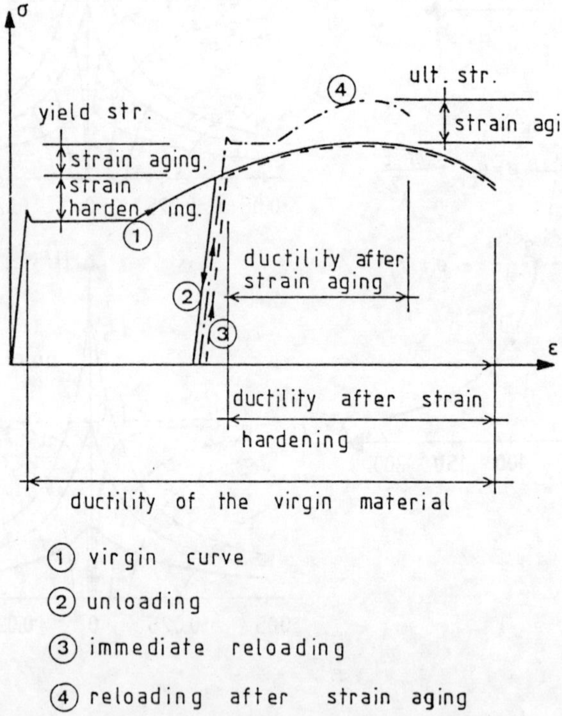

Fig. 2.5. - Effects of strain-hardening and strain-aging on the stress-strain characteristics of a structural steel.

Several authors have made theoretical proposals allowing for a prediction of the yield stress increase during the cold-forming process, with respect to the steel grade, the sheet thickness and the profile geometry (number of folds) [19, 20].

Most existing standards take account at least partly of this increase of the yield stress of the virgin material. More especially, the residual stresses are related to the strain-hardening ; they reach a maximum value in the corners, where the strain-hardening is the most pronounced. Thus, both effects, i.e. residual stresses and yield stress increase, respectively, are a result of the manufacturing process and trend sometimes to compensate each other.

2.2. Global column buckling design equations.

2.2.1. Flexural buckling.

Although interesting observations on the behaviour of columns have been performed by Leonardo da Vinci (1452-1519) and Petrus van Musschenbroek (1692-1761), the first valuable column formula is due to Leonhard Euler and has been published, for the first time, in 1744, in the appendix of his book on variational calculus. But, as shown by Lucien Navier, in 1826, in the first edition of his book on strength of materials, the Euler formula gives a load, called critical load, which is an upper bound of the ultimate load of an actual imperfect column.

At present time, nearly all the recommendations on the design of steel structures are based on the concept of the "imperfect bar", i.e. a bar fitted with geometrical imperfections, residual stresses and material inhomogeneities.

In Europa, these imperfections are taken into account by means of a global imperfection parameter used in connection with an Ayrton-Perry type equation.

If one neglects the second order term in equation (2.2.), the amplification of the central deflection during the flexural buckling writes :

$$f_t = \frac{1}{1-N/N_{cr}} f_o \tag{2.6.}$$

and equation (2.4.) leads, after rearrangements of the terms, to :

$$(\sigma_{crF} - \sigma)\ (f_y - \sigma) = f_o \frac{A}{W} \cdot \sigma_{crF} \cdot \sigma = \eta \cdot \sigma_{crF} \cdot \sigma \tag{2.7.}$$

where W is the elastic flexural modulus and η is the imperfection parameter which must cover the effect of all the imperfections.

By introducing reduced coordinates :

$$\bar{N} = \sigma/f_y \tag{2.8.}$$

and

$$\bar{\lambda} = \sqrt{f_y/\sigma_{crF}} = \frac{1}{i\pi\ \sqrt{E/f_y}} \tag{2.9.}$$

equation (2.8.) takes the dimensionless expression :

$$(1-\overline{N})(1-\overline{N}\,\overline{\lambda}^2) = \eta\,\overline{N} \qquad (2.10.)$$

Rondal and Maquoi [21] have shown that the form :

$$\eta = \alpha(\overline{\lambda}-\overline{\lambda}_o) \qquad (2.11.)$$

of the imperfection parameter agrees very well with the tests and the numerical simulations performed by the European Convention for Constructional Steelwork [14].

Account taken of (2.11.), equation (2.10.) can be solved in terms of $\overline{N}$ versus reduced slenderness $\overline{\lambda}$:

$$\overline{N} = \frac{1+\alpha(\overline{\lambda}-\overline{\lambda}_o)+\overline{\lambda}^2}{2\,\overline{\lambda}^2} - \frac{1}{2\overline{\lambda}^2}\sqrt{[1+\alpha(\overline{\lambda}-\overline{\lambda}_o)+\overline{\lambda}^2]^2 - 4\overline{\lambda}^2} \ngtr 1 \qquad (2.12.)$$

Values of the global imperfection α and the length of the yield plateau $\overline{\lambda}_o$, are given by Table 2.5. for the main flexural buckling curves of the European Recommendations for Steel Construction [14]. Curve "c" must be used for cold-formed profiles and curve "a" is devoted to hollow sections having received a stress relief treatment. These profiles are called "hot finished" in practice.

Fig. 2.6. shows the european column flexural buckling curves.

$\overline{\lambda}_o = 0.2$			
Curve	a	b	c
α	0.21	0.34	0.49

Table 2.5. - Values of parameters $\overline{\lambda}_o$ and α for the european column flexural buckling curves.

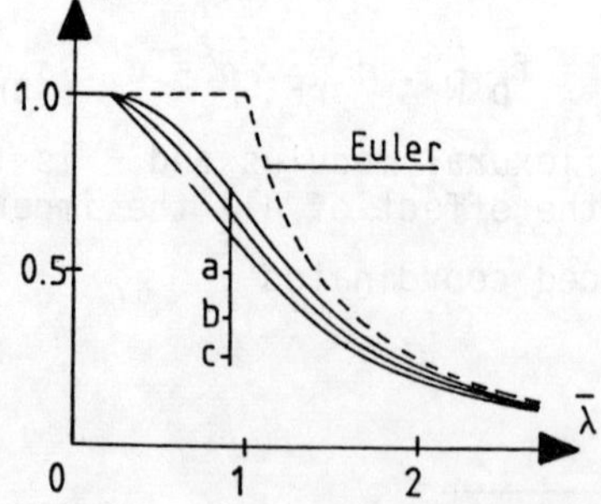

Figure 2.6. - European column flexural buckling curves.

2.2.2. Torsional buckling and flexural-torsional buckling.

As shown in [5] , the coefficient of amplification of the torsional angle during torsional buckling has the same form than the coefficient of amplification in flexural buckling (equation 2.6.) :

$$\theta_t = \frac{1}{1-N/N_{cr}}\,\theta_o \qquad (2.13.)$$

But, as mentioned in section 2.1.1., the initial twist of profiles is rather small in practice. Numerical simulations and experimental tests have shown that the erosion due to these imperfections is only of 1 or 2 % of the critical load and can be disregarded. The torsional buckling load is therefore given by the well-known Wagner solution [22], account taken of the yield stress :

$$\overline{N} = \frac{\sigma}{f_y} = \frac{N_{crT}}{A.\,f_y} \ngtr 1 \qquad (2.14.)$$

with :

$$N_{crT} = \frac{A}{I_p}\left(C + \pi^2 \frac{C_w}{l^2}\right) \qquad (2.15.)$$

where I_p is the polar inertia, C the torsional rigidity and C_w the warping rigidity of the cross-section.

For cross-sections which are monosymmetrical or unsymmetrical, the shear center and the centroid do not coincide any more and the buckling can arise from an interaction between bending and torsion. As, in practice, the difference between this ultimate load and the ultimate load calculated for torsional buckling is small, it is usually neglected for sake of simplicity of the design equations.

Figure 2.7. illustrates above conclusions. Column flexural buckling curve "c", critical flexural buckling load for the weak axis N_f, critical flexural buckling load for the strong axis N_F, critical torsional buckling load N_T and flexural-torsional buckling load N_{F+T} are drawn on the figure, as well as tests results on angles. The picture shows clearly that the experimental values are in good correlation with the critical torsional buckling load N_T for short columns and with the flexural buckling curve "c" for long columns.

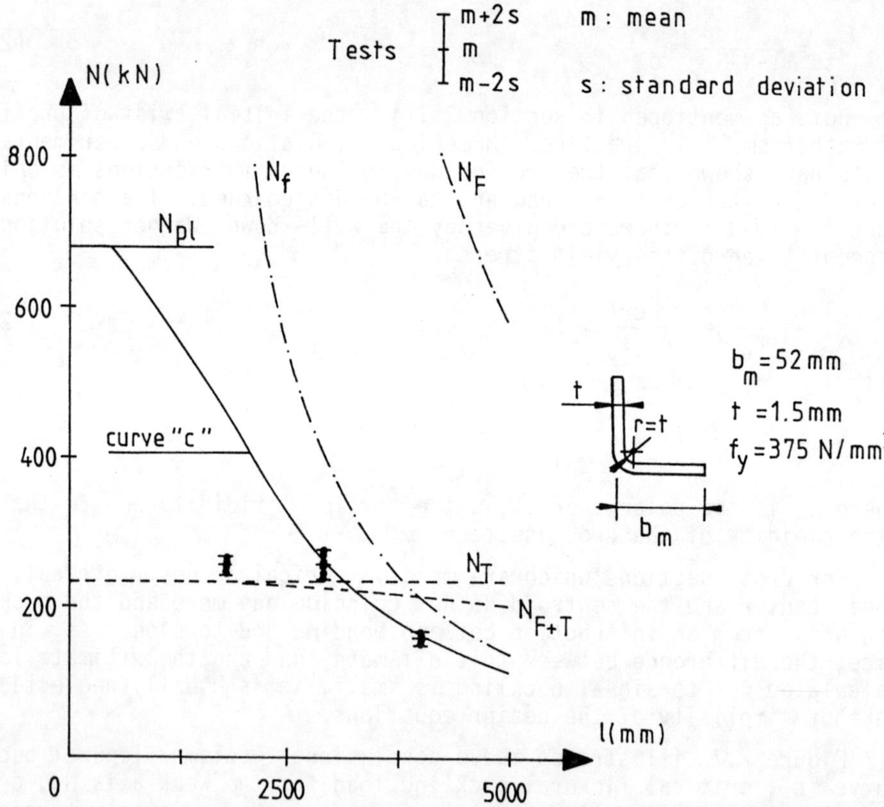

Figure 2.7. - Buckling of an angle.

2.2.3. Lateral buckling.

As shown in [5], an Ayrton-Perry type equation can also be used for the calculation of the lateral-torsional buckling ultimate load.

In agreement with the European Recommendations for Steel Construction [14] and with tests performed by Fukumoto and Kubo [24], equation (2.12.) can be used with the values of parameters given below :

- $\bar{\lambda}_o$ = 0.4
- curve a, hot-rolled sections and stress relief hollow sections : α = 0.32

- curve b, cold-formed sections : $\alpha = 0.78$.

Figure 2.8. shows these two buckling curves where the reduced slenderness is given by :

$$\bar{\lambda} = \sqrt{f_y/\sigma_{crL}} \tag{2.16.}$$

where, for a simply supported beam subject to pure bending, the critical lateral torsional buckling stress is given by the well-known Prandl solution [25] :

$$\sigma_{crL} = \frac{\pi\sqrt{EIC}}{l\,W}\sqrt{1 + \frac{C_w}{C}\frac{\pi^2}{l^2}} \tag{2.17.}$$

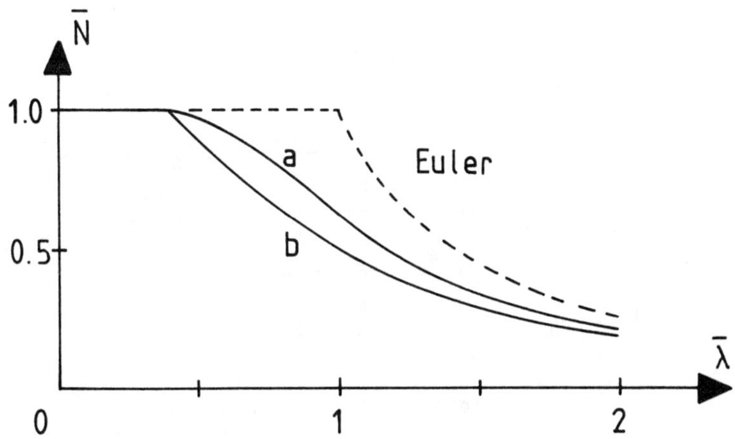

Figure 2.8. - Lateral buckling curves.

2.3. Local plate buckling design equations.

It has been pointed out, in section 2.1.1., that in practice, post-critical effects are of limited interest for the column behaviour. This is quite the reverse for compressed plates which exhibits, at least for stiffened compression elements, an important post-critical behaviour.

In cold-formed sections, the plate elements which constitute the profiles are generally classified in stiffened compression elements and unstiffened compression elements [2]. A stiffened element is a flat compression element both edges parallel to the direction of stress of which are stiffened by web elements, flanges or edge stiffeners of sufficient rigidity (Fig.2.9.a). An unstiffened element is a flat compression element that is stiffened at only one edge parallel to the direction of stress (Fig. 2.9.b).

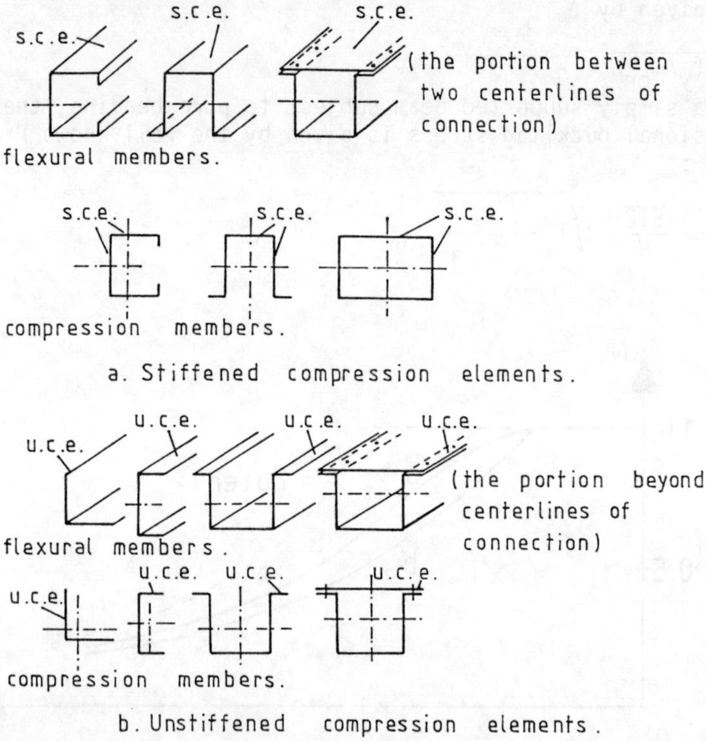

Figure 2.9. - Stiffened and unstiffened compression elements.

Post-buckling strength ability of both kinds of compression elements can be easily understood physically by means of the grid model proposed by Winter [26]. By isolating one element and replacing it by a grid model yields the situation of Fig. 2.10. for the stiffened case ; the rigid edge guides represent the action of the webs which hold the two flange edges straight.

If the five compression struts were to act independently, they would all collapse at the same stress: the plate would fail immediately when it starts to buckle. A plate, however, is a two-dimensional thing and the cross-ties roughly represent its action perpendicular to the direction of compression. It is evident that when the struts start buckling (bending), tension and bending are induced in the cross-ties which, thereby, counteract further buckling of the struts. They do so more effectively for those struts (portions of the plate) closer to the stiffened edges. In consequence, the grid (plate) is enable to carry additional, increasing load, but it will be primarily those portions which are close to the edges which resist increasing stress. The distribution of compressive stresses is, therefore, non-uniform in this post-buckling range.

The plate refuses to carry further additional load and begins to fail only when the most highly stressed portions reach the yield stress of the material. It is thus seen that the stresses in the cross-ties, representing the membrane stresses in the compressed plate, are the cause of post-buckling strength and of the difference between column and plate buckling. Membrane stresses are not only of the tension or compression variety, but also consist of shear stresses in the plane of the plate. For this reason, to make the grid model more representative of the actual plate, diagonals should be added in all panels. These have been omitted for the sake of graphical clarity.

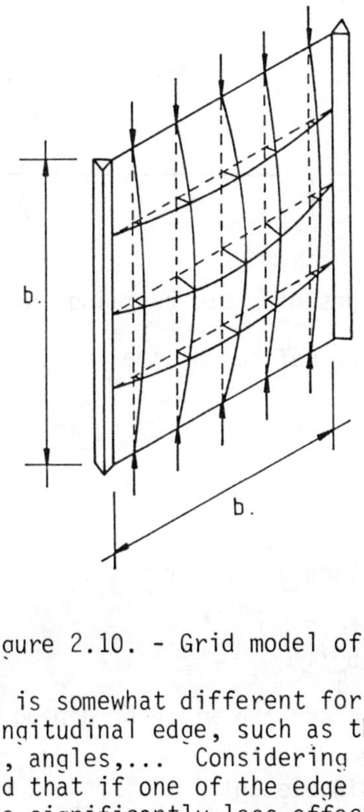

Figure 2.10. - Grid model of post-buckling strength.

The situation is somewhat different for compressed plates stiffened only along one longitudinal edge, such as the outstanding flange projections of channels, angles,... Considering the model of Fig. 2.10., it is easily realized that if one of the edge stiffeners is removed, the cross-ties will be significantly less effective. That is, the membrane stresses in such a plate in the post-buckling range, though present, are smaller ; corresponding, they will permit larger buckling deformations to occur and will not raise the ultimate strength as much above the incipient buckling load.

These differences in the plate buckling behaviour leads to plate buckling curves which are associated with the von Karman ultimate load for sections the main element of which is stiffened (hollow sections, C, U, Z,...) and with the Euler critical load for sections the elements of which are unstiffened (angles, X,...). Figure 2.11. shows these two types of plate buckling curves.

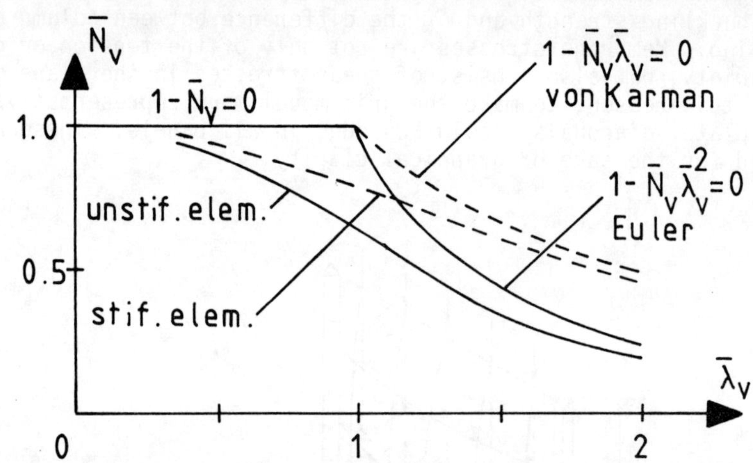

Figure 2.11. - The two types of plate buckling curves.

The reduced coordinates are defined by the relations :

$$\bar{N}_v = \frac{\sigma}{f_y} \tag{2.18.}$$

and

$$\bar{\lambda}_v = \sqrt{f_y/\sigma_{crv}} \tag{2.19.}$$

where the critical plate buckling stress writes :

$$\sigma_{crv} = k \frac{\pi^2 E}{12\,(1-\nu^2)} \left(\frac{t}{b_m}\right)^2 \tag{2.20.}$$

The buckling coefficient k and the width b_m depend of the type of section.

For sections where the main element is stiffened, b_m is defined in Fig. 2.12. and k is given, with $\phi = c_m/b_m$ by [27] :

- for square and rectangular hollow sections (fig. 2.12.a) :

$$k = 6.56 - 5.77\,\phi + 8.56\,\phi^2 - 5.36\,\phi^3 \tag{2.21.}$$

- for [, ⅃⅂ and ⊏ sections (Fig. 2.12.b), with $0.1 \leq \frac{d_m}{b_m} \leq 0.3$:

$$k = 6.8 - 5.8\,\phi + 0.2\,\phi^2 - 6.0\,\phi^3 \tag{2.22.}$$

- for [and ∫ sections (Fig. 2.12.c), with $\phi \geq 0.5$:

$$k = 9.03 - 16.10\,\phi + 7.94\,\phi^2 \qquad (2.23.)$$

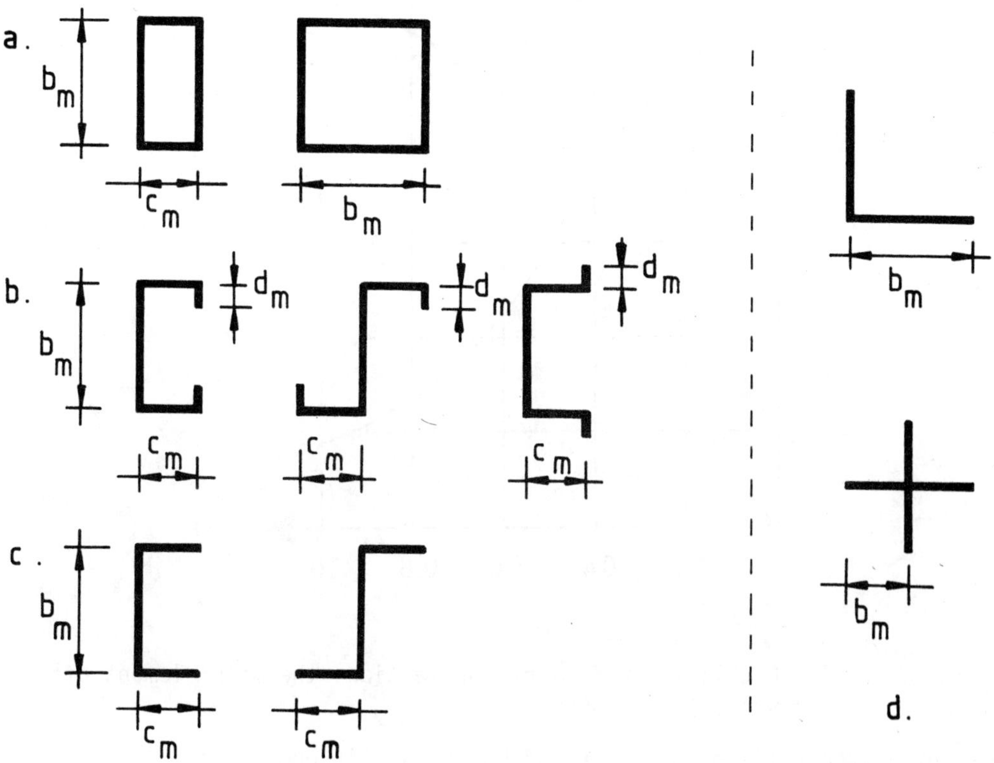

Figure 2.12. - Width taken into account for the calculation of the critical stress.

Figure 2.13. illustrates these relations.

For sections with only unstiffened elements (fig. 2.12.d) the buckling coefficient is equal to the classical value :

$$k = 0.425 \qquad (2.24.)$$

The local plate buckling curves for the two types of sections can be represented by an Ayrton-Perry equation calibrated by means of numerical simulations and experimental tests :

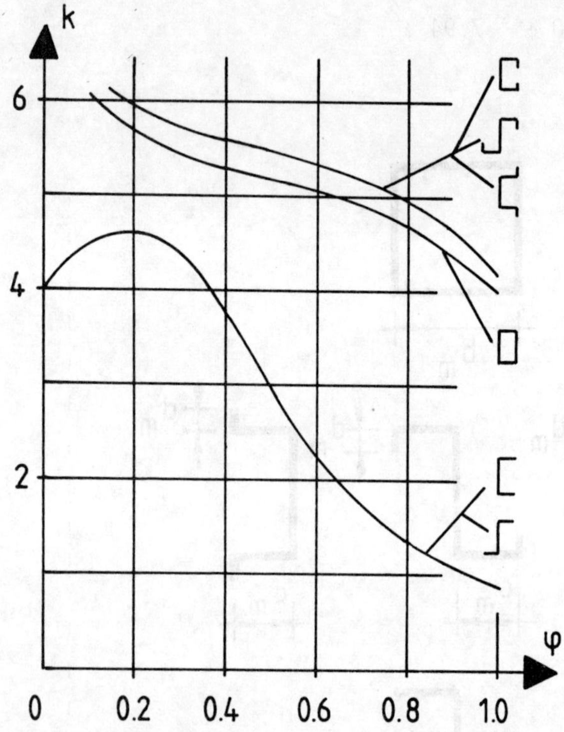

Figure 2.13. - Buckling coefficient for sections the main element of which is stiffened.

- for sections the main element of which is stiffened [28] :

$$(1 - \overline{N}_v)(1 - \overline{N}_v\,\overline{\lambda}_v) = \beta\,(\overline{\lambda}_v - \overline{\lambda}_{vo})\,\overline{N}_v \tag{2.25.}$$

which leads to :

$$\overline{N}_v = \frac{1 + \beta\,(\overline{\lambda}_v - \overline{\lambda}_{vo}) + \overline{\lambda}_v}{2\,\overline{\lambda}_v} - \frac{1}{2\overline{\lambda}_v}\sqrt{[1+\beta(\overline{\lambda}_v-\overline{\lambda}_{vo})+\overline{\lambda}_v]^2 - 4\overline{\lambda}_v} \not> 1 \tag{2.26.}$$

with :

- $\overline{\lambda}_{vo} = 0.8$;
- curve a, stress relief hollow sections : $\beta = 0.35$;
- curve b, cold formed sections : $\beta = 0.67$;

- for sections with only unstiffened elements [23] :

$$(1 - \overline{N}_v)\,(1 - \overline{N}_v\,\overline{\lambda}_v^2) = \beta\,(\overline{\lambda}_v - \overline{\lambda}_{vo})\,\overline{N}_v \tag{2.27.}$$

which leads to :

$$\bar{N}_v = \frac{1 + \beta(\bar{\lambda}_v - \bar{\lambda}_{vo}) + \bar{\lambda}_v^2}{2\,\bar{\lambda}_v^2} - \frac{1}{2\bar{\lambda}_v^2}\sqrt{[1+\beta(\bar{\lambda}_v-\bar{\lambda}_{vo})+\bar{\lambda}_v^2]^2 - 4\bar{\lambda}_v^2} \ngtr 1 \quad (2.28.)$$

with :

- $\bar{\lambda}_{vo} = 0.8$;
- curve c : $\beta = 0.15$.

Figure 2.14. shows these three local plate buckling curves.

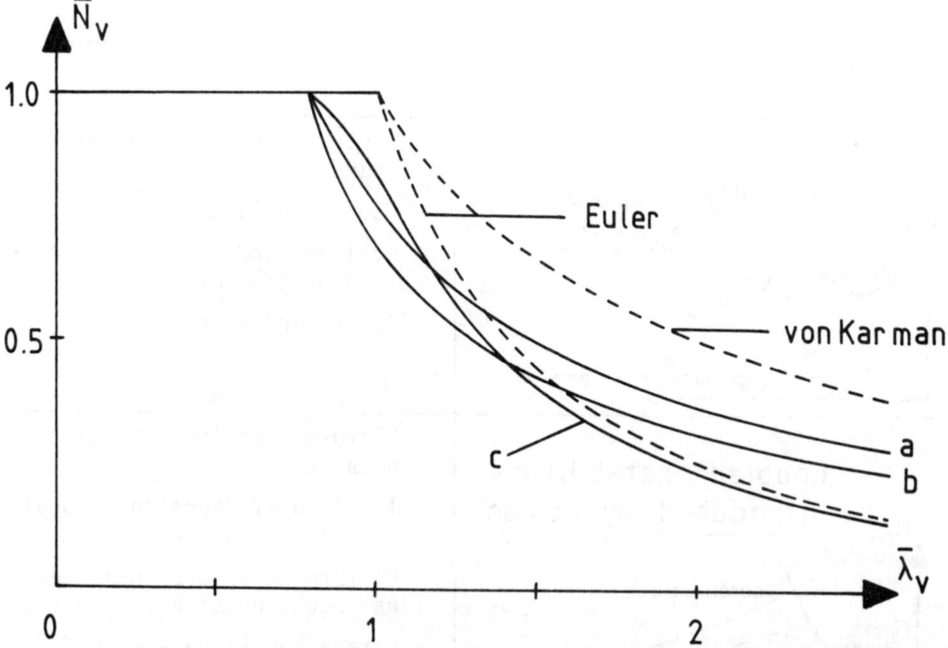

Figure 2.14. - Local plate buckling curves.

3. MODE INTERACTION IN THIN-WALLED BARS

3.1. General

It was shown, in section 2.1.1., that the plastic behaviour of steel leads to an interaction between stability and strength in compressed bars. This interaction can result in a decrease of the maximum load with respect to the case of the undefinite elastic material.

Sometimes, several stability constraints govern the problem ; the coupling of these stability modes must then be examined with caution because the sensitivity to imperfections in case of simultaneous bifurcations is significantly enhanced as compared to simple bifurcation.

These coupled instabilities may be of two kinds [29] :

- naturally coupled instabilities resulting from "Garland" curves ; two or more forms of instability being possible at the point of interaction of the curve ;
- instabilities where the coupling is due to the design, i.e. the geometric dimensions are chosen such that two forms of instabilities are possible.

This last case is of very great importance in optimum design under stability constraints.

Table 3.1. presents some examples of these two types of coupled instabilities.

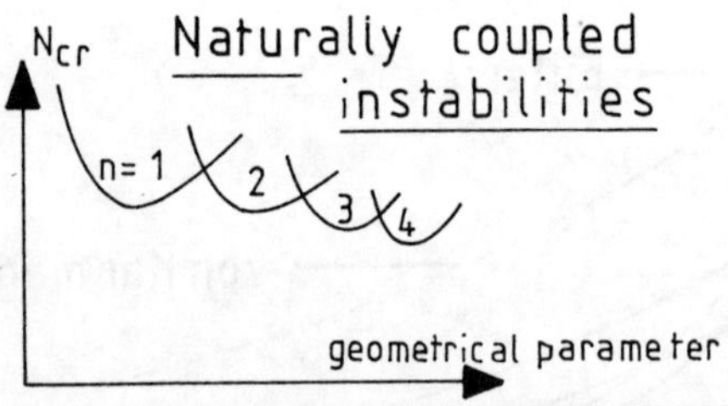	- Bar, arch and ring on elastic foundation - Compressed plate - Cylinder under axial compression or pressure - Sphere under pressure
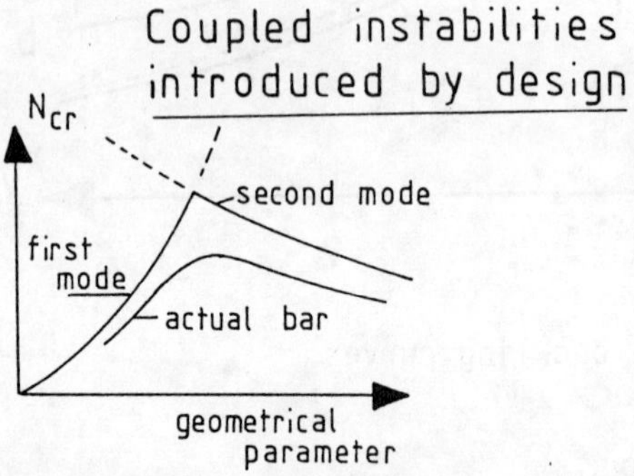	- Flexural and local buckling of a bar - Torsional-flexural and local buckling of a bar - Flexural and torsional-flexural buckling of a bar - Lateral buckling and local buckling of a bar - General and member buckling of a laced column - Individual and general buckling of columns in a frame - General and individual buckling of a guyed mast - Buckling of stiffeners and of a stiffened plate or shell

Table 3.1. - Example of coupled instabilities.

Investigations into the post-critical behaviour of coupled bifurcations, or nearly coupled bifurcations, and into the influence of imperfections have been carried out by Thompson and Hunt [8]. They have shown that the interaction of two stable buckling modes may generate an unstable coupled mode, making the system highly imperfection sensitive.

For bars, the two main problems - interaction between global flexural buckling and local plate buckling, on the one hand, and lateral buckling and local plate buckling, on the other hand - are characterized by an overall mode with a long half wave length, of the order of magnitude of the length of the member, associated with a local mode of short wave. The half wave length of this local buckling mode is of the order of magnitude of the width of the member.

In the last years, several analytical and numerical methods have been proposed to solve these problems. They can be classified in three categories [30] :

- exact methods ; they are called "exact" in the frame of the general theory of elastic stability. Due to the mathematical complexities of this approach, only some interactive buckling problems have been solved up to now with simplified approximations, e.g. residual stresses are disregarded and the material has a perfectly elastic behaviour ;
- numerical methods ; the application of general numerical methods - such as for instance, the finite element method - to interactive buckling in thin-walled members is not very convenient, at least from an economic point of view. A satisfactory result can only be expected by using an assembly of thin-shell elements with a fine mesh, with the result of a great number of degrees of freedom. Simplified approaches, based on the Rayleigh-Ritz method or finite-strips, have also been adopted by several authors.
- approximate methods ; they are generally based on reduced stiffness approaches to take account of the effect of the local buckling.

In reduced stiffness approaches, the effect of local buckling is classically taken into consideration by introducing an effective cross-section, i.e. a section where only the effective widths of the plate elements are considered. As the geometrical cross-section of each plate is constant throughout the length of the bar, the effective width of the plate depends only on the state of stress ; however the latter varies, as a result of the flexure of the bar, from plate to plate over the cross-section and along the length of the bar. This problem is thus very complicated and can only be solved by a step-by-step procedure [31]. Some authors have introduced an important simplification in the process by considering effective widths which are constant throughout the length of the member and calculated on base of the state of stress which appears at mid-length of the bar. This assumption gives a good approximation of the maximum load but overestimates the displacements near the collapse.

Effective width formulae have been proposed by numerous authors since the original proposal of von Karman et al [32] . Recently published A.I.S.I. (American Iron and Steel Institute) and E.C.C.S. (European Convention for Constructional Steelwork) recommendations give effective width formulae for stiffened and unstiffened elements under uniform compression or stress gradient [33, 34].

Some authors have also proposed, for the interaction between flexural global buckling and local plate buckling, direct methods based on modifications of global buckling curves to take into account the influence of the local plate buckling [28, 35, 36] . One of these proposals, which is of peculiar interest in the optimum design of thin-walled bars, is developed in section 3.2.

3.2. A direct method for the interaction between flexural global buckling and local plate buckling.

In the practical buckling design of steel columns, a prominent role is played by the stub-column, i.e. and element that is sufficiently short to avoid flexural global buckling. The ultimate load N_y of this stub-column is indeed a reference value of the buckling curves.

For compact sections, the load N_v is equal to the squash load N_{pl} since yielding occurs prior to plate buckling and the effective yield stress f_y defined as the ratio N_{pl}/A, allows one to take implicit account of nonuniform values of the yield stress across the section, which follows from the manufacturing process (cold-forming, welding, rolling, cooling, etc...). On the other hand, for thin-walled sections, which are affected by plate buckling, the ratio N_y/A represents the mean plate buckling stress f_v of the thin-walled section.

As explained in section 2.3., f_v is given by :

$$f_v = \overline{N}_v . f_y \tag{3.1.}$$

where $\overline{N}_v$ is the reduced load.

It was suggested in references [28, 37] that the ultimate load of a compressed thin-walled member may be determined using column buckling curves in which the plate buckling stress f_v could be substituted for the effective yield stress f_y. The validity of this very simple design method has been demonstrated on the basis of experimental results obtained by Rondal, on square and rectangular hollow sections [28], by Batista, on U and C sections [27] and by Costa Ferreira, on angles [5].

Table 3.1. summarizes this design procedure.

Local plate buckling

$$\bar{\lambda}_v = \sqrt{f_y/\sigma_{crv}}$$

$$\sigma_{crv} = k \frac{\pi^2 E}{12(1-\nu^2)} \left(\frac{t}{b_m}\right)^2$$

with :

- hollow sections; k:equation (2.21.)
-  ; k:equation (2.22.)
-  ; k:equation (2.23.)
- angles ; k = 0.425

$$f_v = \bar{N}_v f_y$$

with :

- main element stiffened ; $\bar{N}_v$:equation(2.26.)
- only unstiffened elements ; $\bar{N}_v$:equation(2.28.)

Interaction between local and overall buckling

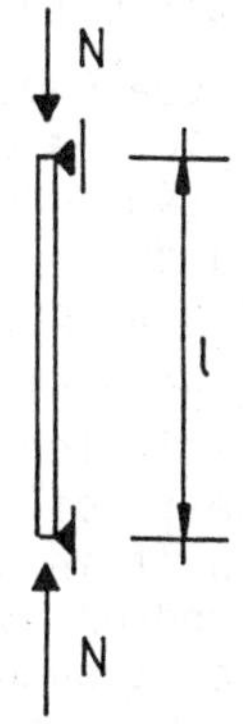

$$\bar{\lambda}' = \sqrt{f_v/\sigma_{cr}} = (1/\pi i)\sqrt{f_v/E}$$

$$\bar{N}' = (1/2\,\bar{\lambda}'^2)\,\{1+\,(\bar{\lambda}'-0.2)+\,\bar{\lambda}'^2 - \sqrt{[1+\alpha(\bar{\lambda}'-0.2)+\bar{\lambda}'^2]^2 - 4\bar{\lambda}'^2}\,\}$$

$$N = \bar{N}\,A\,f_v$$

Manufacturing process	α
Stress relief hollow sections	0.21
Cold-formed sections	0.49

Notation :

A cross - sectional area

N ultimate load

i radius of gyration

f_y yield stress

l buckling length

Table 3.1. - Summary of the design method.

3.3. Optimum design of perfect and imperfect bars.

According to the classical linear buckling theory, the optimum design of thin-walled bars requires equal or nearly equal critical stresses for the global and local modes in order to achieve the highest structural efficiency. This concept, called "simultaneous mode design" is generally attributed to Bleich [38].

But, "Redde Caesari que sunt caesaris", as mentioned by Shanley [39], this principle was proposed for the first time by Tuckerman, Petrenko and Johnson [40] who called it "the one-hoss shay" principle in accordance to the famous poem of Oliver Wendell Holmes : "The Deacon's Masterpiece or, the Wonderful One-Hoss Shay".

One cannot resist to the temptation to reproduce the beginning and the end of this logical story :

"Have you heard of the wonderful one-hoss shay,
That was built in such a logical way
It ran a hundred years to a day,
And then, of a sudden, it-ah, but stay,
I'll tell you what happened without delay ;
\- - -
What do you think the parson found,
When he got up and stared around ?
The poor old chaise in a heap or mound,
As if it had been to the mill and ground !
You see, of course, if you're not a dunce,
How it went to pieces all at once, -
All at once, and nothing first, -
Just as bubbles do when they burst.

End of the wonderful one-hoss shay.
Logic is logic. That's all I say".

If the simultaneous mode design principle is a pertinent concept for the optimum design of perfect systems, it is seriously questionable for real structures which exhibit unavoidable imperfections. Thompson and Hunt [41], in a paper which has won a great renown , have drew the attention on the dangers of structural optimization under stability constraints, which can leads to an "explosive" instability. This important point is discussed carefully, for compressed bars, in section 4.

4. EFFICIENCY CHARTS

4.1. General

Having a simple and accurate design method for interactive buckling of thin-walled bars, it is highly attractive to use it extensively in order to determine the optimum dimensions of the cross-section. As an example, thin-walled compressed columns are considered in this section.

Though the weight is generally not a satisfactory criterion for the optimization of an engineering structure, it can constitute a valuable reference for a single compression member. Minimization of the weight of a column with a specified length and subject to a specified compressive force may be also regarded as a maximization problem of the ultimate strength of a column, the weight of which is specified. This latter aspect is used here.

The aim of this chapter is to answer the three following basic questions :

- Is there a relationship between the optimal solution for a perfect column and that for an actual imperfect column ?
- What is, for the optimal solution, the erosion of the ultimate strength which has to be expected because of the interaction between buckling modes ?

and, last but not least :

- Is there some benefit in using sections the wall thinness of which is such that local plate buckling occurs ?

An efficiency chart due to Thompson and Lewis [42] and another one suggested by Usami and Fukumoto [35] will serve as supports to the investigations.

The first efficiency chart has been used, for the first time, by Thompson and Lewis [42] to optimize the model cross-section used by Van der Neut [43] for his analytical investigation on the interaction of local buckling and column failure in thin-walled compression members.

The idealized section treated by Van der Neut [43] is shown in Fig. 4.1. and comprises two load-carrying flange-plates with an unspecified web which simply serves to maintain the structural integrity of the strut without contributing to the transmission of axial stresses (web rigid in shear and laterally but which has no longitudinal stiffness).

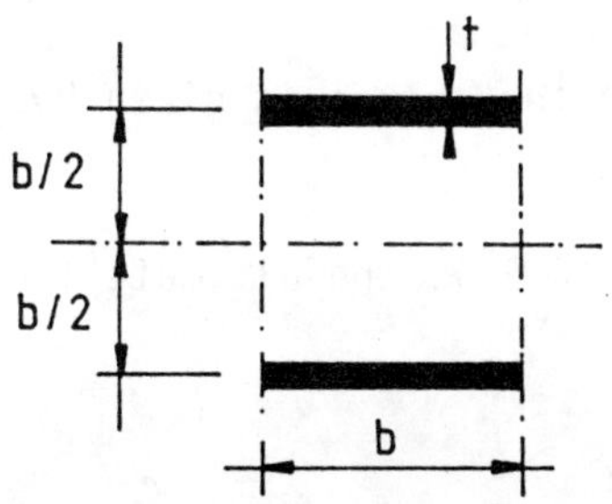

Figure 4.1. - van der Neut's model.

Unfortunately, the consideration of this idealized model and of column imperfections which are considered as quite mild as compared to the effect of flange imperfections, have led Thompson and Lewis to erroneous conclusions. The Thompson efficiency chart is used here for the optimization of a square hollow section and gives interesting answers at the two first questions.

Fukumoto's efficiency chart is used afterwards to answer the third question on the benefit of using thin-walled sections. It is applied to close and some open cross-sections.

4.2. Thompson's efficiency chart.

Rondal and Maquoi [44] have considered a square cross-section with an average width b_m and a wall thickness t, neglecting the rounded corners (fig. 4.2.). The critical buckling stresses are, respectively :

- for overall column buckling :

$$\sigma_{cr} = \frac{\pi^2 EI}{A l^2} = \frac{\pi^2 E b_m^2}{6 l^2} \qquad (4.1.)$$

- for local plate buckling :

$$\sigma_{crv} = \frac{\pi^2 E}{3(1-\nu^2)} \left(\frac{t}{b_m}\right)^2 \qquad (4.2.)$$

where E is the modulus of elasticity and ν is the Poisson ratio, because the cross-section area A and the moment of inertia I are approximated as follows :

$$A = 4b_m t, \qquad (4.3.)$$

$$I = \frac{2}{3} b_m^3 t. \qquad (4.4.)$$

In accordance with Thompson and Lewis [42] , two characteristic ratios are introduced :

$$x = \frac{\sigma_{cr}}{\sigma_{crv}} = 8(1-\nu^2) \frac{b_m^6}{l^2 A^2}, \qquad (4.5.)$$

$$y = \sigma/\sigma_{crv}, \qquad (4.6.)$$

where the ultimate stress of the column is given by :

$$\sigma = N/A. \qquad (4.7.)$$

From (4.2.), (4.6.) and (4.7.), the ultimate load of the column writes :

$$N = K.y.x^{-2/3} \qquad (4.8.)$$

where the numerical factor :

$$K = \frac{\pi^2 EA^{5/3} l^{-4/3}}{12(1-\nu^2)^{1/3}}, \qquad (4.9.)$$

is constant for specified values of A and l. As a consequence of (4.8.) the efficiency function :

$$\rho(x) = yx^{-2/3} \qquad (4.10.)$$

can replace N in the analysis, and be plotted against x.

For the perfect structure, the optimal solution is defined by the relation (see chapter 2) :

$$\sigma_{cr} = \sigma_{crv} \tag{4.11.}$$

which leads to the following optimum :

$$x = \rho(x) = 1 \tag{4.12.}$$

In the Thompson efficiency chart, the analysis is characterized as follows (Fig. 4.2.) :

a) perfect column :
   - overall column buckling : in equation (4.7.) N is replaced by :

$$N_{cr} = \frac{\pi^2 EI}{l^2} \quad ;$$

   - local plate buckling : $y = 1$;

b) imperfect column, without interactive buckling :
   - overall column buckling : in equation (4.7.), N is replaced by $\overline{N}.A.f_y$, where $\overline{N}$ is given by expression (2.12.) ;
   - local plate buckling : in equation (4.7.), N is replaced by $\overline{N}_v.A.f_y$ where $\overline{N}_v$ is given by expression (2.26.) ;

c) imperfect column, with interactive buckling : in equation (4.7.), N is replaced by the value calculated in accordance with section 3.2.

Rondal and Maquoi [44] have performed an extensive analysis on base of the Thompson efficiency chart with a peculiar attention paid to a large range of values of the length, the area of the cross-section and the yield stress which cover all the cases encountered in practice.

A first conclusion deals with the location of the actual optimum with respect to the optimum for the perfect column, which lies at $x = 1$ in the Thompson efficiency chart. Fig. 4.3. shows that when the cross-sectional area increases - all the other parameters remaining constant - the location of the true optimum moves. For small values of A, the optimum occurs for high values of x. With increasing A, abscissa x of the optimum first decreases, becoming lower than 1, and then increases infinitely. This conclusion, drawn from Fig. 4.3. for a specified set of data, is quite general, as is demonstrated by the analysis of an extended range of data values. Fig. 4.4. shows, for example, a similar move of the location of the optimum for different lengths and two steel grades.

It may be concluded that the maximum efficiency of a section can be obtained as well for $x > 1$ as for $x < 1$, according to the values of the different parameters : there is thus no specified relationship between the optimum for a perfect column and that for an actual one. This conclusion is in agreement with the results - less extended than these - obtained by Maquoi and Massonnet [45] but disagrees with the conclusion drawn by Thompson and Lewis [42] according to which the optimum would appear in any case for $x < 1$. This example shows clearly the imperative necessity to use realistic models in the optimum design of structures.

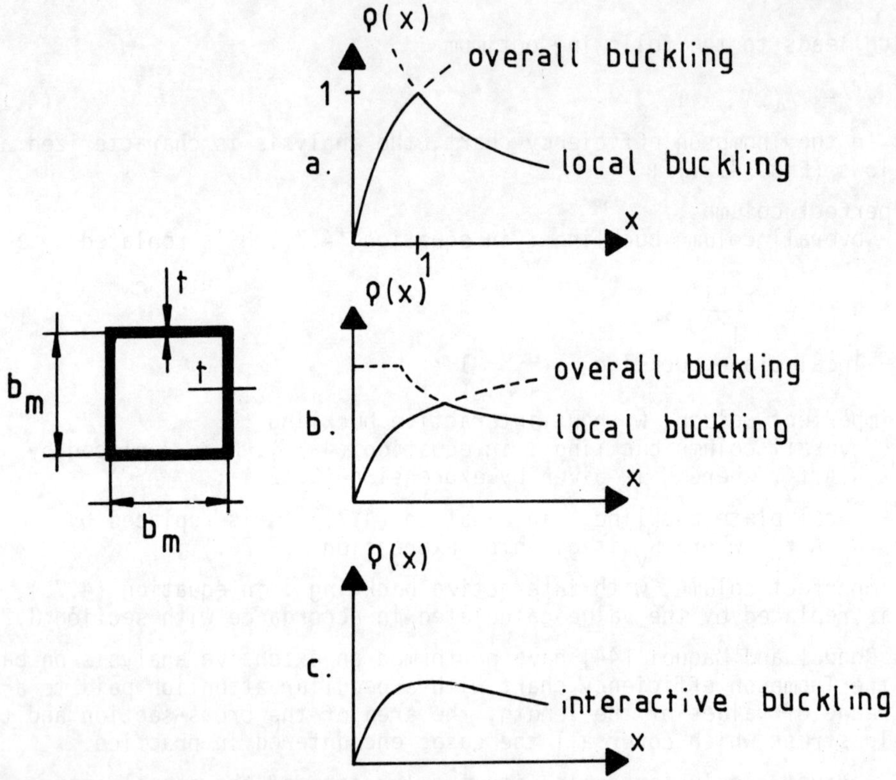

Figure 4.2. - Thompson's efficiency chart : (a) perfect column, (b) imperfect column (without interactive buckling), (c) imperfect column (with interactive buckling).

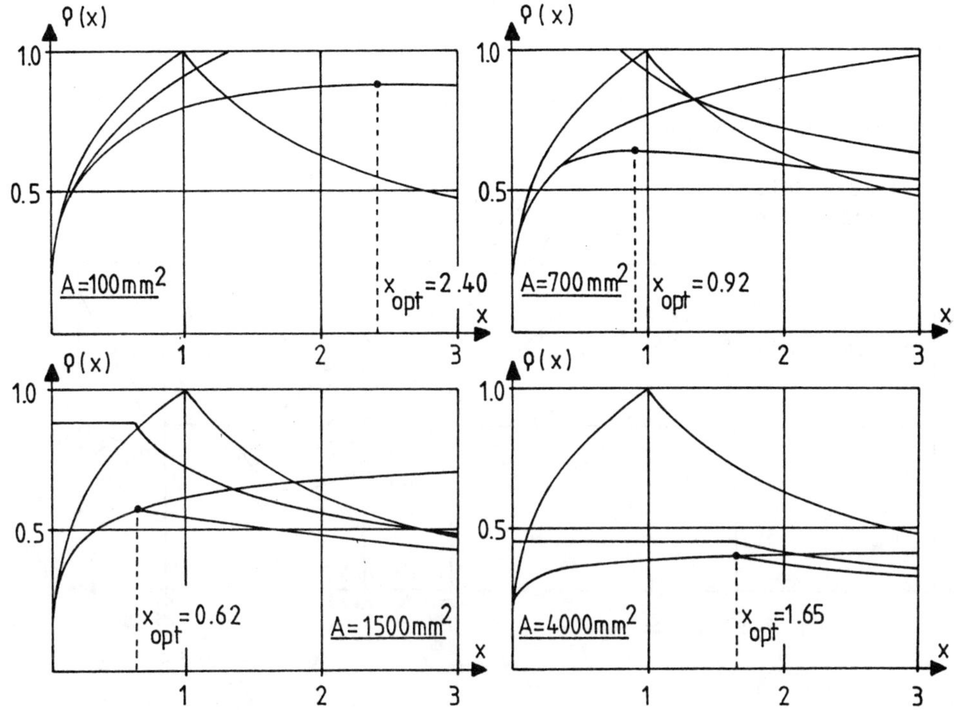

Figure 4.3. - Hot-finished square hollow section (l = 5000 mm ; f_y = 235 N/mm^2 ; α = 0.21 ; $\overline{\lambda}_o$ = 0.2 ; β = 0.35 ; $\overline{\lambda}_{vo}$ =0.8).

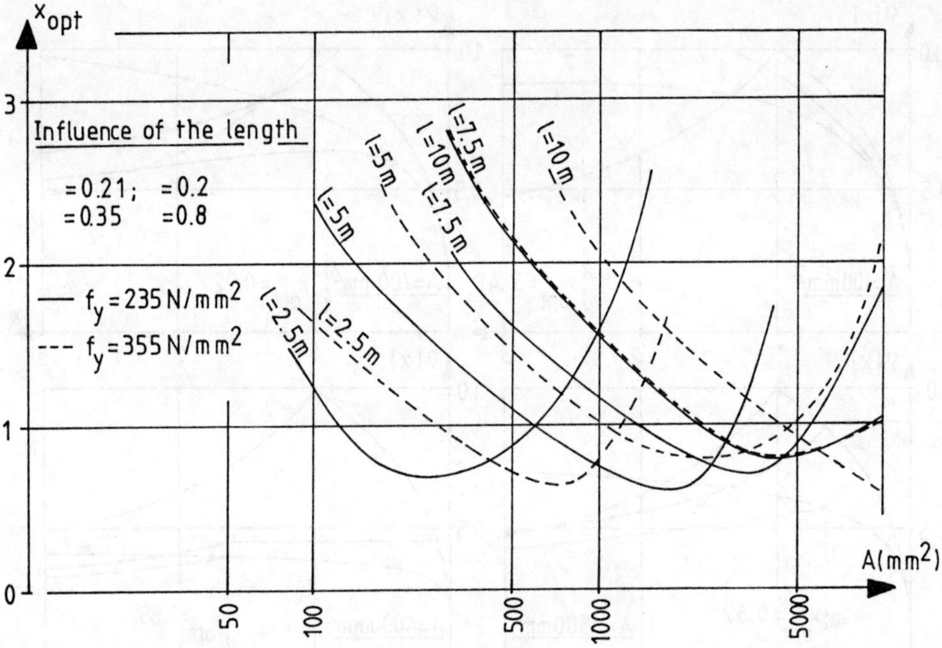

Figure 4.4. - Location of the optimum.

As mentioned in section 3.3., Thompson and Hunt [41] drew attention to the danger of an optimization based on simultaneous occurrence of different buckling modes. They pointed out the possibility of an appreciable decrease in strength because of the increasing coupling of stability modes and unavoidable imperfections. Because they based their study on the difference in strength Δ_{ac} between a perfect structure, on the one hand, and an imperfect one with interaction, on the other (Fig.4.5.), such a conclusion has a restricted validity. Indeed, the design philosophy in the field of structural engineering refers nowadays to imperfect structures. In this respect, the actual erosion of strength due to interactive buckling modes is closer to the practical erosion Δ_{pr} than to the academic one Δ_{ac}, as shown in Fig. 4.5. Both kinds of erosion - Δ_{ac} and Δ_{pr} - are plotted on Fig. 4.6. for a square section, by means of their dimensionless gap :

$$g_{ac}(\%) = 100\,\frac{1-\rho_{interaction}}{1} \qquad (4.13.)$$

$$g_{pr}(\%) = 100\,\frac{\rho_{without\ interaction}-\rho_{interaction}}{\rho_{without\ interaction}} \qquad (4.14.)$$

The two values vary in different ways; when area A isincreasing, the academic gap becomes rather large while the practical gap decreases and becomes even negligible in some cases.

As a summary, it is observed that :

- when A $\nearrow$ - g_{ac} $\nearrow$ and g_{pr} $\searrow$;
- when l $\nearrow$ - g_{ac} $\searrow$ and g_{pr} $\nearrow$;
- when $\dot{f}_y$ $\nearrow$ -g_{ac} $\searrow$ and g_{pr} $\nearrow$.

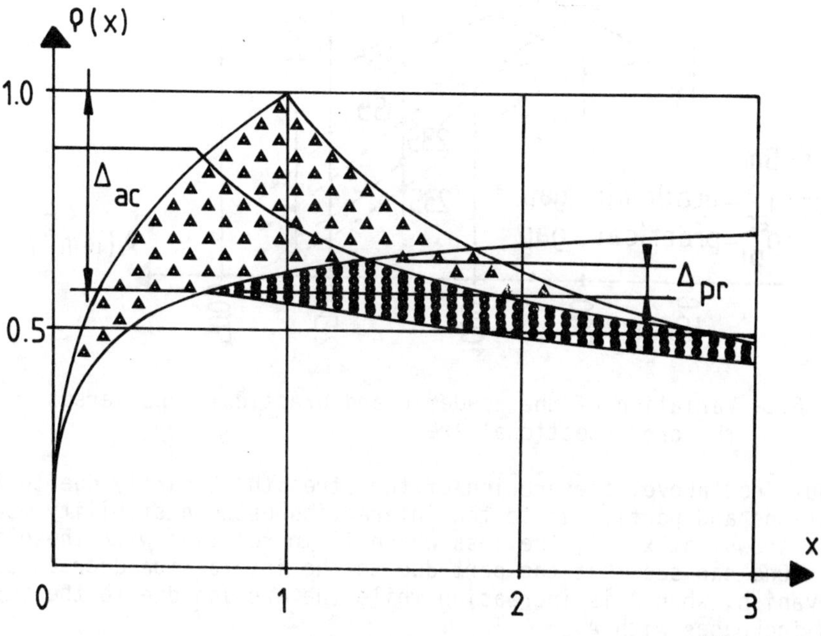

Figure 4.5. - Definition of the academic erosion and of the practical erosion of the strength.

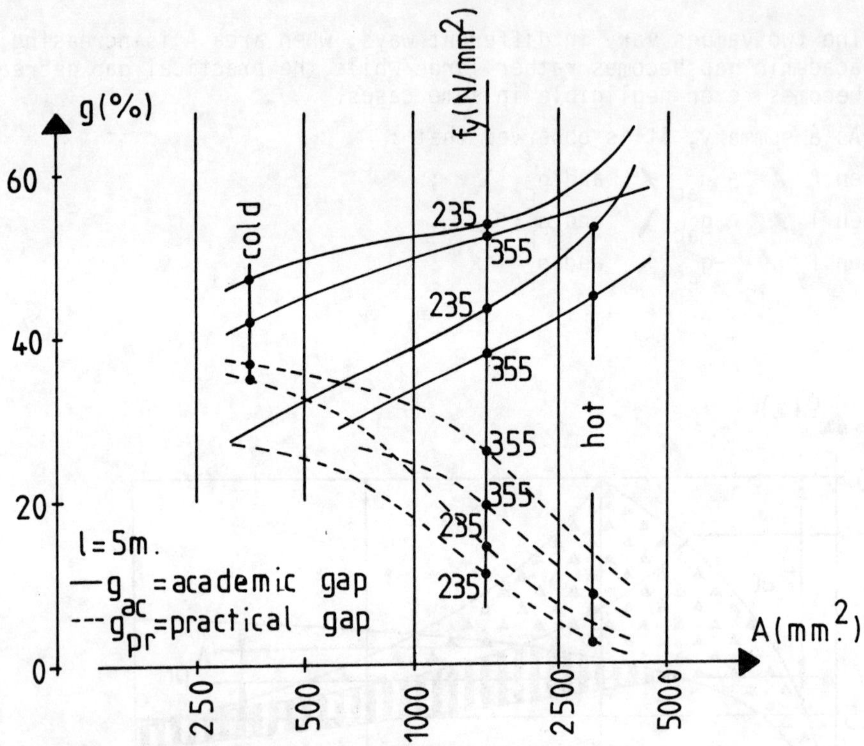

Figure 4.6. - Variation of the academic and practical gaps versus the cross-sectional area.

As outlined above, the erosion of the strength is partly due to the imperfections and partly due to the interaction between stability modes. Fig. 4.7. shows, at x = 1, the loss which is in relation with these two factors. One can see that the part due to the interaction decreases, and can vanish, when A is increasing while the erosion due to the imperfections increases with A.

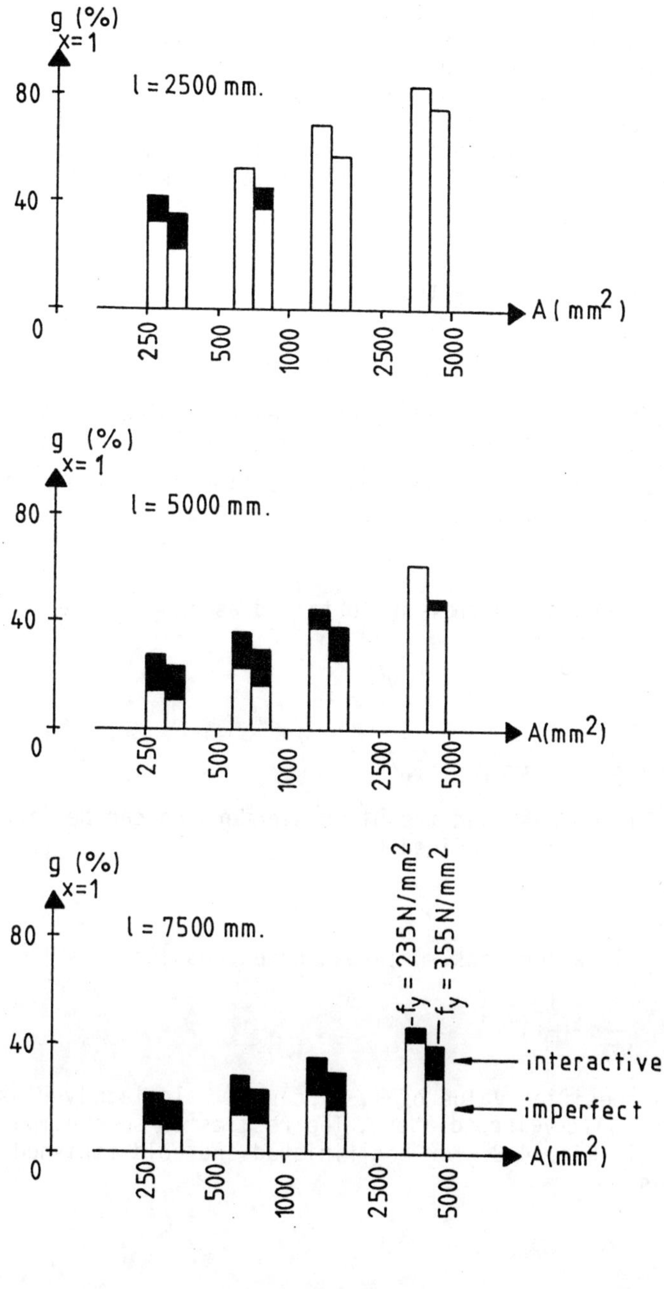

Figure 4.7. - The two components of the erosion of the strength at x = 1 (Hot-finished square hollow section).

4.3. Fukumoto's efficiency chart.

Consideration of Thompson's efficiency chart allows us to answer two of the three questions asked above. However, it does not satisfactorily answer the third, concerning the benefit of using sections the wall thickness of which is such that plate buckling occurs. In this respect, another efficiency chart suggested by Usami and Fukumoto [35] is more adequate. The efficiency function ρ^*, equal to the ratio of the ultimate stress:

$$\sigma = \frac{N}{A} \tag{4.15.}$$

to the yield stress f_y, writes thus :

$$\rho^* = \frac{\sigma}{f_y} \tag{4.16.}$$

In (4.15.), the collapse load is computed as indicated in chapter 3, with account taken of the interaction between local and global buckling.

For the square hollow section used in section 4.2., the radius of gyration is given by :

$$i = b_m/\sqrt{6} \tag{4.17.}$$

and the dimensionless slenderness are obtained as :

$$\bar{\lambda} = \sqrt{f_y/\sigma_{cr}} = 0.78 \frac{1}{b_m} \sqrt{f_y/E} \tag{4.18.}$$

and

$$\bar{\lambda}_v = \sqrt{f_y/\sigma_{crv}} = \frac{b_m}{1.9t} \sqrt{f_y/E} \tag{4.19.}$$

The column slenderness and the plate slenderness can be linked by a the dimensionless factor γ defined by :

$$\gamma = \bar{\lambda} \sqrt{\bar{\lambda}_v} \tag{4.20.}$$

For square hollow sections, γ takes the expression :

$$\gamma = 1.13 \frac{1}{\sqrt{A}} \left(\frac{f_y}{E}\right)^{3/4} \tag{4.21.}$$

Thus, for a specified value of γ, ρ^* can be calculatedversus $\bar{\lambda}_v$ by means of the expressions developped above. The results of such computations are represented in Fig. 4.8. for hot-finished and cold-finished square hollow sections.

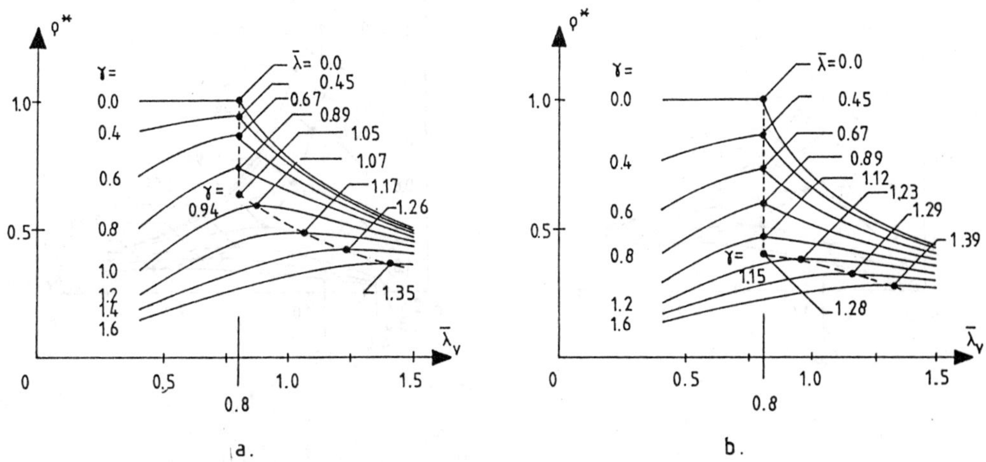

Figure 4.8. - Fukumoto's efficiency chart for square hollows sections :
a. hot-finished profiles ;
b. cold-finished profiles.

Interesting conclusions can be drawn from Fig.4.8. :

- the optimum efficiency is never reached for a thick-walled tube, i.e. for $\bar{\lambda}_v < 0.8$;
- for small values of γ , the optimal efficiency corresponds to the limiting plate slenderness $\bar{\lambda}_v = 0.8$ adopted for plate buckling curves (see chapter 2) ;
- for large values of γ , the optimal efficiency corresponds to thin-walled sections.

The Fukumoto efficiency chart can also be drawn for open sections. For instance, charts for U profiles and for angles are respectively given in Fig. 4.9.a and 4.9.b. The expressions which are used to build the charts are given in table 4.1. One can mention that, as explained in section 2, the buckling curves are the same for cold-finished hollow sections and for cold-formed U sections. For this reason, the charts for these two types of profiles are also identicals.

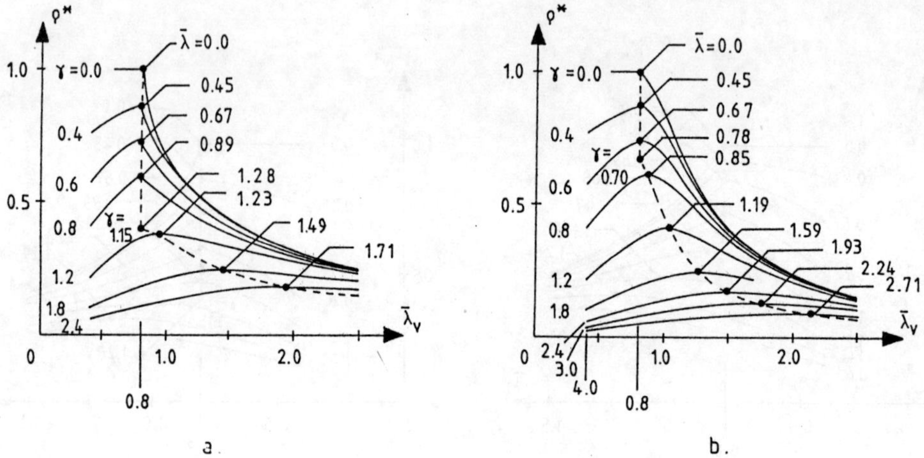

Figure 4.9. - Fukumoto's efficiency chart for cold-formed sections :
a. square hollow and U profiles ;
b. angles.

It is interesting to compare, for the three types of cold-formed profiles studied above, the benefits provided by the effective optimal solution as compared to the optimal situation when plate buckling is not allowed, i.e., for $\bar{\lambda}_v = 0.8$.

Fig. 4.10. gives, for these three types of profiles, the efficiency ratio :

$$\rho^{**} = \frac{\rho^{*}_{optimal}}{\rho^{*}_{(\bar{\lambda}_v=0.8)}} \tag{4.22.}$$

with respect to the part of γ which is independent of the type of profiles.

Symbols	U profiles	Angles
Notations	$\phi = \frac{c_m}{b_m}$ 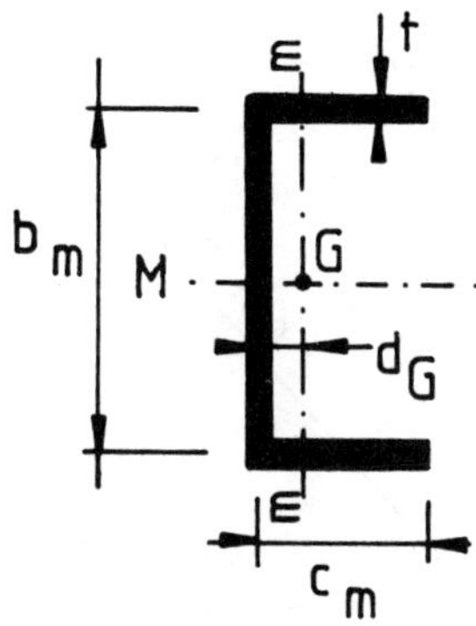	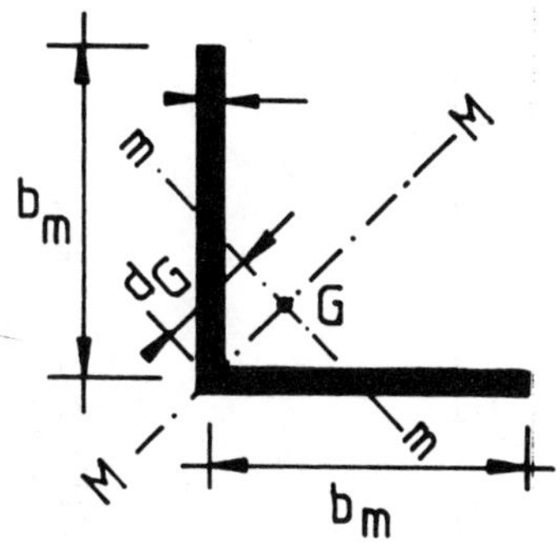
d_G	$\frac{\phi^2}{1+2\phi} . b_m$	$b_m/2\sqrt{2}$
A	$(1+2\phi).b_m.t$	$2\ b_m.t$
I	$\frac{2+\phi}{3(1+2\phi)} . b_m^3 . t$	$\frac{1}{12} b_m^3.t$
i	$\sqrt{\frac{2+\phi}{3}} . \frac{b_m}{1+2\phi}$	$\frac{b_m}{2\sqrt{6}}$
$\bar{\lambda}=\sqrt{f_y/\sigma_{cr}}$	$\frac{1+2\phi}{\pi\sqrt{\frac{2+\phi}{3}}} . \frac{1}{b_m} . \sqrt{\frac{f_y}{E}}$	$1.56\ \frac{1}{b_m} . \sqrt{\frac{f_y}{E}}$
$\bar{\lambda}_v = \sqrt{f_y/\sigma_{crv}}$	$\frac{1}{\sqrt{\psi}} . \frac{b_m}{t} . \sqrt{\frac{f_y}{E}}$ with $\psi = \frac{\pi^2 . k}{12(1-\nu^2)}$	$1.61\ \frac{b_m}{t} . \sqrt{\frac{f_y}{E}}$
$\gamma = \bar{\lambda}\sqrt{\bar{\lambda}_v}$	$\frac{(1+2\phi)^{3/2}}{\pi(\psi)^{1/4}\sqrt{\frac{2+\phi}{3}}} . \frac{1}{\sqrt{A}} . (\frac{f_y}{E})^{3/4}$	$2.80\ \frac{1}{\sqrt{A}} . (\frac{f_y}{E})^{3/4}$

Table 4.1.- Main relations for U sections and angles (weak axis buckling).

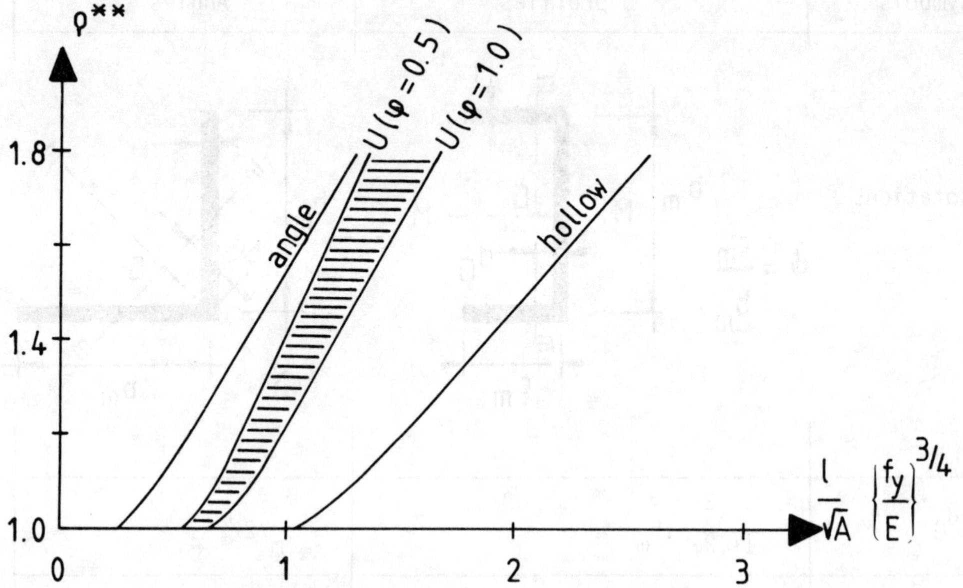

Figure 4.10. - Benefit due to the optimal thinness for cold-formed profiles .

This figure shows that an important benefit can be obtained when the length l increases, or the cross-sectional area A decreases, or the yield stress f_y increases. One can see also that this benefit is more quickly obtained for open sections that for hollow profiles. Let us add that because high strength steels are currently used in practice (f_y = 350 - 400 N/mm^2 in Europe and up to 700 N/mm^2 in Japan), the optimization of thin-walled steel sections is the more up to date.

5. OPTIMAL DESIGN AND MANUFACTURABILITY.

In optimal design, the use of the correct problem model is essential when realistic results are expected quickly. It must be stressed that structural optimization is a two stage process [46]. First, a real problem must be modelled in a mathematical form and, second, this mathematical problem must be solved. Often the first stage is the more difficult but it is vitally important to get the right relationship between the real-world structure and its mathematical idealization. The well-known optimal fixed-ended beam of constant depth demonstrates this point very well (Fig. 5.1.). The result is optimal but totally impracticable. Here, the idealization is incorrect and constraints to ensure a practical solution should be included [47, 48].

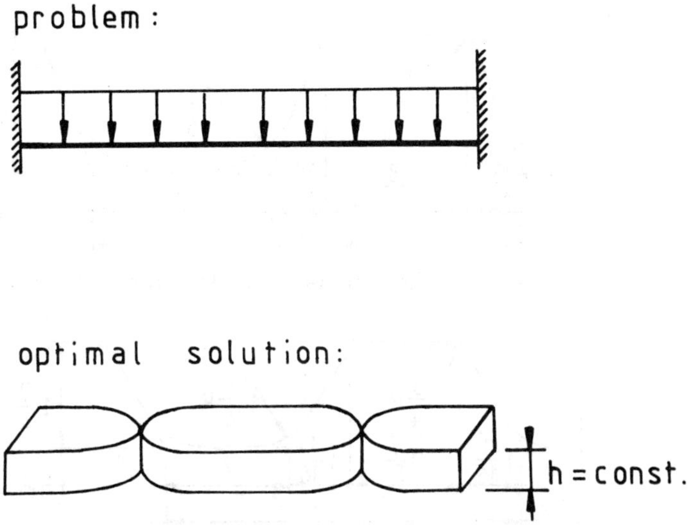

Figure 5.1. - Naive optimal design of a fixed-ended beam.

These remarks are also of interest in the optimal design of thin-walled bars which must be adequate not only for strength and stability but must also fulfil all the constraints related to the manufacturability.

As pointed out by Halmos [49]:
"There are millions of product designs made around the world every year but only a part of them are considered a success.
...
Quite frequently the balance between success and failure has nothing to do with the calculation of the structural strength ."

Fig. 5.2. shows that additional constraints such that a minimum thickness coming from connection requirements or corrosion (Fig.5.2.b), or a maximum corner radius which is necessary to stabilize the geometry of the profile (Fig. 5.2.c.), or a limitation of section height due to the roll-forming line (Fig. 5.2.c), can, sometimes, highly increase the weight of the optimal solution [50].

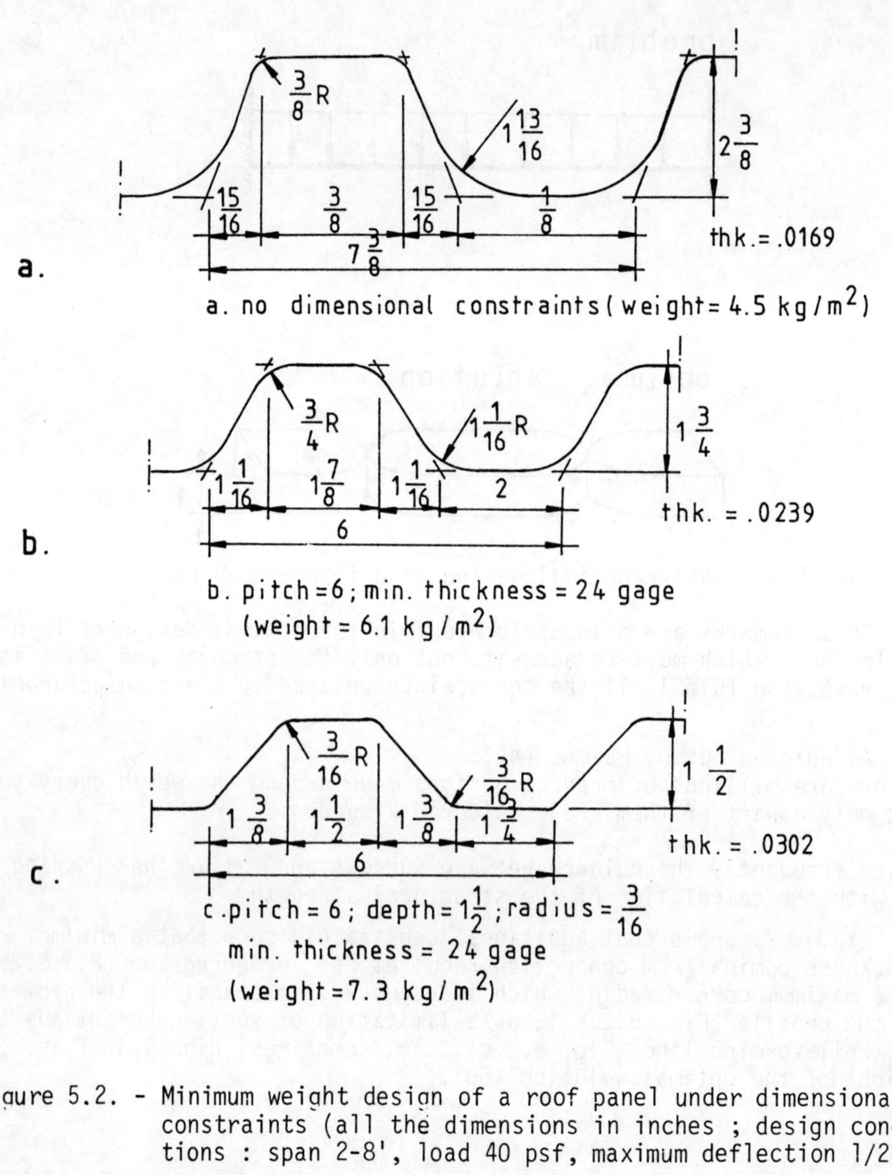

Figure 5.2. - Minimum weight design of a roof panel under dimensional constraints (all the dimensions in inches ; design conditions : span 2-8', load 40 psf, maximum deflection l/240).

6. OPTIMAL RANGES OF BARS.

6.1. General.

Optimization of a steel structure is largely dependent on the sections that are available on the market. Indeed, mathematical procedures are existing and can be used when the sections are tailored as, for instance, the welded beam shown in section 7. On the contrary, for usual shapes, optimization leads to choose, among the production series, the section which is just complying with the geometrical and mechanical requirements as obtained from the global optimization of the structure itself. Therefore, the production ranges have to be well formed, especially for elements subjected to major compression load because, for this case, the material distribution across the section is governing.

Present section is aimed to present guidelines in defining optimal production ranges. Hollow sections are considered as an example but similar procedure can be applied to other shapes.

6.2. Optimal ranges of columns with square hollow sections.

A range of square hollow sections is completely defined by a fan of thickness values for each specified width b. In order to guarantee a nearly optimal character to the investigated ranges, following procedure can be used [51]:

- choice of a discrete series of values of width b, for example those adopted in the specifications of the International Organization for Standardization (ISO 657/14 for hot finished sections and ISO 4019.2 for cold-finished sections) ;
- computation, for each specified width b, on the one hand, of the optimum thickness t_{opt}, and on the other hand, of the efficiency ratio ρ^* - i.e. the ratio between the ultimate stress σ of the section and the yield stress f_y - for various values of the thickness t close to the optimal value. σ is computed by means of the procedure given in section 3.2. ;
- choice of lower and upper bounds of thickness t, enabling one to comply with the condition:

$$\rho^* \geqslant 0.95\, \rho^*_{opt} \qquad (6.1.)$$

which preserves the nearly optimal character of the profile series.

Fig. 6.1. gives the results obtained for four different widths and a yield stress equal to 355/N/mm^2. In agreement with the conclusions of section 4.3., the optimum is reached for a thickness lower than the limiting value $\bar{\lambda}_v = 0.8$ for small widths (e.g. for b = 20 mm) and for a thickness which corresponds to $\bar{\lambda}_v = 0.8$ for larger values of the width b.

Fig. 6.2. gives an example of the optimal ranges obtained for cold finished sections with a steel grade of 355 N/mm^2. It is seen on Fig. 6.1. that the loss of efficiency is less pronounced with regard to the optimal thickness for an increase of thickness than for a decrease. This explains that, in Fig. 6.2., the thickness range is substantially larger above the line $\bar{\lambda}_v = 0.8$ than below it.

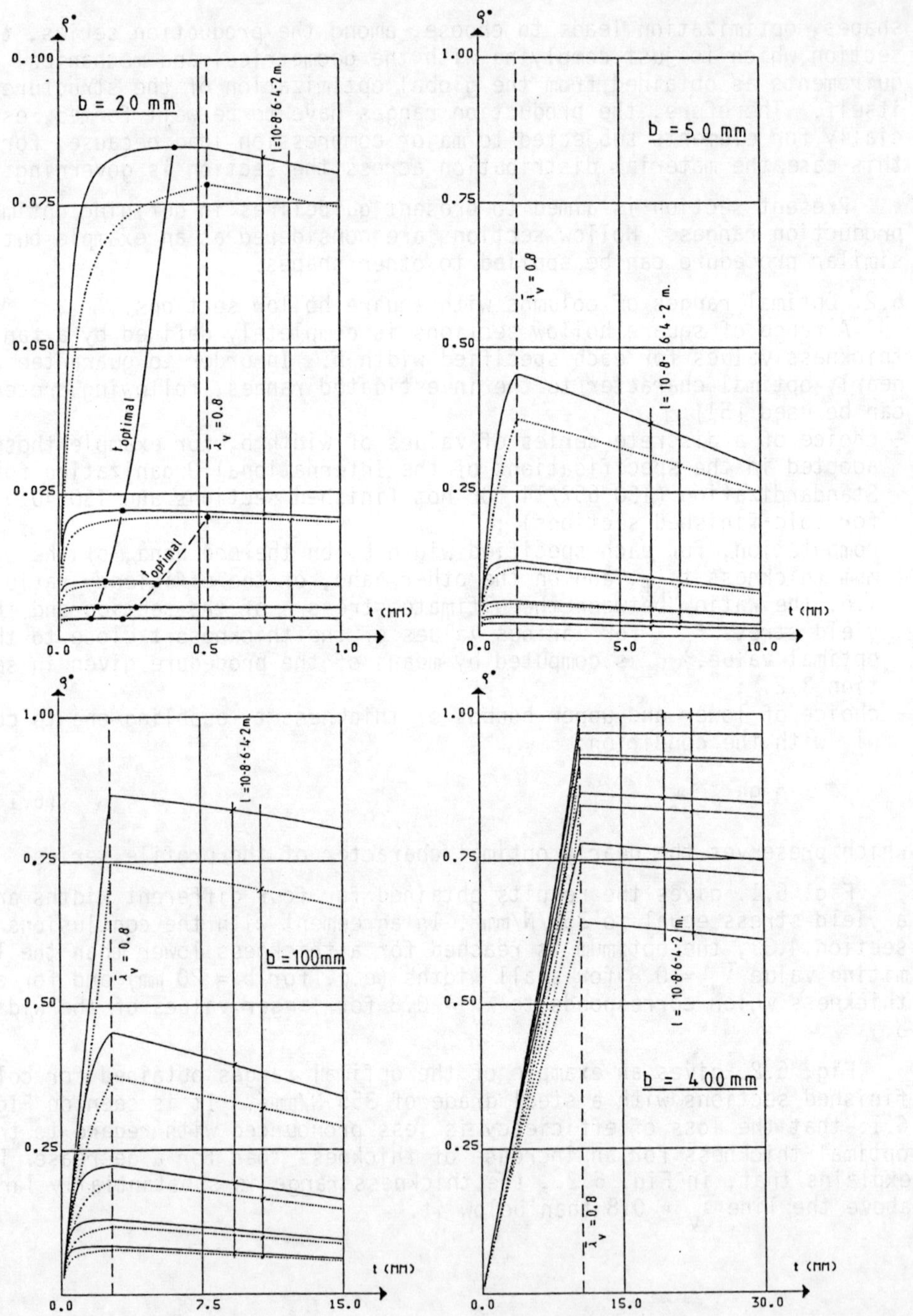

Figure 6.1. - Optimal thickness (f_y = 355 N/mm^2)
—— hot finished sections ---- cold finished sections.

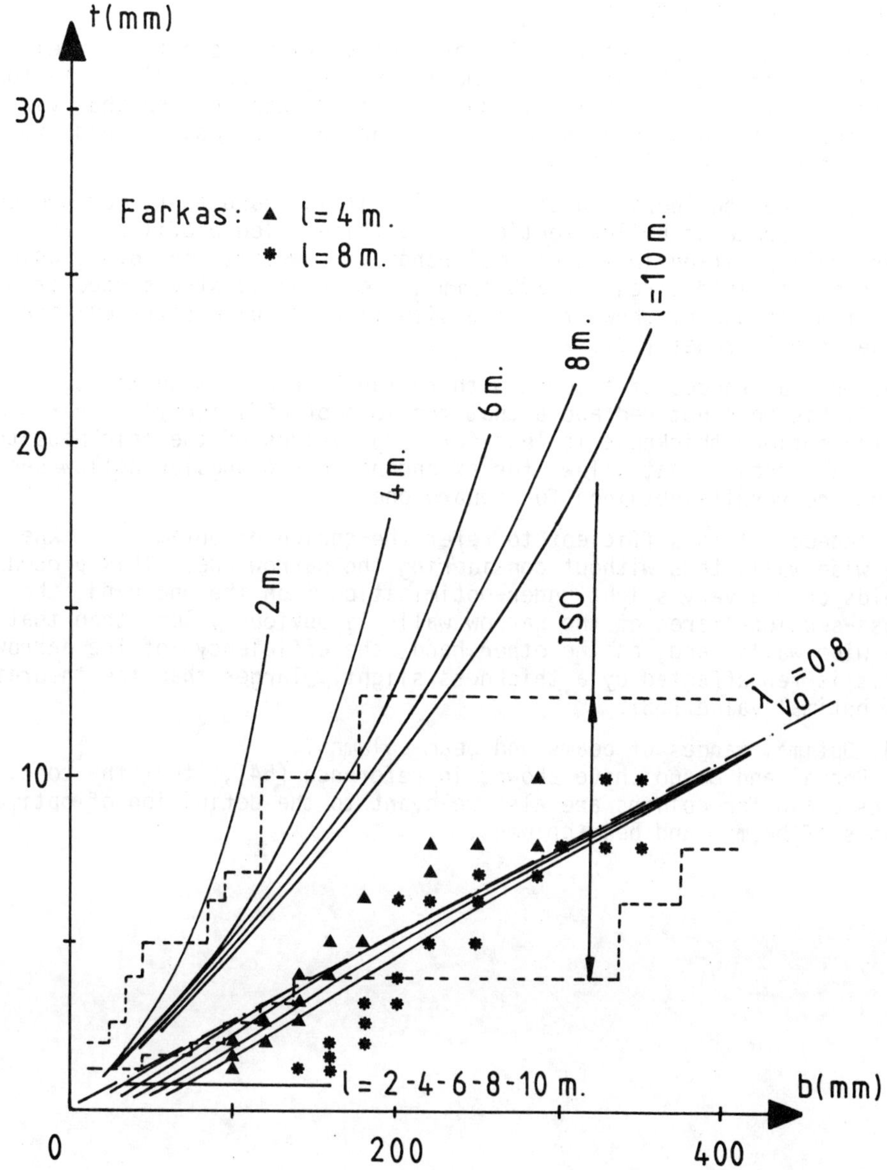

Figure 6.2. - Optimal ranges of cold finished profiles (f_y = 355 N/mm^2).

Results obtained by Farkas [52], by backtrack programming, have also been drawn on Fig. 6.2. They are in good agreement with the results given by above procedure.

Fig. 6.2. shows that the ISO range of profiles is not very well calibrated ; the thickness is too small for large size sections and too large for small size sections. But, one must keep in mind that various problems, i.e. requirements for connections and corrosion, lead to walls which are not too thin.

From the consideration of Fig. 6.2., it is found that, for the common steel grade of hollow sections, the thin-walled profiles have a rather small position in the optimal ranges. However, Fig. 6.3., which is drawn for a yield stress of 700 N/mm^2, i.e. a yield stress used at present time in Japan, shows that the situation is quite different for higher steel grades [53].

6.3. Optimal ranges of columns with rectangular hollow sections.

It has been noticed above that the loss of efficiency, with respect to the optimum thickness is less for large values of the thickness than for small ones. That allows for extending to rectangular hollow sections the results obtained for square ones.

Indeed, it is sufficient to refer the choice of optimal thickness to the wide wall, thus without considering the narrow one. This procedure yields only a very slight under-optimization ; on the one hand, the cross-sectional area of the narrow walls is obviously less than that of the wide walls, and, on the other hand, the efficiency of the narrow walls is few affected by a thickness slightly larger than the theoretical optimal value [53].

6.4. Optimal ranges of beams and beam-columns.

Rondal and Maquoi have shown, in reference [54], that the conclusions drawn for columns are also relevant to the definition of optimal ranges of beams and beam-columns.

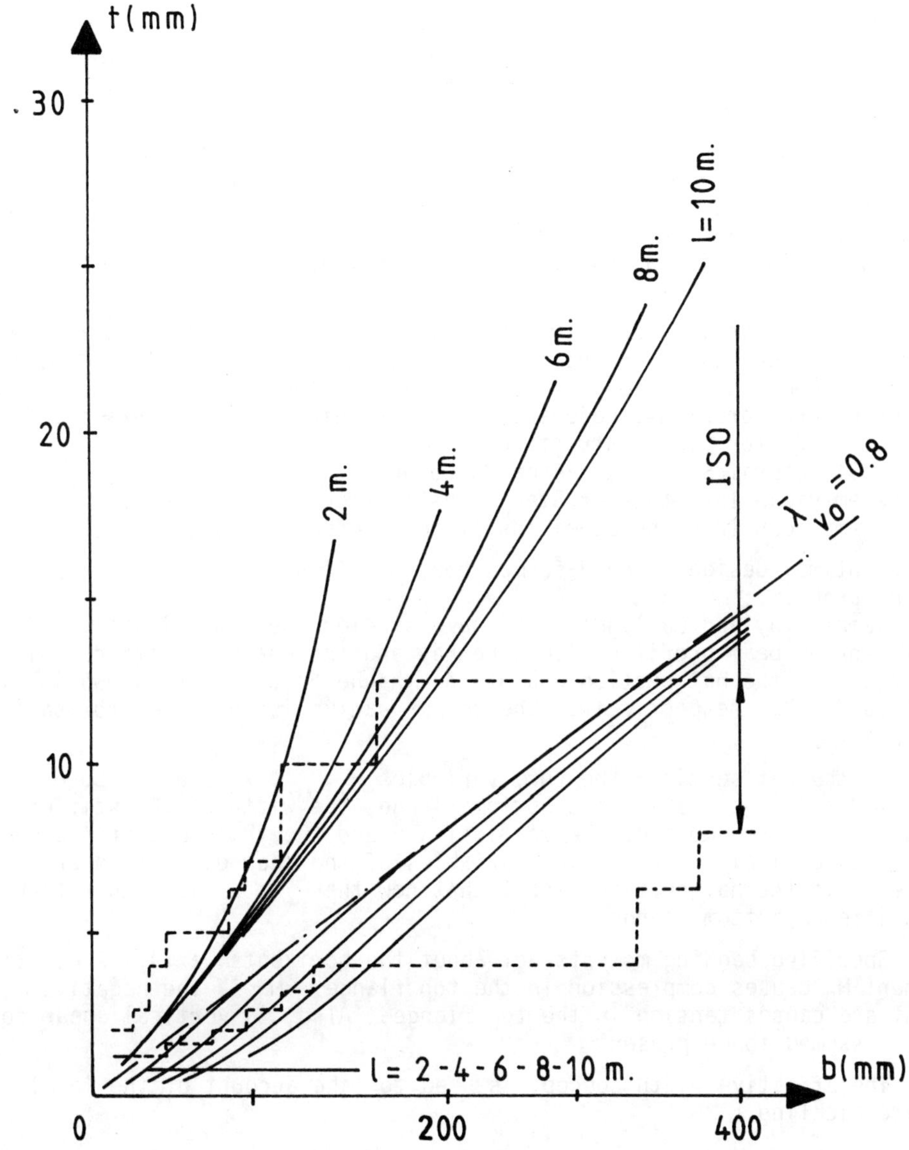

Figure 6.3. - Optimal ranges of cold finished profiles (f_y = 700 N/mm^2).

7. OPTIMAL DESIGN AND MATHEMATICAL PROGRAMMING METHODS.

7.1. General

Mathematical programming methods can, of course, be used for the optimal design of thin-walled bars. This section is only dedicated to some case studies in the field. More extensive references are given by Gajeswski and Zyczkowski [55].

Examples given in this section does not include mathematical exercises which are only of interest for academicians but are, in fact, totally impracticable. One must keep in mind that the minimum weight - an objective function largely used in optimal design - represents generally a grossly simplified approximation and can lead, sometimes, to a crazy design [56].

Using realistic stability constraints is also of vital importance. These equations can be found in recent recommendations published by :

- the European Convention for Constructional Steelwork : reference [14] for general problems, reference [34] for thin-walled members and reference [57] for plated structures ;
- the Structural Stability Research Council : reference [58] ;
- the American Institute of Steel Construction : reference [59] ;
- the American Iron and Steel Institute : reference [33].

7.2. Optimal design of cold-formed beam profiles.

- The problem :

Ramamurthy and Gallagher [60] have studied the optimal design of two cold-formed beam profiles, i.e. the hat and the channel sections. The geometry of the hat section and of the channel section are shown in Fig. 7.1. and 7.2., respectively. The corner radius is neglected for sake of simplicity.

In the hat section, the form variables x_1, x_2, x_3 and x_4 represent the width of the unstiffened bottom flange, the width of the stiffened top flange, the depth of the vertical web and the thickness of the member, respectively. The design variables of the channel section are the same as in the hat section except that now there is a lip rather than an unstiffened bottom flange.

Specified bending moments act about the horizontal axis. A positive moment M_p causes compression in the top flange. M_n is the negative moment and causes tension in the top flange. Also, a vertical shear force V is assumed to be present.

The effective width concept is used to take account of the local plate buckling.

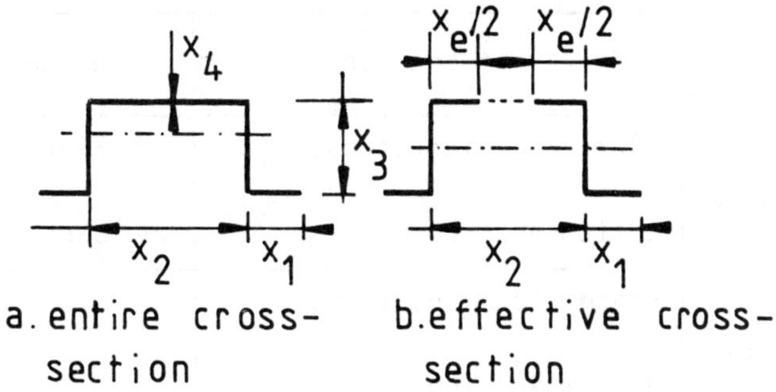

Figure 7.1. - Hat section geometry.

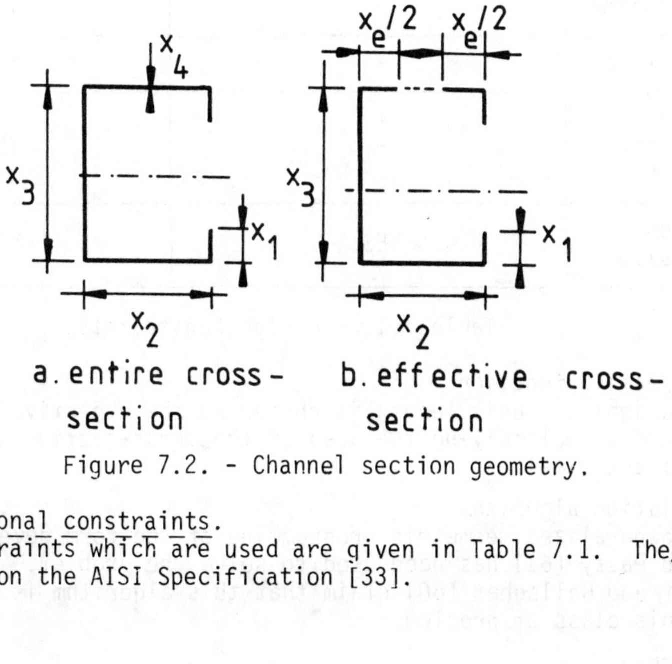

Figure 7.2. - Channel section geometry.

- <u>The operational constraints</u>.

The constraints which are used are given in Table 7.1. They are mainly based on the AISI Specification [33].

Constraint Condition	Hat Section	Channel Section
M_p (Pos. Bending Moment)	Flange Bending Stresses ; Bottom and Top	Top Flange Bending and Web Bending Stresses
M_N (Neg. Bending Moment)	Flange Bending Stress ; Bottom and Top	-
V (Shear)	YES	YES
Web Crippling	-	YES
Lip Stiffener	-	YES
Combined Bending and Shear-Web	-	YES
Limits on Member sizes	YES	YES

Table 7.1. - Design Constraints.

- The objective function.

The weight per unit length is chosen as the objective function. It is obtained by multiplying the area of the cross-section by the unit weight of the steel.

- The solution algorithm.

The generalized geometric programming of the form developed by Avriel, Dembo and Passy [61] has been used to solve the problem. Ramamurthy and Gallagher [60] claim that this algorithm is especially suited to this class of problem.

- An example.

The design conditions chosen for the case of the hat section are adapted from a problem described by Wei-Wen Yu [2] where a positive bending moment M_p = 15.4 kN.m is specified. A negative bending moment M_N = 7.4 kN.m and a shear force of 90 kN have been added. With a yield stress f_y of 278 N/mm^2, Wei-Wen Yu's design was x_1 = 59.7 mm, x_2 = 368 mm, x_3 = 241 mm and x_4 = 2.65 mm giving a weight per unit length w of 20.329 kg/m.

Taking a convergence tolerance of 0.001 on the value of the objective function, the final weight converges to 12.668 kg/m in ten iterations.

Fig. 7.3. and 7.4. show, in normalized form, the optimal values of the objective function and of the variables x_1 and x_2 at every iteration. The results of the computation have shown that the values of x_3 and x_4 did not change after the first iteration.

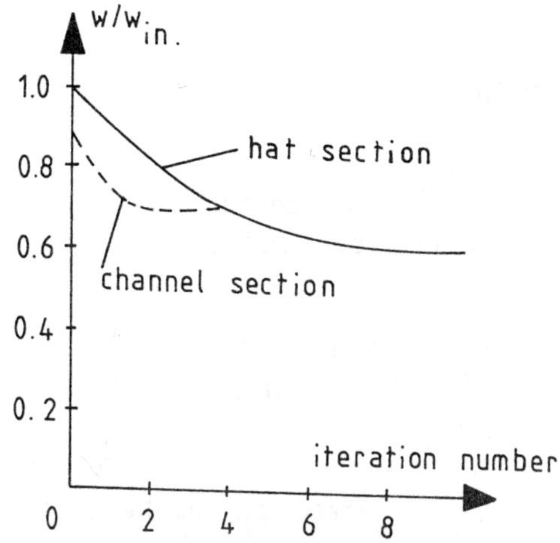

Figure 7.3. - Convergence of objective functions.

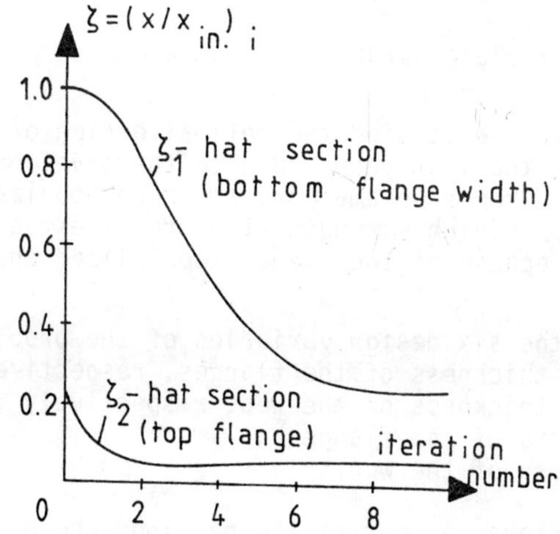

Figure 7.4. - Convergence of design variables (hat section)

The channel section has been optimaly designed to carry a uniformly distributed load of 2.65 kN/m over three spans of 7.6 m length each. The yield stress f_y is assumed equal to 312 N/mm^2. The initial values of the design variables was x_1 = 28.7 mm, x_2 = 139.7 mm, x_3 = 216 mm and x_4 = 3.02 mm giving a weight per unit length w of 13.102 kg/m.

Fig. 7.3. shows that the weight has converged to a value of 9.163 kg/m in four iterations and Fig. 7.5. gives the optimal values of the variables x_1 and x_2 at every iteration.

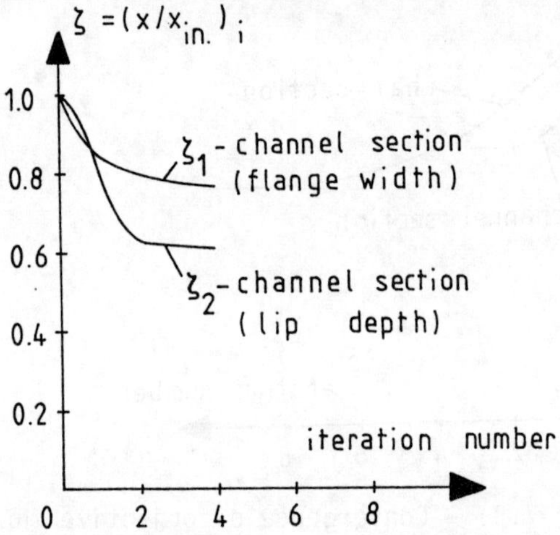

Figure 7.5. - Convergence of design variables (channel section).

7.3. Optimal design of plate girders.

- The problem.

Maquoi and Rondal have studied the optimal design of hybrid plate girders [62, 63]. In the last years, due to the development of automatic welding, steel plate girders became more and more popular in buildings and bridges. The use of high strength steel for the flanges enables an interesting economy because of the highly specialized character of this type of girder.

Fig. 7.6. shows the six design variables of the problem which are :

- b_f, t_f : width and thickness of the flanges, respectively ;
- h_w, t_w : depth and thickness of the web, respectively ;
- $f_{y,f}$: yield stress of the flanges ;
- $f_{y,w}$: yield stress of the web.

The web is unstiffened except at the bearings where it is vertically stiffened

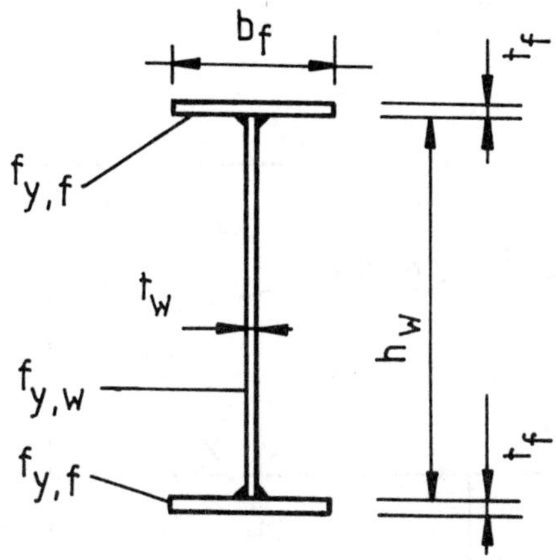

Figure 7.6. - Design variables.

- The operational constraints.

The operational constraints are related to strength, stability and deflection requirements and can be summarized as follows :

- strength of the flanges for the bending moment ; with account taken of the partial yielding of the web in the hybrid cross-section ;
- strength of the web under shear ;
- lateral buckling of the compressed flange ;
- plate buckling of the web ;
- web crippling ;
- web buckling under concentrated loads ;
- lateral torsional buckling of the beam ;
- deflection under service load.

- The objective function.

As the steel grades of web and flanges differ, the cost of the girder must be considered as the objective function.

One can used the following function, which is proportional to the cost of the girder :

$$F = h_w . t_w + 2 \frac{c_f}{c_w} . b_f . f_f \tag{7.1.}$$

For determining the analytical expression of the unit price c versus the steel grade, it has been refered to the price list edicted by the European Community for Coal and Steel and a parabolic law has been obtained by means of the least square method (Fig. 7.7.). In Fig. 7.7., the unit price and the yield stress have been normalized with respect to steel Fe 235 (f_y = 235 N/mm^2).

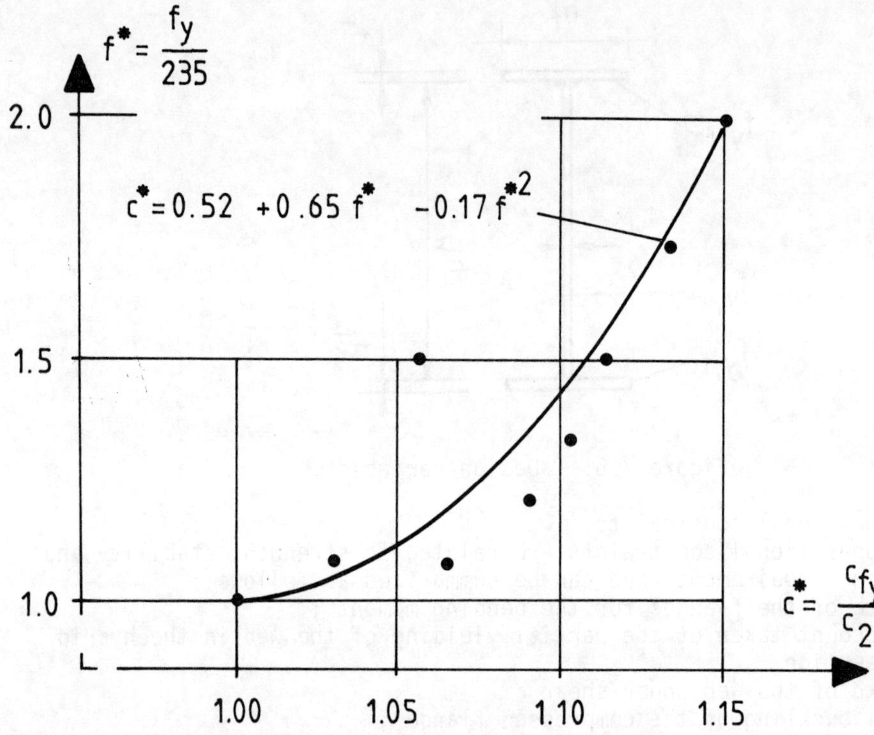

Figure 7.7. - Unit price versus steel grade.

- The solution algorithm.

The flexible tolerance method of the form developed by Himmelblau has been used to solve the problem [64].

- An example.

The results are illustrated by means of a simply supported beam of 24 m length with a uniformly distributed load of 13.6. kN/m. The yield stress of the web material is $f_{y,w} = 235$ N/mm^2 (current mild steel) and the yield stress of the flanges is limited to 470 N/mm^2.

Fig. 7.8. shows the results obtained. One can see that the most economical girder has a flange yield stress of 470 N/mm^2, which is the upper bound of this variable, and leads to an economy of 5 % of the price and a saving of 13 % in weight with respect to the homogeneous optimized girder in steel Fe 235.

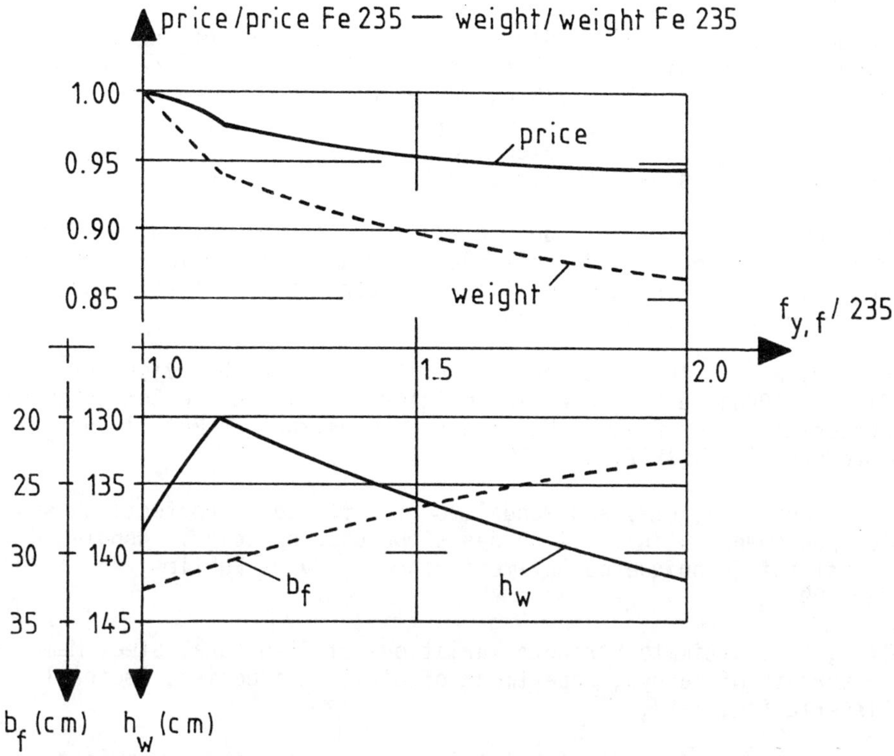

Figure 7.8. - Price, weight and design variables.

REFERENCES.

1. Rondal, J. : Thin-Walled Structures. Proceedings of the Second Regional Colloquium on Stability of Steel Structures, Final Report, Hungary September 25-26, 1986, 269 - 285.

2. Wei-Wen Yu : Cold-Formed Steel Design. John Wiley and sons, New-York, 1985.

3. Skaloud, M. : Le critère de l'état limite des plaques et systèmes de plaques. Proceedings du Colloque International sur le Comportement Postcritique des Plaques utilisées en Construction Métallique, Liège Novembre 12-13, 1962, 41 - 63.

4. Koiter, W.T. : Introduction to the Post-Buckling Behaviour of Flat Plates. Proceedings du Colloque International sur le Comportement Postcritique des Plaques Utilisées en Construction Métallique, Liège Novembre 12-13, 1962, 17 - 35.

5. Costa Ferreira, C.M. and Rondal, J. : Effet des imperfections sur les phénomènes d'instabilité des structures en acier. Annales de l'Institut Technique du Bâtiment et des Travaux Publics, 451 (1987), 78 - 99.

6. Itoh, Y. : Ultimate Strength Variations of Structural Steel Members. University of Nagoya, Department of Civil Engineering, Doctoral Dissertation, 1984.

7. Koiter, W.T. : Over de stabiliteit van het elastisch evenwicht. Technological University of Delft, Department of Civil Engineering, Doctoral Dissertation, 1945.

8. Thompson, J.M.T. and Hunt, G.W. : A General Theory of Elastic Stability. John Wiley and sons, London, 1973.

9. Koiter, W.T. : Post-Buckling of a Simple Two-Bar Frame, in : Recent Progress in Applied Mechanics. Almqvist and Wiksell, Stockholm, 1967, 337 - 354.

10. Roorda, J. : Buckling of Elastic Structures. University of Waterloo, Department of Civil Engineering, 1980.

11. Gioncu, V. and Ivan, M. : Bazele calculului structurilor la stabilitate. Editura Facla, Timisoara, 1983.

12. Roorda, J. : The Buckling Behaviour of Imperfect Structural Systems. Journal of the Mechanics and Physics of Solids, 13 (1965), 267 - 280.

13. Ayrton, W.E. and Perry, J. : On Struts. The Engineer, December 10, (1886), 464-465 and December 24 (1886), 513-515.

14. European Convention for Constructional Steelwork : European Recommendations for Steel Construction. ECCS, EG 77-1E, 1977.

15. Rondal, J. : Residual Stresses in Cold-Rolled Profiles. Construction and Building Materials, 1 (1987), 150 - 164.

16. Costa Ferreira, C.M. and Rondal, J. : Influence of Flexural Residual Stresses on the Stability of Compressed Angles. Proceedings of the International Conference on Steel Structures, Budva September 28 - October 1, 1986, 147-155.

17. Karren, K.W. and Winter, G. : Effects of Cold-Forming on Light-Gage Steel Members. Journal of the Structural Division, 93 (1967), 433 - 469.

18. Chajes, A., Britvec, S.J. and Winter, G. : Effects of Cold-Straining on Structural Sheet Steels. Journal of the Structural Division, 89, (1963), 1 - 32.

19. Karren, K.W. : Corner Properties of Cold-Formed Steel Shapes. Journal of the Structural Division, 93 (1967), 401-432.

20. Lind, N.C. and Schroff, D.K. : Utilization of Cold Work in Cold-Formed Steel. Journal of the Structural Division, 101 (1975), 67-78.

21. Rondal, J. and Maquoi, R. : Formulations d'Ayrton-Perry pour le flambement des barres métalliques. Construction Métallique, 4(1979) 41-53.

22. Wagner, H. : Verdrehung und Knickung von offenen Profilen. Festschrift 25. Jahre Technische Hochschule Danzig, 1929.

23. Costa Ferreira, C.M. and Rondal, J. : Flambement des cornières à parois minces. Annales des Travaux Publics de Belgique, 2 (1986), 101-121.

24. Fukumoto, Y. and Kubo, M. : A Survey of Tests on Lateral Buckling Strength of Beams. Proceedings of the Second International Colloquium on Stability of Steel Structures, Preliminary Report, Liège, April 13-15, 1977, 233-240.

25. Prandtl, L. : Kipperscheinungen. Dissertation, München, 1899.

26. Winter, G. : Cold-Formed Light-Gage Steel Construction. Journal of the Structural Division, 85 (1959), 151-173.

27. Batista, E. : Etude de la stabilité des profils à parois minces et section ouverte de types U et C. University of Liège, Department of Civil Engineering, Doctoral Dissertation, 1988.

28. Grimault, J.P., Maquoi, R., Mouty, J., Plumier, A. and Rondal, J. : Stabilité des poutres-poteaux en profils creux à parois minces. Construction Métallique, 4 (1984), 33 - 41.

29. Gioncu, V. : New Conceptions, Trends and Perspectives in the Theory of Postcritical Behaviour of Structures. Proceedings of the Third International Colloquium on Stability, Timisoara October 16, 1982, 3 - 26.

30. Reis, A.J. : Interactive Buckling in Thin-Walled Structures, in : Developments in Thin-Walled Structures, Vol. 3 (Ed. J. Rhodes and A.C. Walker), Elsevier, London, 1987, 237 - 279.

31. Djubek, J., Kodnar, R. and Skaloud, M. : Limit State of the Plate Elements of Steel Structures, Veda, Bratislava, 1983.

32. von Karman, T., Sechler, E.E. and Donnel, L.H. : Strength of Thin-Plates in Compression. Transactions of the American Society of Mechanical Engineers, Journal of Applied Mechanics, 54 (1932),53-57.

33. American Iron and Steel Institute : Specification for the Design of Cold-Formed Steel Structural Members. AISI, August 19, 1986 Edition.

34. European Convention for Constructional Steelwork: European Recommendations for the Design of Light Gauge Steel Members. ECCS, 49, 1987.

35. Usami, T. and Fukumoto, Y. : Local and Overall Buckling of Welded Box Columns. Journal of the Structural Division, 108 (1982), 525 - 542.

36. Hasegawa, A., Abo, H., Mauroof, M. and Nishino, F. : A Simplified Analysis and Optimality on the Steel Column Behavior with Local Buckling. Proceedings of the Japan Society of Civil Engineers, Structural Engineering and Earthquake Engineering, 3 (1986),195-204.

37. Braham, M., Grimault, J.P., Massonnet, C., Mouty, J. and Rondal, J.: Buckling of Thin-Walled Hollow Sections - Cases of Axially Loaded Rectangular Sections. Acier-Stahl-Steel, 5 (1980), 30-36.

38. Bleich, F. : Buckling Strength of Metal Structures, McGraw-Hill, New-York, 1952.

39. Shanley, F.R. : Optimum Design of Eccentrically Loaded Columns. Journal of the Structural Division, 93 (1967), 201 - 226.

40. Tuckerman, L.B., Petrenko, S.N. and Johnson, C.D.: Strength of Tubing under Combined Axial and Transverse Loading. National Advisory Committee for Aeronautics, Technical Note 307, 1929.

41. Thompson, J.M.T. and Hunt, G.W. : Dangers of Structural Optimization. Engineering Optimization, 1 (1974), 99 - 110.

42. Thompson, J.M.T. and Lewis, G.M. : On the Optimum Design of Thin-Walled Compression Members. Journal of the Mechanics and Physics of Solids, 20 (1972), 101 - 109.

43. Van der Neut, A. : The Interaction of Local Buckling and Column Failure of Thin-Walled Compression Members. Technological University of Delft, Department of Aeronautical Engineering, Report VTH 149, 1968.

44. Rondal, J. and Maquoi, R. : On the Optimum Design of Square Hollow Compression Members. Proceedings of the IUTAM Symposium on Collapse-The Buckling of Structures in Theory and Practice, London August 31 - September 3, 1982, 333 - 344.

45. Maquoi, R. and Massonnet, C. : Interaction between Local Plate Buckling and Overall Buckling in Thin-Walled Compression Members - Theories and Experiments. Proceedings of the IUTAM Symposium on Buckling of Structures, Cambridge June 17-21, 1974, 365-382.

46. Templeman, A.B. : Synthesis and Conclusions on the Theme : Progress in Structural Optimization. Proceedings of the Tenth IABSE Congress, Final Report, Tokyo September 6-11, 1976, 155-156.

47. Hartmann, D. : Uber die Grundlagen und Methoden der Optimierung. Proceedings of the Tenth IABSE Congress, Final Report, Tokyo September 6-11, 1976, 111-114.

48. Maquoi, R. and Rondal, J. : Preponderance of Idealization in Structural Optimization. Proceedings of the Tenth IABSE Congress, Final Report, Tokyo September 6-11, 1976, 95 - 98.

49. Halmos, G.T. : Design for Manufacturability. Proceedings of the Fifth International Specialty Conference on Cold-Formed Steel Structures, St. Louis November 18-19, 1980, 727 - 754.

50. Seaburg, P.A. and Salmon, C.G. : Minimum Weight Design of Light Gage Steel Members. Journal of the Structural Division, 97 (1971), 203 - 222.

51. Rondal, J. and Maquoi, R. : Etude d'une gamme optimale de profils creux carrés et rectangulaires. Annales de l'Institut Technique du Bâtiment et des Travaux Publics, 409 (1982), 61 - 72.

52. Farkas, J. : Optimum Design of Metal Structures. Ellis Horwood, Chichester, 1984.

53. Maquoi, R., Massonnet, C. and Rondal, J. : Promoting the Use of Thin-Walled Hollow Rectangular Sections by Increasing Steel Grade, in : Behaviour of Thin-Walled Structures. (Ed. J. Rhodes and J. Spence), Elsevier, London, 1984.

54. Rondal, J. and Maquoi, R. : Optimal Ranges of Beam-Columns with Square or Rectangular Hollow Sections. Proceedings of the Second Regional Colloquium on Stability of Steel Structures. Volume II, Hungary September 25-26, 1986, 285 - 291.

55. Gajeswski, A. and Zyczkowski, M. : Optimal Structural Design under Stability Constraints, Martinus Nijhoff, Dordrecht, 1987.

56. Turner, H.K. and Plant, R.H. : Optimal Design for Stability under Multiple Loads. Journal of the Structural Division, 106 (1980), 1365 - 1382.

57. European Convention for Constructional Steelwork : Behaviour and Design of Steel Plated Structures. (Ed. P. Dubas and E. Gehri). ECCS, 44, 1986.

58. Structural Stability Research Council : Guide to Stability Design Criteria for Metal Structures (Ed. B.G. Johnston), Third Edition, Wiley, New-York, 1976.

59. American Institute of Steel Construction: Stability of Metal Structures - A World View, (Ed. D. Sfintesco, L.S. Beedle, G.W. Schulz and R. Zandonini). AISC, 1982.

60. Ramamurthy, S. and Gallagher, R.H. : Generalized Geometric Programming in Cold-Formed Steel Design. Proceedings of the Fourth International Specialty Conference on Cold-Formed Steel Structures, St. Louis June 1-2, 1978, 41 - 72.

61. Avriel, M., Dembo, R. and Passy, U. : Solution of Generalized Geometric Programs. International Journal for Numerical Methods in Engineering, 9 (1975), 149 - 168.

62. Maquoi, R. and Rondal, J. : Optimum Cross-Sectional Properties for Unstiffened Plate Girders. Proceedings of the Second International Colloquium on Stability of Steel Structures, Final Report, Liège April 13-15, 1977, 155 - 156.

63. Rondal, J. and Maquoi, R. : Optimization of Unstiffened Hybrid I Beams with Stability Constraints. Proceedings of the Regional Colloquium on Stability of Steel Structures, Hungary October 19-21, 1977, 373 - 382.

64. Himmelblau, D.M. : Applied Nonlinear Programming, McGraw-Hill, New-York, 1972.

PART V

A. P. Seyranian
USSR Academy of Sciences, Moscow, USSR

ABSTRACT

Part V deals with sensitivity analysis and structural optimization in stability and vibration problems. Chapter 1 is an introduction. Chapter 2 is devoted to qualitative and quantitative sensitivity analysis of vibrational frequencies of mechanical systems with respect to problem parameters. As an example linear gyroscopic system is considered. In chapters 3 and 4 sensitivity analysis for nonconservative problems of elastic stability is given and discussed. Discrete as well as distributed structures are considered. Chapter 5 is devoted to optimization of critical loads of columns subjected to follower forces. Chapter 6 is devoted to optimization of aeroelastic stability of panels in supersonic gas flow. Both static and dynamic forms of the loss of stability are considered. In chapters 7 and 8 bending-torsional flutter problem of a wing in incompressible flow is considered. Influence of mass and stiffness distributions on aeroelastic stability characteristics is studied. Optimization problem how to maximize the critical speed at which aeroelastic stability is lost is stated and solved numerically.

1. INTRODUCTION

In mechanics as well as in physics one of the most general and important thing is the study of the dependence of characteristics of a physical process on problem parameters. For stability and vibration problems it is interesting and important to obtain qualitative and also quantitative information about the dependence of vibrational frequencies or critical loads on discrete and distributed parameters of the problem. Since these quantities are not explicit functions on problem parameters we have to find the implicit derivatives of mentioned quantities in different functional spaces. This is what we call the sensitivity analysis. It is shown below that the solutions to the main and the adjoint eigenvalue problems provide all the necessary information for evaluating the sensitivities and to choose the most essential parameters. Thus, information on sensitivities demands rather few calculations because all the necessary data are available from the ordinary stability analysis. Nevertheless, this information is very important for the designer because it allows to present the whole picture of dependence on problem parameters and to use the most simple and effective tools to improve the stability and vibration characteristics of the system.

One major application of sensitivity analysis lies within optimal structural design, because sensitivities provide all the necessary information for numerical gradient optimization procedures. It should be also noted that the necessary optimality conditions are usually formulated in terms of derivatives and gradient functions, i.e. sensitivities.

Sensitivity analysis and optimization problems for discrete as well as distributed structures like gyroscopic systems, elastic columns loaded by follower forces, panels in supersonic gas flow and slender wing under bending-torsional flutter conditions are studied below.

The subject of this paper is based on the author's papers [1-5] and joint papers with A.V.Sharanyuk [6-8] and P.Pedersen [9].

2. SENSITIVITY ANALYSIS OF VIBRATIONAL FREQUENCIES OF MECHANICAL SYSTEMS

2.1 General case

Let's consider vibrations of a linear system of general form [8]

$$A\ddot{q} + \Gamma\dot{q} + Cq = 0 \tag{2.1}$$

in which A, Γ, C, are $m \times m$ matrices with real coefficients, q is a vector of dimension m. It is assumed that the elements of A, Γ and C are smooth functions of the vector of parameters h of dimension n. Parameters h_1, h_2, ..., h_n are assumed to be independent variables.

We shall find a solution of (2.1) in the form $X \cdot \exp(\lambda t)$, where t is time. As a result we arrive at the generalized eigenvalue problem, λ is an eigenvalue, while X is an eigenvector of dimension m

$$[A\lambda^2 + \Gamma\lambda + C]X = 0 \tag{2.2}$$

In addition to the right eigenvector X we also consider left eigenvector Y satisfying the equation

$$Y^T[A\lambda^2 + \Gamma\lambda + C] = 0 \tag{2.3}$$

The index T denotes transposition. This equation can also be written in the form $[A^T\lambda^2 + \Gamma^T\lambda + C^T]Y = 0$.

Let us consider variations of eigenvalues λ with respect to changes in the vector of parameters h, assuming λ to be a simple eigevalue. For this purpose, we give vector h an increment εk, where ε is a small positive number, k is an arbitrary vector of dimension n, $|k| = 1$.

Then, matrices A, Γ and C obtain the increments

$$A(h+\varepsilon k) = A(h) + \varepsilon A_1(k) + \cdots \tag{2.4}$$

$$\Gamma(h+\varepsilon k) = \Gamma(h) + \varepsilon \Gamma_1(k) + \cdots$$

$$C(h+\varepsilon k) = C(h) + \varepsilon C_1(k) + \cdots$$

$$A_1 = \|(\nabla a_{ij}, k)\|, \quad \Gamma_1 = \|(\nabla \gamma_{ij}, k)\|, \quad C_1 = \|(\nabla c_{ij}, k)\|,$$

$$\nabla = (\partial/\partial h_1, \partial/\partial h_2, \cdots, \partial/\partial h_n)$$

As a result of perturbation of parameters h , the eigenvalues λ and eigenvectors X receive the increments

$$\lambda(h+\varepsilon k)=\lambda(h)+\varepsilon\lambda_1+\varepsilon^2\lambda_2+\cdots \tag{2.5}$$
$$X(h+\varepsilon k)=X(h)+\varepsilon X_1+\varepsilon^2 X_2+\cdots$$

Substituting expansions (2.4) and (2.5) into (2.2) in first approximation of ε we obtain

$$\lambda_1[2A\lambda+\Gamma]X+[A_1\lambda^2+\Gamma_1\lambda+C_1]X+[A\lambda^2+\Gamma\lambda+C]X_1=0$$

We multiply this equation from the left side by Y^T and note that the last term of the resultant equation is zero due to (2.3). Ultimately we obtain an expression for the increment of the eigenvalue λ_1

$$\lambda_1=-\frac{Y^T[A_1\lambda^2+\Gamma_1\lambda+C_1]X}{Y^T[2A\lambda+\Gamma]X} \tag{2.6}$$

This is the derivative of eigenvalue λ with respect to direction k . In view of the continuity of the right hand side of (2.6) an ordinary derivative also exists. Multiplying both sides of (2.6) by ε and using notations $\delta h=\varepsilon k$, $\delta\lambda=\varepsilon\lambda_1$, we can write (2.6) in the form [8]

$$\delta\lambda=(\nabla\lambda,\delta h), \tag{2.7}$$
$$\nabla\lambda=-\frac{Y^T[\nabla A\lambda^2+\nabla\Gamma\lambda+\nabla C]X}{Y^T[2A\lambda+\Gamma]X},$$
$$\nabla A=\|\nabla a_{ij}\|,\ \nabla\Gamma=\|\nabla\gamma_{ij}\|,\ \nabla C=\|\nabla c_{ij}\|$$

The variation $\delta\lambda$ is a complex number. Separating the real and imaginary parts, we obtain expressions for the variations of the attenuation decrement and the vibrational frequency respectively.

Thus, to estimate the sensitivity of a simple (nonmultiple) eigenvalue λ with respect to variation of the vector of parameters δh , we need to know the left and right eigenvectors Y and X corresponding to this eigenvalue, to calculate the gradients of matrices ∇A, $\nabla \Gamma$, ∇C for given h (this operation is performed analytically), and then to find the gradient $\nabla\lambda$ from (1.7) and obtain the desired unknown $\delta\lambda$. The effectiveness of the sensitivity analysis increases as the dimension n of the vector of parameters and the system dimension m increase.

2.2 Linear gyroscopic system

Let us consider the case of a linear gyroscopic system. Assume that A and C are symmetrical $m \times m$ matrices with real elements, where A is a positive definite mass matrix, while C - stiffness matrix - is a nonnegative one, gyroscopic matrix Γ is a skew-symmetric matrix with real elements of the same dimension. In accordance with (2.2) we consider the following eigenvalue problem

$$[A\lambda^2 + \Gamma\lambda + C]X = O \tag{2.8}$$

It is known that the eigenvalues λ of the system (2.8) are purely imaginary: $\lambda = i\omega$, ω are natural oscillation frequencies, while i is the square root of -1.

In view of the fact that A, Γ and C are real matrices, we note that if λ is an eigenvalue of (2.8) corresponding to eigenvector X, then $\bar{\lambda} = -i\omega$ is also an eigenvalue with eigenvector $\bar{X}$. Therefore

$$[-A\omega^2 - i\omega\Gamma + C]\bar{X} = 0 \tag{2.9}$$

Since $A^T = A$, $C^T = C$, $\Gamma^T = -\Gamma$, it follows that $\bar{X}^T[-A\omega^2 + i\omega\Gamma + C] = O$. Thus, in the case of gyroscopic system (2.8), if X is a right eigenvector, then $\bar{X}$ is the left eigenvector corresponding to the same frequency ω.

From (2.7) we obtain an expression for the variation of frequency, using $Y = \bar{X}$, $\lambda = i\omega$, $\delta\lambda = i\delta\omega$

$$\delta\omega = (\nabla\omega, \delta h) \tag{2.10}$$

$$\nabla\omega = \frac{\bar{X}^T[-\omega^2\nabla A + i\omega\nabla\Gamma + \nabla C]X}{\bar{X}^T[2\omega A - i\Gamma]X}$$

We multiply the numerator and denominator of the last expression by ω and, performing simple manipulations, obtain a final expression for the gradient of frequency [8]

$$\nabla\omega = \frac{\omega(\bar{X}^T\nabla C X) + i\omega^2(\bar{X}^T\nabla\Gamma X) - \omega^3(\bar{X}^T\nabla A X)}{\omega^2(\bar{X}^T A X) + (\bar{X}^T C X)} \qquad (2.11)$$

All the terms in parantheses in (2.11) are real quantities except for ($\bar{X}^T\nabla\Gamma X$) which is purely imaginary. Thus, all the terms in the numerator and denominator of (2.11) are real. If we write vector X in the form $X = X_1 + iX_2$, then we see that, for example, the expression $\bar{X}^T A X$ is positive for any X, $|X| \neq 0$

$$\bar{X}^T A X = X_1^T A X_1 + X_2^T A X_2 > 0$$

because A is a positive definite matrix.

On the basis of (2.10), (2.11) we can draw some qualitative conclusions of a general nature.

Let us consider first the case when $\delta A = \nabla A \delta h$ is a nonnegative matrix, while $\delta\Gamma = \nabla\Gamma\delta h$ and $\delta C = \nabla C \delta h$ are zero matrices. Physically, this case corresponds to an increase of the mass of the system $(\bar{X}^T \delta A X) \geqslant 0$, while the stiffness and gyroscopic characteristics remain constant. Then expressions (2.10), (2.11) imply $\delta\omega \leqslant 0$. Thus, we have proved a theorem.

Theorem 1. Vibrational frequencies of the gyroscopic system do not increase when mass is increased $(\bar{X}^T\delta A X) \geqslant 0$.

We can similarly consider another case, in which $\delta C = \nabla C\, \delta h$ is a nonnegative matrix, while δA and $\delta\Gamma$ are zero matrices. Then expressions (2.10) and (2.11) imply $\delta\omega \geqslant 0$. So, we have proved the validity of the theorem

Theorem 2. Vibrational frequencies of the gyroscopic system do not decrease when stiffness is increased $(\bar{X}^T\delta C X) \geqslant 0$.

The requirement that matrices $A, C, \delta A, \delta C$ be strictly positive definite leads to the strict inequalities $\delta\omega < 0$, $\delta\omega > 0$ in the first and second cases respectively.

The theorems we have proved are local analogs of the theorem by Zhuravlev in the case of simple eigenvalues. The case of multiple eigenvalues is more complicated. Nevertheless, the qualitative results regarding the influence of stiffness and mass characteristics on vibrational frequencies of a gyroscopic system, proved above for simple eigenvalues ω , remain valid in the multiple case too [8].

We should point out that local relations (2.7), (2.10), (2.11) enable us to make not only qualitative but also quatitative estimates of the sensitivity of vibrational frequencies of gyroscopic systems.

2.3 Numerical example

As an example let us consider the equations of small oscillations of a uniaxial gyrostabilizer with allowance for elastic compliance of the elements [8]

$$\begin{aligned} I\ddot{\alpha} + H\dot{\beta} &= K(\psi - \alpha) \\ I\ddot{\beta} - H\dot{\alpha} &= 0 \\ \Psi\ddot{\psi} &= K(\alpha - \psi) + N(\theta - \psi) \\ \Theta\ddot{\theta} &= N(\psi - \theta) \end{aligned} \qquad (2.12)$$

Here I , Ψ , Θ are moments of inertia, they characterize mass properties of the system, K and N are stiffness parameters, H is angular momentum of gyroscope rotor, α , β , ψ and θ are angles. Matrices A , Γ and C from previous paragraph have the form

$$A = \begin{Vmatrix} I & 0 & 0 & 0 \\ 0 & I & 0 & 0 \\ 0 & 0 & \Psi & 0 \\ 0 & 0 & 0 & \Theta \end{Vmatrix}, \quad \Gamma = \begin{Vmatrix} 0 & H & 0 & 0 \\ -H & 0 & 0 & 0 \\ 0 & 0 & 0 & 0 \\ 0 & 0 & 0 & 0 \end{Vmatrix}, \quad C = \begin{Vmatrix} K & 0 & -K & 0 \\ 0 & 0 & 0 & 0 \\ -K & 0 & K+N & -N \\ 0 & 0 & -N & N \end{Vmatrix}.$$

Thus, gyroscopic system (2.12) is of dimension $m = 4$, and the vector of design parameters is of dimension $n = 6$ $h = (I, \Psi, \Theta, H, K, N)$. Matrix A is positive definite for positive values of the parameters I , Ψ , Θ , while matrix C is nonnegative for positive values of K and N .

In accordance with (2.11) the gradient of vibrational frequency ω becomes equal to

$$\nabla\omega = D^{-1}[-\omega^3(\bar{X}^T M_1 X), -\omega^3(\bar{X}^T M_2 X), \qquad (2.13)$$

$$-\omega^3(\bar{X}^T M_3 X),\ i\omega^2(\bar{X}^T M_4 X),$$
$$\omega(\bar{X}^T M_5 X),\ \omega(\bar{X}^T M_6 X)].$$

Here D is a constant, $D=\omega^2(\bar{X}^T A X)+(X^T C X)$, while M_1, M_2, ..., M_6 are matrices of the following structure

$$M_1=\left\|\begin{array}{cc|c} 1 & 0 & \\ 0 & 1 & O \\ \hline & & \\ \multicolumn{2}{c|}{O} & O \end{array}\right\|,\quad M_2=\left\|\begin{array}{c|cc} O & \multicolumn{2}{c}{O} \\ \hline & 1 & 0 \\ O & 0 & 0 \end{array}\right\|,\quad M_3=\left\|\begin{array}{c|cc} O & \multicolumn{2}{c}{O} \\ \hline & 0 & 0 \\ O & 0 & 1 \end{array}\right\|$$

$$M_4=\left\|\begin{array}{cc|c} 0 & 1 & \\ -1 & 0 & O \\ \hline & & \\ \multicolumn{2}{c|}{O} & O \end{array}\right\|,\quad M_5=\left\|\begin{array}{cc|cc} 1 & 0 & -1 & 0 \\ 0 & 0 & 0 & 0 \\ \hline -1 & 0 & 1 & 0 \\ 0 & 0 & 0 & 0 \end{array}\right\|,\quad M_6=\left\|\begin{array}{c|cc} O & \multicolumn{2}{c}{O} \\ \hline & 1 & -1 \\ O & -1 & 1 \end{array}\right\|$$

Matrices M_1, M_2, M_3, M_5, M_6 are nonnegative. We can determine the increment $\delta\omega$ from (2.10), (2.11)

$$\delta\omega=(\nabla\omega,\delta h), \tag{2.14}$$
$$\delta h=(\delta I,\delta\Psi,\delta\Theta,\delta H,\delta K,\delta N)$$

Calculations were performed for the following values of parameters [8] : $H/I = 200\ \text{sec}^{-1}$, $K/I = 10^4\ \text{sec}^{-2}$, $K/\Psi = 1600\ \text{sec}^{-2}$, $N/\Theta = 6400\ \text{sec}^{-2}$, $\Theta/\Psi = 0{,}2$, $I = 10\ \text{g.cm.sec}^2$.

System (2.12) has three nonzero vibrational frequencies and one zero frequency. A computer was employed to calculate the vibrational frequences ω, the corresponding eigenvectors X and the gradients $\nabla\omega$ from (2.13). The results of calculations are given below

$$\omega_1 = 32.067\ , \quad \nabla\omega_1 = (-0.331, -0.199, -0.282, 0.332\cdot 10^{-2},$$
$$0.122\cdot 10^{-3},\ 0.708\cdot 10^{-5}),$$

$$\omega_2 = 88.957\ , \quad \nabla\omega_2 = (-0.053, -0.155, -2.78, 0.443\cdot 10^{-3},$$
$$0.114\cdot 10^{-4},\ 0.536\cdot 10^{-3}),$$

$$\omega_3 = 224.36\ , \quad \nabla\omega_3 = (-20, -0.0128, -0.271\cdot 10^{-3},$$
$$0.885\cdot 10^{-1}, 0.236\cdot 10^{-3},\ 0,333\cdot 10^{-6}).$$

Thus, for example, the increment of the third frequency in accordance with (2.14) is

$$\delta\omega_3 = -20\delta I - 0.0128\delta\Psi - 0.271\cdot 10^{-3}\delta\Theta +$$
$$+ 0.0885\delta H + 0.236\cdot 10^{-3}\delta K + 0.333\cdot 10^{-6}\delta N$$

As is to be expected, an increase in the mass characteristics I , Ψ , Θ corresponds to a decrease in the frequencies ω_1 , ω_2 and ω_3 , while an increase in the stiffness parameters K and N corresponds to an increase in these frequencies. Note that an increase in the angular momentum H (for given values of parameters) corresponds to an increase in all three frequencies.

The frequency ω_3 is most greatly affected by the parameter I. The first frequency ω_1 is affected by the parameters I , Ψ and Θ in roughly the same degree.

As for ω_2 , it is most affected by the parameter Θ , while the angular momentum H has an appreciable effect on the third frequency. The effect of the stiffness parameters K and N is relatively slight. The parameter N exerts the greatest effect on ω_2 .

3. SENSITIVITY ANALYSIS FOR NONCONSERVATIVE PROBLEMS OF ELASTIC STABILITY (DISCRETE CASE)

3.1 Introduction

In this and the next chapters we consider problems of elastic stability of nonconservative systems [10-12] . These problems are connected with the question of stability of vibrations. Dealing with nonconservative problems we have to be aware that instability may occur either statically (divergence) or dynamically (flutter). Note that static form of instability is a special case of dynamic form at the frequency equal to zero. Therefore, dynamic method is a general method of the study of stability of nonconservative systems [10-12]. From mathmatical point of view nonconservative problems of elastic stability lead to eigenvalue problems with nonselfadjoint operators.

Boundaries between stable and unstable domains naturally depend on the parameters of the problem, i.e. on the design, on the boundary conditions, mass and stiffness, load distribution etc. It is very interesting and important to obtain qualitative and also quantitave information about the dependence of stability on the discrete and distributed parameters of the problem. This is what we mean talking about sensitivity analysis. Study of the dependence of boundaries between stable and unstable domains on problem parameters is one of the most important and general questions of the theory of elastic stability.

Our inte ntion is to present detailed sensitivity analysis by means of derivatives of stability characteristics with respect to discrete parameters and gradient functions for the case when independent parameter is a function. We shall show that calculation of sensitivities requires the solution to the main stability problem (i.e. determination of the critical load, the critical frequency and the form of the loss of stability), and the solution to the so called adjoint stability problem (more precisely, only the form of the loss of stability is required). With the use of the solutions to these two stability problems it is possible to get all the necessary information about variation of critical loads with respect to all discrete and distributed design parameters and to choose the most essential parameters affecting the stability domain.

Thus, information on sensitivities demands rather few calculations because all the necessary data are available from the ordinary stability analysis. Nevertheless, this information is very important for the designer because it allows to see the whole picture of dependence on parameters, and to use most simple and effective tools to improve the stability characteristics of the system.

It is interesting that the sensitivity analysis with respect to an increment of the load parameter clarifies the condition of instability, and thus makes possible a more rigorous definition of terms like critical load, flutter load and divergence load.

One major application of sensitivity analysis lies within optimal structural design, because sensitivities provide all the necessary information for gradient optimization algorithms. It should be also noted that necessary optimality conditions are usually formulated in terms of gradients, i.e. sensitivities of functions and functionals.

In the chapters 3 and 4 discrete and continuous systems subjected to instability phenomena like flutter and divergence are considered. Sensitivity analysis is presented in general discrete case and with the use of mechanical examples-elastic columns loaded by following forces.

3.2 Sensitivity analysis in matrix formulation

Let us consider nonconservative system with the finite degrees of freedom m subjected to dynamic instability [1, 3, 7, 9]. It is assumed that the system is characterized by the design parameters h_1, h_2, ..., h_n. As design parameters various geometric, stiffness and mass parameters may be considered.

Separating the time t by the exponential function $e^{\lambda t}$ in the linearized equations of motion of the system we get homogeneous matrix equation

$$L(P, \lambda, h)\xi = 0 \tag{3.1}$$

where the matrix L depends on the real load parameter P, on the complex eigenvalue $\lambda = \alpha + i\omega$, and on the real parameters h_i, $i = 1, 2, ..., n$. The complex eigenvector ξ is a vector of generalized displacements. We may write the matrix L as a linear function of real matrices, specifying the dependence on λ explicitly by

$$L = S + Q + \lambda^2 M + \lambda C \tag{3.2}$$

where the symmetric stiffness matrix S and the symmetric mass matrix M depend on h_i but not on P. The nonsymmetric load matrix Q and the damping matrix C depend on P as well as on h_i. It is assumed that all the matrices considered are smooth functions of the parameters.

The dynamic stability of the system is determined by the eigenvalues λ_i, $i = 1, 2, \ldots, 2m$. The system is considered as stable if all the real parts of the eigenvalues satisfy the inequality

$$Re\,\lambda_i \leq -\varepsilon_\circ , \quad i = 1, 2, \cdots, 2m \qquad (3.3)$$

This inequality specifies the degree of stability, $\varepsilon_\circ$ is a small positive constant. When $\varepsilon_\circ = 0$ then the given definition of stability (3.3) coincides with the usual [7].

The condition (3.3) defines the critical load P_c and the critical eigenvalue $\lambda_c = \alpha_c + i\omega_c$. We assume that it is simple eigenvalue

$$\begin{aligned} &Re\,\lambda_c = -\varepsilon_\circ \qquad \text{and} \qquad Re\lambda_i < -\varepsilon_\circ \quad i \neq c,\ i = 1, 2, \cdots, 2m \\ &Re\,\lambda_{c,P} > 0 \qquad\qquad\qquad \text{at} \quad P = P_c \end{aligned} \qquad (3.4)$$

The subindex with the preceding comma means partial differentiation with respect to this index. The condition (3.4) determines the boundary of the region of stability. It means that when P becomes greater than P_c then the stability domain $P < P_c$ is changed by instability domain $P \geqslant P_c$. The derivative $Re\,\lambda_{c,P}$ shows how fast with respect to the increment ΔP the instability develops, i.e. how dangerous is flutter.

Let us find first derivatives of critical load P_c with respect to design parameters $h_1, h_2, \ldots, h_n$ and the derivative $Re\,\lambda_{,P}$. For this purpose we introduce the adjoint to (3.1) problem

$$L^T (P, \lambda, h)\eta = 0 \qquad (3.5)$$

where L^T means transposed matrix, η is an eigenvector of transposed system. The equation (3.5) can be rewritten in the form $\eta^T L = 0$, it means that η is a left eigenvector. At given value P the eigenvalues λ in adjoint systems (3.5) and (3.1) are the same because the determinants of the matrices L and L^T are the same.

Let us take to parameter h_i a variation δh_i. Then the quantities λ, P, ξ obtain the increments $\delta\lambda$, δP, $\delta\xi$. The variational equation has the form

$$L_{,P}\,\xi\,\delta P + L_{,\lambda}\,\xi\,\delta\lambda + L_{,h_i}\,\xi\,\delta h_i + L\,\delta\xi = 0 \quad (3.6)$$

We multiply this equation from the left by η^T and obtain

$$(\eta^T L_{,P}\,\xi)\,\delta P + (\eta^T L_{,\lambda}\,\xi)\,\delta\lambda + (\eta^T L_{,h_i}\,\xi)\,\delta h_i = 0 \quad (3.7)$$

The term ($\eta^T L\,\delta\xi$) vanishes due to (3.5). Especially at a critical load $P = P_c$ we have by (3.4) $\lambda_c = -\varepsilon_0 + i\omega_c$, $\delta\lambda_c = i\,\delta\omega_c$, $\delta P = \delta P_c$. Dividing (3.7) on $(\eta^T L_{,\lambda}\xi)$ and taking the real part we get the expression [1, 3, 7, 9]

$$\frac{\partial P_c}{\partial h_i} = -\,Re\left(\frac{\eta^T L_{,h_i}\,\xi}{\eta^T L_{,\lambda}\,\xi}\right) \Big/ Re\left(\frac{\eta^T L_{,P}\,\xi}{\eta^T L_{,\lambda}\,\xi}\right) \quad (3.8)$$

Similarly, dividing (3.7) on ($\eta^T L_{,P}\,\xi$) and taking the imagine part we obtain

$$\frac{\partial \omega_c}{\partial h_i} = -\,Im\left(\frac{\eta^T L_{,h_i}\,\xi}{\eta^T L_{,P}\,\xi}\right) \Big/ Re\left(\frac{\eta^T L_{,\lambda}\,\xi}{\eta^T L_{,P}\,\xi}\right) \quad (3.9)$$

Thus, to calculate first derivatives of critical load P_c and critical frequency ω_c with respect to design parameters h_i we have to find the derivatives of matrix L $L_{,h_i}$ $L_{,P}$, $L_{,\lambda}$, calculate critical values P_c, λ_c, and eigenvectors ξ and η. Due to explicit dependence of L on parameters (3.2) we find

$$L_{,h_i} = S_{,h_i} + Q_{,h_i} + \lambda^2 M_{,h_i} + \lambda C_{,h_i}\,,$$

$$L_{,P} = \lambda C_{,P} + Q_{,P},$$
$$L_{,\lambda} = 2\lambda M + C.$$

Having obtained the derivatives (3.8), (3.9) we can find finite increments

$$\Delta P_c = \sum_{i=1}^{n} \frac{\partial P_c}{\partial h_i} \Delta h_i + o(|\Delta h|)$$

$$\Delta \omega_c = \sum_{i=1}^{n} \frac{\partial \omega_c}{\partial h_i} \Delta h_i + o(|\Delta h|)$$

If the load is kept unchanged we get from (3.7) with $\delta P = 0$

$$\frac{\partial \lambda}{\partial h_i} = \frac{\partial \alpha}{\partial h_i} + i\,\frac{\partial \omega}{\partial h_i} = -\,\frac{\eta^T L_{,h_i} \xi}{\eta^T L_{,\lambda} \xi} \tag{3.10}$$

On the other hand, if the design parameters are kept unchanged then from (3.7) we get with $\delta h_i = 0$

$$\frac{\partial \lambda}{\partial P} = -\,\frac{\eta^T L_{,P} \xi}{\eta^T L_{,\lambda} \xi} \tag{3.11}$$

Thus, the instability criterion (3.4) may therefore be written as

$$\operatorname{Re} \lambda_c = -\varepsilon_0$$
$$\text{at } P = P_c$$
$$\operatorname{Re}\left(\frac{\eta^T L_{,P} \xi}{\eta^T L_{,\lambda} \xi} \right) < 0$$

Calculation of the quantity $Re\,\lambda_{,P}$ at $P=P_c$ is important for stability analysis of the system, because it characterizes the measure of instability when P increases.

Now we consider systems without damping. For these systems the results of the previous section can be considerably simplified. The system matrix L reduces to

$$L = S + Q + \lambda^2 M \tag{3.12}$$

The definition of stability (3.3) may be used only with $\varepsilon_o = 0$. The solutions to (3.1) are either real λ^2 with corresponding real ξ ($\lambda^2 > 0$ giving divergence instability and $\lambda^2 < 0$ - stable harmonic vibrations), or complex conjugate solutions λ^2 with corresponding complex conjugate eigenvectors ξ. The flutter instability in this case occurs when two eigenvalues $\lambda^2 < 0$ merge at $P = P_c$ (i.e. become double), and then with the increase of P become complex conjugate. The static instability (divergence) occurs when λ is equal to zero.

Let us consider (3.1), (3.5) with (3.12). At $P=P_c$ λ^2 is double eigenvalue with the single eigenvectors ξ and η. In this point the so called "flutter condition" takes place

$$\eta^T M \xi = 0 \tag{3.13}$$

It means that at $P=P_c$ we have from (3.12), (3.13)

$$(\eta^T L_{,\lambda} \xi) = 2\lambda\,(\eta^T M \xi) = 0 \tag{3.14}$$

For divergence instability (3.14) is also valid because $\lambda = 0$. Using (3.14) we find from equation (3.7)

$$\frac{\partial P_c}{\partial h_i} = -\frac{\eta^T L_{,h_i} \xi}{\eta^T L_{,P} \xi} \tag{3.15}$$

We can also find the derivatives of ω taking in (3.12) $\lambda^2 = -\omega^2$

$$\frac{\partial\omega}{\partial h_i} = \frac{\eta^T L_{,h_i}\xi}{2\omega(\eta^T M\xi)}, \quad \frac{\partial\omega}{\partial P} = \frac{\eta^T L_{,P}\xi}{2\omega(\eta^T M\xi)} \tag{3.16}$$

The expressions (3.16) are valid at $P < P_c$ and $P > P_c$
But they are invalid at the critical point due to double eigenvalue at $P = P_c$

4. SENSITIVITY ANALYSIS FOR NONCONSERVATIVE PROBLEMS OF ELASTIC STABILITY (DISTRIBUTED CASE)

4.1 Extended Beck's problem

Let us consider an extension of well known Beck's problem [10-12] of dynamic stability of elastic column loaded by follower force. The extended problem includes a linear elastic support, a concentrated mass in addition to the distributed mass, and a partial follower force, $0.5 \leq \eta \leq 1.5$ Fig. 1. The stability analysis for this problem was done by P.Pedersen, here we shall concentrate our attention on sensitivity analysis with respect to distributed and discrete parameters of the problem [9]. The equation of small vibrations of a column and boundary conditions after separating the time by $\exp(\lambda t)$ lead to the expressions written in nondimensional form [9].

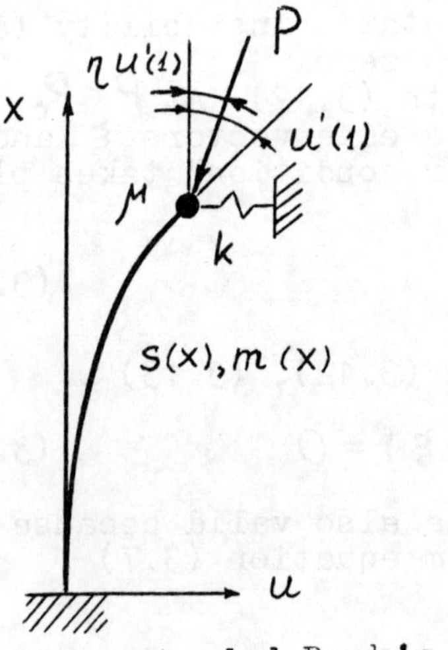

Fig.1, Extended Beck's column

$$Lu \equiv (su'')'' + pu'' + \lambda^2 mu = 0 \tag{4.1}$$

$$u(0)=u'(0)=0\,, \qquad su''(1)=0\,,$$

$$lu \equiv [(su'')' + (1-\eta)pu' - (\mu\lambda^2+k)u\,]_{x=1} = 0$$

In these equations $s(x)$ and $m(x)$ are stiffness and mass distributions respectively, discrete parameters k, μ and η are rigidity of support, concentrated mass and angle of follower force, Fig. 1. The equations (4.1) represent an eigenvalue problem with nonselfadjoint differential operator, λ^2 is an eigenvalue, $u(x)$ is an eigenfunction (deflection of the column), p is a load parameter. Note that eigenvalue as well as load p and parameters μ, k, η are involved in the boundary condition.

For sensitivity analysis we introduce the adjoint to (4.1) problem

$$\begin{aligned}
&(sv'')'' + pv'' + \lambda^2 mv = 0 \\
&v(0)=v'(0)=0\,, \quad (sv''+\eta pv)_{x=1} = 0 \\
&[(sv'')' + pv' - (\mu\lambda^2+k)v\,]_{x=1} = 0
\end{aligned} \tag{4.2}$$

If the quantities s, m, p, η, k, μ_2 are the same in (4.1) and (4.2) then the eigenvalues λ^2 and their multiplicities coincide with each other. When system is unloaded $p=0$ the problems (4.1) and (4.2) become selfadjoint $u\equiv v$. In this case eigenvalues λ^2 are real and negative due to expression

$$\lambda^2 = -\frac{\int_0^1 su''^2dx + k\,u^2(1)}{\int_0^1 mu^2dx + \mu\,u^2(1)}$$

k and μ are assumed to be positive. Therefore we may take $\lambda^2 = -\omega^2$, ω is a frequency of harmonic vibrations. With the increase of load p two of the eigenvalues may merge, become double at $p = p_c$ and then complex conjugate quantities, that means instability.

First of all we obtain "the flutter condition" which is similar the condition (3.13). For this purpose we consider (4.1) with the eigenvalue λ_i^2 and appropriate eigenfunction u_i and (4.2) with the other eigenvalue λ_j^2 and eigenfunction v_j. We multiply (4.1) on v_j and (4.2) on u_i and integrate over $[0,1]$. Taking partial integration with the use of the boundary conditions we obtain

$$\int_0^1 (s u_i'' v_j'' - p u_i' v_j')\, dx + [\eta p u_i' v_j + (\mu \lambda_i^2 + k) u_i v_j]_{x=1} + \lambda_i^2 \int_0^1 m u_i v_j\, dx = 0$$

$$\int_0^1 (s u_i'' v_j'' - p u_i' v_j')\, dx + [\eta p u_i' v_j + (\mu \lambda_j^2 + k) u_i v_j]_{x=1} + \lambda_j^2 \int_0^1 m u_i v_j\, dx = 0$$

Subtracting these equations we get

$$(\lambda_i^2 - \lambda_j^2)\Big[\int_0^1 m u_i v_j\, dx + \mu (u_i v_j)_{x=1}\Big] = 0$$

If $\lambda_i \neq \lambda_j$ then the biorthogonality condition takes place

$$\int_0^1 m u_i v_j\, dx + \mu (u_i v_j)_{x=1} = 0$$

At the critical point eigenvalue λ_c is double. Taking the limits $p \to p_c$, $\lambda_i \to \lambda_c$, $\lambda_j \to \lambda_c$, $u_i \to u_c$, $v_j \to v_c$ we obtain the "flutter condition"

$$\int_0^1 m u_c v_c\, dx + \mu (u_c v_c)_{x=1} = 0 \qquad (4.3)$$

Note that this condition includes the boundary term at $x = 1$.

4.2 Sensitivity analysis

Now we derive partial derivatives and functional gradients of critical load p_c with respect to discrete and distributed parameters. For this purpose we consider (4.1) at the critical point taking there $\lambda_c^2 = -\omega_c^2$, $p = p_c$ and $u = u_c$.

Let us take to the parameter η an increment $\delta\eta$. Due to this variation the critical load p_c, the critical frequency ω_c and the eigenfunction u_c obtain the increments δp_c, $\delta\omega_c$, δu_c. The variational equations have the form

$$L\delta u_c + \delta p_c u_c'' - 2\omega_c \delta\omega_c m u_c = 0 \quad (4.4)$$

$$(\delta u_c)_{x=0} = (\delta u_c')_{x=0} = 0, \quad (s\delta u_c'')_{x=1} = 0 \quad (4.5)$$

$$(\ell\delta u_c)_{x=1} + [-\delta\eta\, p_c u_c' + (1-\eta)\delta p_c u_c' + 2\mu\omega_c\delta\omega_c u_c]_{x=1} = 0$$

We multiply (4.4) on the function $v_c(x)$ and integrate over $[0,1]$. Then we get

$$\int_0^1 v_c L\delta u_c dx + \delta p_c \int_0^1 u_c'' v_c dx - 2\omega_c\delta\omega_c \int_0^1 m u_c v_c dx = 0$$

Performing the integration by parts for the first term of the last equation with the use of (4.5) we obtain

$$\begin{aligned} &\int_0^1 \delta u_c L v_c dx + \delta p_c \int_0^1 u_c'' v_c dx + \\ &+ [v_c(p_c u_c' \delta\eta - (1-\eta)\delta p_c u_c')]_{x=1} - \\ &- 2\omega_c\delta\omega_c [\mu (u_c v_c)_{x=1} + \int_0^1 m u_c v_c dx] = 0 \end{aligned} \quad (4.6)$$

Due to (4.2) $L v_c = 0$, and the last term in (4.6) vanishes because of (4.3). Thus, from (4.6) we have [9]

$$\frac{\partial p_c}{\partial\eta} = \frac{p_c (u_c' v_c)_{x=1}}{(1-\eta)(u_c' v_c)_{x=1} - \int_0^1 u_c'' v_c dx} \quad (4.7)$$

Doing calculations similar to (4.4)-(4.6) we obtain the derivatives

$$\frac{\partial p_c}{\partial k} = \frac{p_c\,(u_c' v_c)_{x=1}}{(1-\eta)(u_c' v_c)_{x=1} - \int_0^1 u_c'' v_c\, dx}, \tag{4.8}$$

$$\frac{\partial p_c}{\partial \mu} = -\,\omega_c^2 \frac{\partial p_c}{\partial k} \tag{4.9}$$

Now let us derive the functional gradients of critical load p_c with respect to the distributed parameters s and m. If the variations δs and δm are independent quantities, then variational equations due to variation δs take the form

$$L\delta u_c + \delta p_c u_c'' - 2\omega_c \delta\omega_c m u_c + (\delta s\, u_c'')'' = 0$$

$$(\delta u_c)_{x=0} = (\delta u_c')_{x=0} = 0\,,\quad [s\delta u_c'' + \delta s\, u_c'']_{x=1} = 0$$

$$(\ell\delta u_c)_{x=1} + [(\delta s\, u_c'')' + (1-\eta)\delta p_c u_c' + 2\mu\omega_c\delta\omega_c u_c]_{x=1} = 0$$

Multiplying the first equation on v_c and using the integration by parts with the boundary conditions and (4.3) we obtain

$$\int_0^1 \delta s\, u_c'' v_c''\, dx + \delta p_c\Big[\int_0^1 u_c'' v_c\, dx - (1-\eta)(u_c' v_c)_{x=1}\Big] = 0$$

From this we get [9]

$$\delta p_c = \int_0^1 g_s \delta s\, dx\,, \tag{4.10}$$

$$g_s = \frac{u_c'' v_c''}{(1-\eta)(u_c' v_c)_{x=1} - \int_0^1 u_c'' v_c\, dx}$$

where g_s is the gradient of the functional p_c with respect to $s(x)$. Similarly, we can find the expression for the gradient function of p_c with respect to $m(x)$

$$\delta p_c = \int_0^1 g_m \delta m \, dx, \tag{4.11}$$

$$g_m = -\frac{\omega_c^2 u_c v_c}{(1-\eta)(u_c' v_c)_{x=1} - \int_0^1 u_c'' v_c \, dx}$$

For geometrically similar cross sections $s = m^2$, so the variations δs and δm are connected by the relation $\delta s = 2m\delta m$. For this case the variation of the critical load with respect to mass distribution takes the form [9]

$$\delta p_c = \int_0^1 g \delta m \, dx, \tag{4.12}$$

$$g(x) = \frac{2m u_c'' v_c'' - \omega_c^2 u_c v_c}{(1-\eta)(u_c' v_c)_{x=1} - \int_0^1 u_c'' v_c \, dx}$$

Thus, to calculate the partial derivatives and the gradients (4.7)-(4.12) it is necessary to solve the stability problems (4.1) and (4.2), i.e. to find the quantities p_c, ω_c, $u_c(x)$ and $v_c(x)$. Having received the sensitivities (4.7)-(4.12) we may obtain finite increments of critical load Δp_c when all the parameters are varied

$$\Delta p_c = \frac{\partial p_c}{\partial \mu} \Delta \mu + \frac{\partial p_c}{\partial \eta} \Delta \eta + \frac{\partial p_c}{\partial k} \Delta k + \\ + \int_0^1 g_m \delta m \, dx + \int_0^1 g_s \delta s \, dx$$

In this relation mass and stiffness distributions are regarded as independent quantities. If they are related, for instance, by the formula $s = m^2$, then instead of the last integrals we have to use the integral $\int_0^1 g \delta m \, dx$ from (4.12).

It is interesting to note that if we take in (4.1) and (4.2) $s = m = 1$, $\eta = 1$, $\mu = k = 0$, then (4.1) will describe the stability of a uniform column loaded by follower load (Beck's problem), and (4.2) will describe the stability problem of the same column loaded by the force with the fixed line of action (Reut's problem) [10]. Using the solutions of these two classical problems we may calculate the derivatives and the gradients (4.7)-(4.12) and find the increment (4.13). Therefore, we may evaluate the effect of the addition of small quantities - concentrated mass $\Delta\mu$, rigidity of support Δk, follower angle $\Delta\eta$ and variations δs, δm from the analytical solutions of Beck's and Reut's problems neglecting the above effects [9, 10].

The stability problems (4.1) and (4.2) for the uniform cross section of the column $s = m = 1$ and arbitrary values of μ, η, k also have analytical solutions [9]

$$u(x,p,\omega) = y + f_1 z, \quad v(x,p,\omega) = y + f_2 z$$

where

$$y(x,p,\omega) = \cosh(ax) - \cos(bx),$$

$$z(x,p,\omega) = a\sin(bx) - b\sinh(ax),$$

$$f_1(p,\omega) = \frac{a^2\cosh a + b^2\cos b}{ab(a\sinh a + b\sin b)},$$

$$f_2(p,\omega) = \frac{(a^2+\eta p)\cosh a + (b^2-\eta p)\cos b}{b(a^2+\eta p)\sinh a + a(b^2-\eta p)\sin b},$$

$$a = \sqrt{-\frac{p}{2} + \sqrt{\frac{p^2}{4} + \omega^2}}, \quad b = \sqrt{+\frac{p}{2} + \sqrt{\frac{p^2}{4} + \omega^2}}$$

The relation $\omega(p)$ is defined by the equation $f_2(p,\omega) = f_3(p,\omega)$, where

$$f_3(p,\omega) = \frac{a(a^2+p)\sinh a - b(b^2-p)\sin b + (\mu\omega^2 - k)\cosh a}{ab[(a^2+p)\cosh a + (b^2-p)\cos b}$$

The critical values ρ_c, ω_c for this case and some values of μ, k, η are given in Table 1. These quantities were computed from the transcendental equation [9] .

	$\mu = k = 0$			$\eta = 1,\ k = 0$			$\eta = 1,\ \mu = 0$		
	$\eta=0.5$	$\eta=1$	$\eta=1.5$	$\mu=0$	$\mu=0.5$	$\mu=1$	k=0	k=10	k=20
P_c	16.1	20.05	30.6	20.05	16.1	16.2	20.05	24.5	30.0
ω_c	7.1	11.0	12.0	11.0	7.3	6.3	11.0	10.9	10.0

Table 1, Critical load ρ_c and frequency ω_c

The partial derivatives (4.7)-(4.9) and the gradients (4.12) were computed with the use of the analytical solutions, they are presented in Table 2 and Fig. 2. This information gives qualitative and quantitative analysis of influence of system parameters on critical load. For example, for $\eta = 1$, $\mu = k = 0$ the relation (4.13) gives

$$\Delta \rho_c = 15 \Delta \eta - 40 \Delta \mu + 0.33 \Delta k$$

Therefore, the addition of the concentrated mass at the tip of a column ($\Delta\mu > 0$) leads to the decrease of the critical load. On the other hand, the addtion of the elastic support ($\Delta k > 0$) leads to the increase of the critical load. These effects are not evident from physical point of view.

If we consider only variations of distributed mass with the relation $S = m^2$ then according to (4.12) and Fig. 2 we see that there are regions along the column length in which the addition of mass $\delta m(x) > 0$ leads to the increase or decrease of ρ_c . These regions are described by the relations $g(x) > 0$ and $g(x) < 0$ respectively.

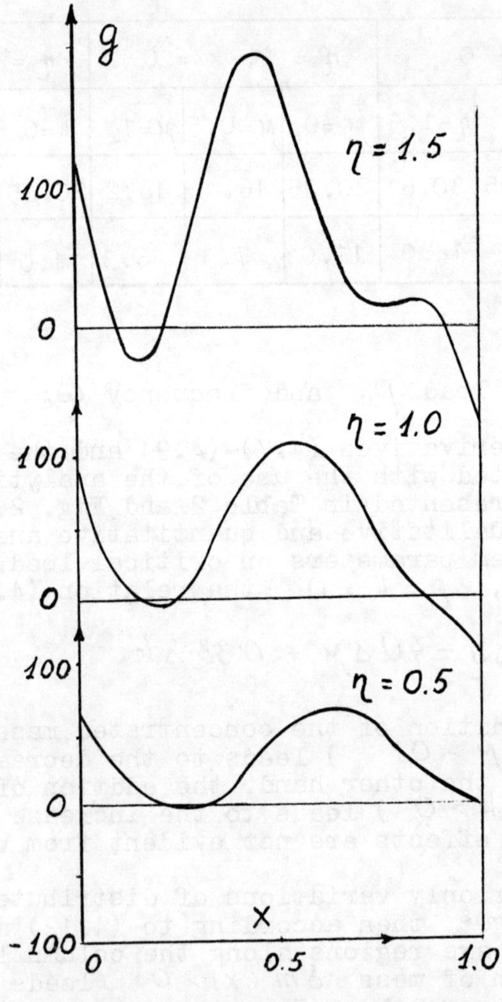

Fig.2, Gradient function for Beck's column with partial follower force (μ=k=0)

	$\mu = k = 0$			$\eta = 1.0,\ k = 0$			$\eta = 1,\ \mu = 0$		
	$\eta=0.5$	$\eta=1$	$\eta=1.5$	$\mu=0$	$\mu=0.5$	$\mu=1.0$	k=0	k=10	k=20
$\partial p_c/\partial \eta$	0.13	15.0	21.0	15.0	0.31	-0.95	15.0	22.0	23.0
$\partial p_c/\partial \mu$	-0.11	-40.0	-66.0	-40.0	-0.28	0.56	-40.0	-61.0	-61.1
$\partial p_c/\partial k$	0.002	0.33	0.46	0.33	0.005	-0.014	0.33	0.51	0.61

Table 2, Sensitivities

5. OPTIMIZATION OF CRITICAL LOADS OF COLUMNS SUBJECTED TO FOLLOWER FORCES

5.1 Introduction

Optimization of critical loads of stability is intimately connected with the sensitivity analysis. Indeed, if expressions of partial derivatives and/or functional gradients of flutter and divergence loads have been drawn and calculated, they can be used through optimization procedures in structural optimization problems. Our opinion is that the gradient methods using sensitivity analysis are very useful for structural optimization problems taking flutter and divergence phenomena into account. This is valid especially when a large finite (or infinite) number of design parameters are considered, because the calculation of the gradient of the critical load only calls for the main and the adjoint flutter problems to be solved once. In contrast to this, a gradient calculation based on numerical differentiation of the critical flutter (or divergence) load would require the stability problem to be solved (N + 1) times if N design parameters are taken into account. Note that flutter problems for itself (the determination of the critical load, the critical frequency and the form of the loss of stability) is rather complicated and expensive computational problem.

5.2 Optimization problem

Let us consider the problem of maximization of the critical load of a column of variable thickness subjected to the action of distributed follower forces, assuming that the total mass of a column is maintained constant, Fig. 3. The design variable is a mass distribution $m(x)$. The loss of stability of a column is described by the equation and the boundary conditions [6].

Fig.3, Distributed follower force

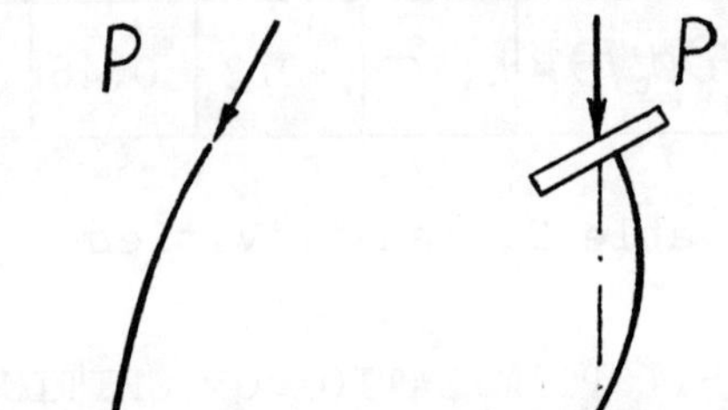

Fig.4, Adjoint problems

$$(m^2 u'')'' + pQu'' - \omega^2 m u = 0, \quad Q = \int_x^1 q\,dx \tag{5.1}$$

$$u(0) = u'(0) = 0, \quad (m^2 u'')_{x=1} = (m^2 u'')'_{x=1} = 0$$

The notations used in this lecture are the same as the notations of the preceding chapter,

The adjoint problem takes the form

$$
\begin{aligned}
&(m^2 v'')'' + p(Qv)'' - \omega^2 m v = 0 \\
&v(0) = v'(0) = 0, \quad [m^2 v'' + pQv]_{x=1} = \\
&= [(m^2 v'')' + p(Qv)']_{x=1} = 0
\end{aligned}
\tag{5.2}
$$

If in eqns (5.1) and (5.2) we take $Q = 1$, then the problem (5.1) will describe the stability problem of a column loaded by the tangential follower force (Beck's problem) and the problem (5.2) will correspond to the stability problem of a column, loaded by the force with the fixed line of action (Reut's problem). This is an example when the main and the adjoint problems of stability have physical meaning, Fig. 4.

In the case of a distributed load $q(X)$ the adjoint problem (5.2) has not explicit physical sense. Using relations $Q(1) = 0$, $Q'(1) = -q(1)$ we write the last two terms in (5.2) in the form

$$(m^2 v'')_{x=1} = [(m^2 v'')' - pqv]_{x=1} = 0$$

If $q(1) = 0$ then all the boundary conditions in the adjoint problems (5.1) and (5.2) are the same.

Due to the absence of damping the loss of stability occurs by merging of two frequencies. The boundary of vibrational stability is defined by double frequencies at the critical point. The flutter condition in our case is

$$\int_0^1 m u_c v_c \, dx = 0 \tag{5.3}$$

Variation of the critical load is defined by the expression [6]

$$
\begin{aligned}
&\delta p_c = \int_0^1 g \delta m \, dx, \\
&g(x) = (\omega_c^2 u_c v_c - 2m u_c'' v_c'') / \int_0^1 Q u_c'' v_c \, dx
\end{aligned}
\tag{5.4}
$$

Function $g(x)$ is the gradient function of critical load with respect to the distribution $m(x)$. This function indicates the sensitivity of critical load with respect to the variation of mass $\delta m(x)$.

The problem of maximization of the critical load of a column with the fixed total mass is formulated below

$$\max_{m\in\Omega} p_c(m), \ \Omega = \left\{ m : \int_0^1 m\,dx = 1,\ m(x) > 0,\ x\in[0,1] \right\} \tag{5.5}$$

It is easy to see from (5.1) and (5.2) immediately that critical parameters p_c, ω_c are homogeneous with respect to m functionals with the degrees $\nu = 2$ and $\nu = 1$ respectively, i.e. $p_c(\gamma m) = \gamma^2 p_c(m)$, $\omega_c(\gamma m) = \gamma\,\omega_c(m)$; $\gamma = const > 0$. This property can be seen also from the expression for p_c

$$p_c(m) = -\frac{\int_0^1 m^2 u_c'' v_c''\,dx}{\int_0^1 Q\,u_c'' v_c\,dx}$$

which is derived from (5.1) and (5.2). The functional of total mass $M = \int_0^1 m\,dx$ is also homogeneous with respect to m with the degree $\nu = 1$.

Using the results of [5] we conclude that the problem of maximization of the critical load p_c at fixed total mass is equivalent to the dual problem of minization of total mass at the fixed critical load $p_c = 1$. The solution of the last problem $m_d(x)$ differs from the solution $m_0(x)$ of the problem (5.5) only by the multiplier [5]

$$m_d(x) = \gamma\, m_0(x),\ \gamma = 1/\sqrt{p_c(m)} \tag{5.6}$$

It is easy to prove that a solution of the optimization problem (5.5) for a column, described by (5.1), is a solution of the same optimization problem for system, described by (5.2), and vice versa, because in the adjoint problems spectrum of eigenvalues at the same p is the same.

Thus, we conclude that the optimization problems for a column, loaded by tangential follower force, and for a column, loaded by the force with the fixed line of action, lead to the same solutions. The solutions of the problems, dual to the mentioned above, are calculated by (5.6). Thus, these four optimization problems are equivalent in the sense that only the solution of one of them is required to describe the solutions of the other three problems. Note that the optimal mass distribution of optimization problem of a column, loaded by follower force, was obtained in [13, 14].

Using the expression of the gradient (5.4) and the gradient of total mass $M=\int_0^1 m\,dx$, which is equal to 1, we may write the expression for improving variation

$$\delta m(x)=\alpha m(x)(g(x)-\mu),\quad \mu=\int_0^1 mg\,dx\Big/\int_0^1 m\,dx \tag{5.7}$$

Here α is a positive constant (step by gradient) chosen by the researcher. In (5.7) instead of $m(x)$ we may use any positive function.

Variation (5.7) satisfies the condition of constant total mass $\int_0^1 \delta m\,dx=0$ and leads to an increase of the critical load because (5.4) with the use of (5.7) can be transformed into

$$\delta p_c=\int_0^1(g-\mu)\delta m\,dx=\alpha\int_0^1 m(g-\mu)^2dx\geq 0 \tag{5.8}$$

The equality sign here is reached only when $g(x)=\mu$, $x\in[0,1]$.

In Fig. 5a the gradient function $g(x)$ and the relation $\omega - p$ are presented for the uniform column $m(x) = 1$, loaded by the distributed follower forces $q(x)=1-x$, $Q=\int_0^1 q\,dx=1/2\,(1-x)^2$.

The critical load is $p_c = 149.8$, and the critical frequency is equal to $\omega_c=\sqrt{131}$.

For the numerical solution of the eigenvalue problems (5.1) and (5.2) they were discretized and transformed to the algebraic eigenvalue problems. The interval [0, 1] was divided to N = 30, 60 subintervals, the double precision computations were performed.

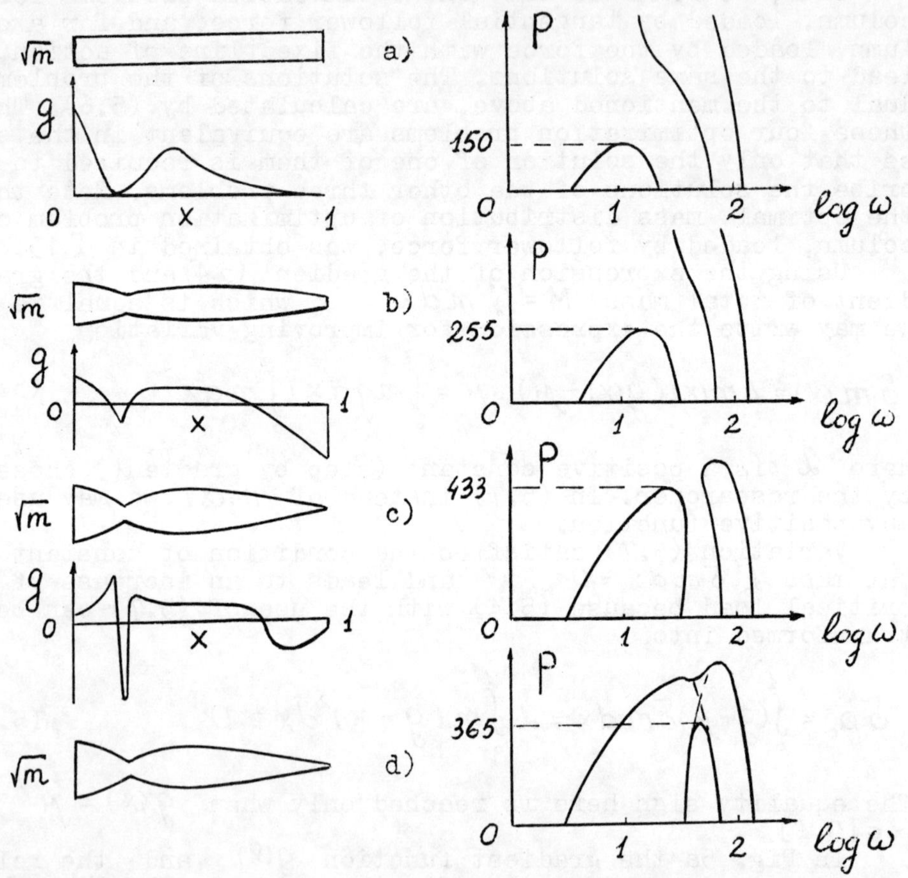

Fig.5, Mass distributions, gradient functions, and characteristic curves during iterations

The results of the successive iterations, done according to (5.7), are presented in Fig. 5. It is seen that when the critical load p_c, caused by the counteraction of the first and the second frequencies, increases, then the critical value p_*, caused by the counteraction of the third and the fourth frequencies, decreases, Fig. 5b,c. When optimization process of p_c is continued, then characteristic curves merge and the catastrophe occurs. It leads to the jump of the critical load and to appearence of additional domains of stability and instability. The qualitative difference is obvious: the loss of stability now occurs due to counteraction of the second and the third frequencies, Fig. 5d.

Through optimization process the critical load increases started from $p_c = 149.8$ till $p_c = 470$, the sensitivity of critical load being increasing too. The character of the gradient function $g(x)$ also changes.

We emphasize the great advantage in maximized functional compared with the advantages in other (static) structural optimization problems. This effect is explained by greater sensitivity of critical flutter load with respect to variations of mass and stiffness distributions.

6. OPTIMIZATION OF AEROELASTIC STABILITY OF PANELS IN SUPERSONIC GAS FLOW

6.1 Introduction

Here the problems of determining the thickness function of solid panels of constant volume having maximal critical speed of aeroelastic stability are considered. The expressions for increments of the critical flutter and divergence speeds with respect to thickness variation are derived. The necessary optimality conditions are established and with their use the optimal solutions for cantilever panel are obtained numerically. Panel flutter optimization problems for other boundary conditions were considered by many authors, see the review paper [15] with the comprehensive list of references.

6.2 Optimization problem

Let's consider small vibrations of thin elastic panel of variable thickness in supersonic gas flow. It is assumed that the panel is symmetric with respect to its neutral plane and that the panel span is much greater than its dimension a in the direction of the flow, Fig. 6.

For the description of aerodynamic forces we use the linearized piston theory formula. Equation of vibrations of the panel in supersonic gas flow is [3,4]

$$\frac{\partial^2}{\partial x^2}\left(D(x)\frac{\partial^2 w}{\partial x^2}\right)+2\rho H(x)\frac{\partial^2 w}{\partial t^2}+g_a=0 \qquad (6.1)$$

$$g_a=\frac{2\rho_0 \varkappa}{c_0}\left(\frac{\partial w}{\partial t}+U\frac{\partial w}{\partial x}\right),\quad D=\frac{2EH^3}{3(1-\nu^2)}$$

Here $D(x), 2H(x), w(x,t)$ are respectively bending stiffness, thickness and deflection functions of the panel. The parameters $\rho, E, \nu, U, \varkappa, c_0, \rho_0$ denote density of panel material, Young's modulus, Poisson's ratio, speed of undisturbed flow, the sound speed in a gas and pressure at infinity.

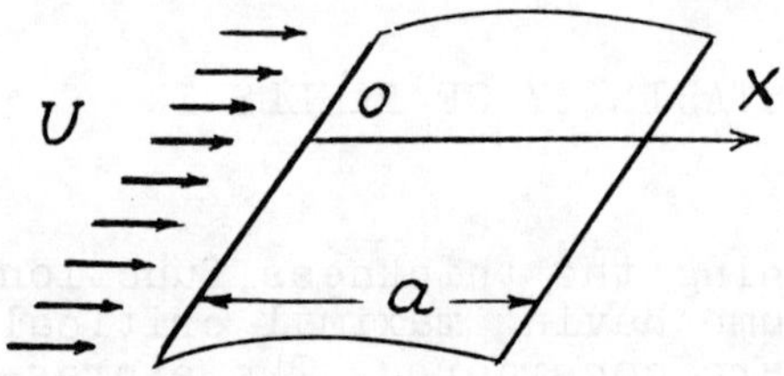

Fig.6, Panel flutter

Let us consider panel deflections expressed by the formula

$$w(x,t)=u(x)e^{st} \qquad (6.2)$$

where $u(x)$ is complex function of real variable x and s is a complex number. With the use of (6.2) the vibrational equation (6.1) in nondimensional variables takes the form

$$(h^3u'')''+\sigma hu+\Lambda\sigma u+\beta u'=0 \qquad (6.3)$$

$$u(0)=u'(0)=0,\ (h^3u'')_{x=1}=(h^3u'')'_{x=1}=0$$

Here the nondimensional quantities are used: $h(x)$ is a thickness function, $u(x)$ is a deflection function, Λ and β are damping coefficient and velocity of the flow, σ is a complex eigenvalue. Primes denote differentiation with respect to x. The boundary conditions correspond to the case of clamped-free panel. Positive as well as negative values of β, i.e. both directions of the speed U, will be considered, Fig. 6.

When $h(x)$, Λ and β are given the equation with the boundary conditions (6.3) determine nonselfadjoint eigenvalue problem in which σ is an eigenvalue. The equilibrium form $u = 0$ of the panel in gas flow is stable when all eigenvalues σ belong to the left half of complex plane, i.e. $\operatorname{Re} \sigma < 0$. If quantities $h(x)$ and Λ are fixed the mentioned equilibrium may become unstable for some values of nondimensional speed β. The critical divergence speed β_d is determined by the condition $\sigma = 0$ and the critical flutter speed is characterized by the relations $\operatorname{Re} \sigma = 0$, $\operatorname{Im} \sigma = \omega \neq 0$.

The nondimensional volume of the panel is

$$V = \int_0^1 h(x)\,dx \tag{6.4}$$

Now we consider the optimization problem: find the thickness function $h_0(x)$ satisfying the constant volume condition $V(h_0) = V_0$ and maximizing the minimal of the critical speeds β_f, β_d.

Quantities Λ and V_0 are the problem parameters. The mathematical formulation of the stated problem is described by the relations

$$\max_{h \in \Omega} \min \left[\beta_f(h), \beta_d(h) \right] = \min \left[\beta_f(h_0), \beta_d(h_0) \right]$$
$$\Omega = \left\{ h(x) : V(h) = \int_0^1 h\,dx = V_0 ,\ h(x) \geq 0 \right\} \tag{6.5}$$

6.3 Sensitivity analysis

Now we calculate the variations of flutter and divergence critical speeds with respect to thickness variation δh. For this purpose we consider the vibrational equation at the boundary of flutter taking in (6.3) $\sigma = i\omega$, $\beta = \beta_f$

$$Lu \equiv (h^3 u'')'' - \omega^2 h u + i\omega\Lambda u + \beta_f u' = 0 ,$$
$$u(0) = u'(0) = 0 ,\ (h^3 u'')_{x=1} = (h^3 u'')'_{x=1} = 0 , \tag{6.6}$$

here i is the imaginary unity $i = \sqrt{-1}$ and ω is flutter frequency.

The adjoint to (6.6) eigenvalue problem is described by the relations

$$L^* v = (h^3 v'')'' - \omega^2 h v + i\omega\Lambda v - \beta_f v' = 0 \tag{6.7}$$

$$v(0) = v'(0) = 0, \quad (h^3 v'')_{x=1} = [(h^3 v'')' - \beta_f v]_{x=1} = 0$$

Note that the eigenvalues and their multiplicities are the same in adjoint problems. To calculate the increment of critical flutter speed we take to the functions h and u variations δh and δu, to the frequency and the critical flutter speed variations $\delta\omega$, $\delta\beta_f$. Using (6.6), (6.7) and integrating by parts we get the variational equation [3]

$$\int_0^1 A\delta h\, dx + B\delta\omega + C\delta\beta_f = 0 \tag{6.8}$$

$$A = 3h^2 u'' v'' - \omega^2 u v, \quad C = \int_0^1 u' v\, dx, \tag{6.9}$$

$$B = -2\omega \int_0^1 h u v\, dx + i\Lambda \int_0^1 u v\, dx$$

Here B and C are complex constants and the function A is a complex function of real variable x.

From (6.8) we obtain the variations [3]

$$\delta\beta_f = \int_0^1 g\delta h\, dx, \quad g = -Im(A\bar{B})/Im(C\bar{B}) \tag{6.10}$$

$$\delta\omega = \int_0^1 t\delta h\, dx, \quad t = -Im(A\bar{C})/Im(B\bar{C})$$

In this expressions $\bar{B}$ and $\bar{C}$ are complex-conjugate quantities to B and C. Thus, to determine the gradients g and t it is necessary to solve the adjoint flutter problems (6.6), (6.7) and calculate complex functions $u(x)$, $v(x)$ and real quantities β_f, ω. Then according to (6.9), (6.10) we get the gradients g and t.

The panel may violate its stability by static form (divergence). Let us determine the variation of the critical divergence speed. Taking in (6.6)-(6.9) $\omega = 0$,

$\delta\omega = 0$ and repeating above calculations we get [3]

$$\delta\beta_d = \int_0^1 e\delta h\,dx\,,\quad e = -3h^2u''v'' / \int_0^1 u'v\,dx \tag{6.11}$$

The function e represents the gradient of the critical divergence speed. The eigenfunctions u and v in this case are real quantities.

Note that if some other boundary conditions in (6.6) are considered, then only the boundary conditions in (6.7) will be changed. All the other quantities in (6.7)-(6.11) remain the same.

6.4 Necessary optimality conditions

Now we derive the necessary optimality conditions for the stated problem (6.5). Because the gradient of the panel volume functional is equal to 1 we obtain optimality conditions of the function $h_0(x)$ in general form [3]

$$\mu_1 e(x) + (1-\mu_1)g(x) + \mu_2 = 0 \tag{6.12}$$

$$\begin{aligned} &\mu_1 = 0 \quad \text{if} \quad \beta_f(h_0) < \beta_d(h_0) \\ &\mu_1 = 1 \quad \text{if} \quad \beta_d(h_0) < \beta_f(h_0) \\ &0 \leq \mu_1 \leq 1 \quad \text{if} \quad \beta_d(h_0) = \beta_f(h_0) \end{aligned} \tag{6.13}$$

The Lagrange's multipliers μ_1 and μ_2 are determined by the isoperimetric condition $V(h_0) = V_0$ and the conditions (6.13). The cases $\mu_1 = 0$ and $\mu_1 = 1$ correspond to the problems of maximization of the flutter and the divergence speeds respectively. The last case $0 \leq \mu_1 \leq 1$ corresponds to the equality condition of the critical speeds for the optimal solution.

Besides the considered cases the other possibilities may take place for the optimal solution. For example, two vibrational forms of the loss of stability, corresponding to the critical speed may appear.

6.5 Divergence

Consider first the case $\beta < 0$. Aeroelastic stability of optimal panel in this case is violated by the divergence. To prove this fact it is necessary to solve the problem of the panel having maximal critical speed of divergence β_d, and calculate for it critical flutter speed β_f. If $|\beta_f| > |\beta_d|$ then the obtained solution realizes the maximal value of critical speed at which the aeroelastic stability is violated.

Thus, we consider the optimal problem for divergence instability. The solution to this problem we denote $h_*(x)$. Numerical results were obtained with the use of the gradient method in the space of control function $h(x)$. The process of successive iterations was stopped when the necessary optimality condition (6.12) with $\mu_1 = 1$ was satisfied.

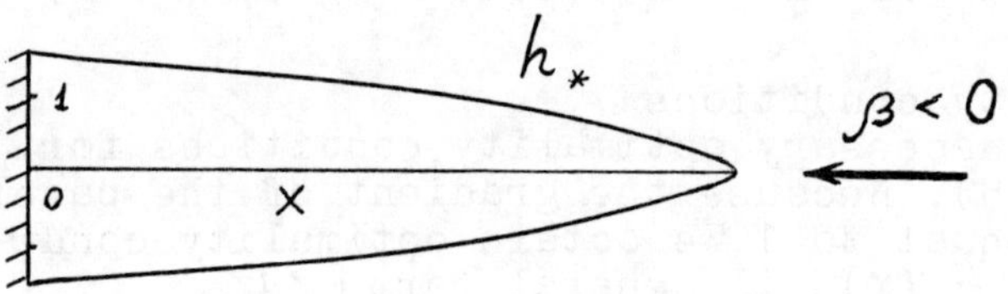

Fig.7, Optimal thickness

The function $h_*(x)$ realizing the maximal value of divergence speed is shown in Fig. 7. The critical value β_d for this panel is equal to $\beta_d = -11.8$ [4]. The critical divergence speed for the panel of constant thickness ($h = 1$) is equal to $\beta_d = -6.33$ [10]. Therefore, the divergence speed for the optimal panel exceeds that of the panel of constant thickness and the same volume in 1.86 times. The study of stability for this panel shows that the critical flutter speed is much greater than the critical divergence speed. Thus, we conclude that for $\beta < 0$ and $V_0 = 1$ the function $h_*(x)$ realizes the maximal value of the critical speed at which aeroelastic stability is violated. Due to homogeneity property of the considered functionals [5] the solution to optimization problem with $V(h) = V_0$ is $h_0(x) = V_0 h_*(x)$

6.6 Flutter

Consider now the case $\beta > 0$ with the same boundary conditions and zero damping coefficient $\Lambda = 0$. It is known that the stability of the panel of constant thickness $h(x) = 1$ is violated by flutter with $\beta_f = 144$ [6]. The numerical results of successive maximization iterations of the critical flutter speed are shown in Fig. 8. It follows from the presented results that critical flutter speed is very sensitive with respect to mass and stiffness distribu-

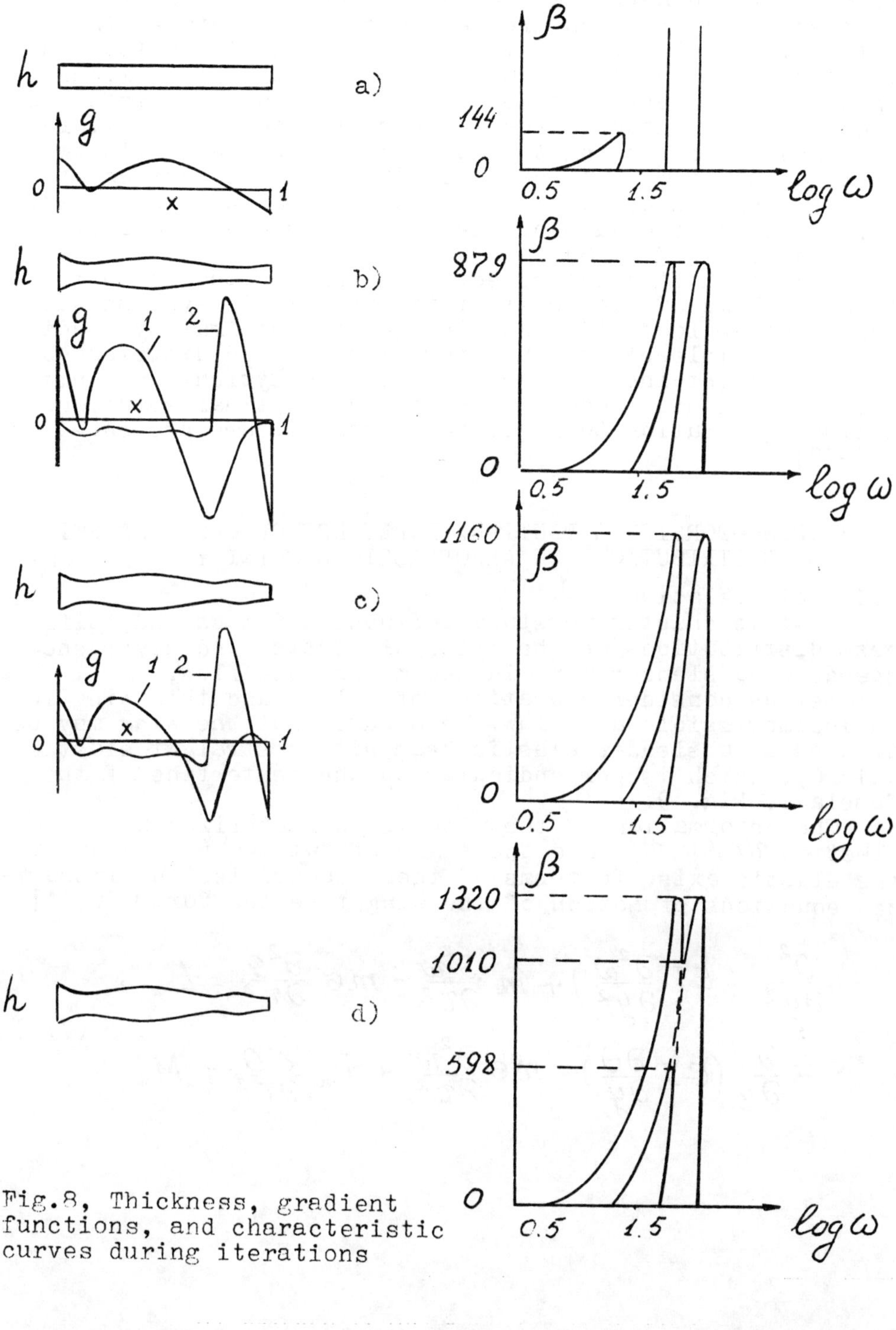

Fig.8, Thickness, gradient functions, and characteristic curves during iterations

tions of the panel. The possible advantages in critical flutter speed are much greater than the advantages in other (static) structural optimization problems. It is interesting that through the optimization process at some iteration the bimodal effect takes place, Fig. 8b,c. It means that the critical flutter speed is characterized by two vibrational modes. Continuing the optimization process with two critical modes we see that the characteristic curves of the frequencies are approached and then cross each other. This leads to the jump of the critical speed of aeroelastic stability, Fig. 8d.

Similar multimodal effects take place also in optimization problems of static stability of elastic systems with conservative loads [16-18]. The general optimality conditions of multiple eigenvalues for discrete and distributed cases were obtained in [19-21]. Some analytical and numerical examples of bimodal and multimodal optimal solutions, including solution to Lagrange's problem, are presented in [22-29].

7. BENDING-TORSIONAL FLUTTER. INFLUENCE OF MASS AND STIFFNESS DISTRIBUTIONS ON AEROELASTIC STABILITY

7.1 Basic relations

In this chapter we study influence of mass and stiffness distributions on the critical flutter and divergence speeds of a slender wing in incompressible flow.

Let us consider vibrations of a long and thin wing in an incompressible air flow. We assume that the wing may be treated as a slender elastic beam with a straight elastic axis Oy, which is perpendicular to the centerline of the fuselage, Fig. 9.

The deformation of the wing is characterized by the deflection $w(y,\tau)$ and the angle of twist $\theta(y,\tau)$ about the elastic axis. In terms of these quantities the linearized equations of motion of the wing take the form [30,31]

$$\begin{aligned} &\frac{\partial^2}{\partial y^2}\left(EI\frac{\partial^2 w}{\partial y^2}\right)+m\frac{\partial^2 w}{\partial \tau^2}-m\sigma\frac{\partial^2\theta}{\partial\tau^2}=L_a,\\ &-\frac{\partial}{\partial y}\left(GJ\frac{\partial\theta}{\partial y}\right)-m\sigma\frac{\partial^2 w}{\partial\tau^2}+I_m\frac{\partial^2\theta}{\partial\tau^2}=M_a \end{aligned} \tag{7.1}$$

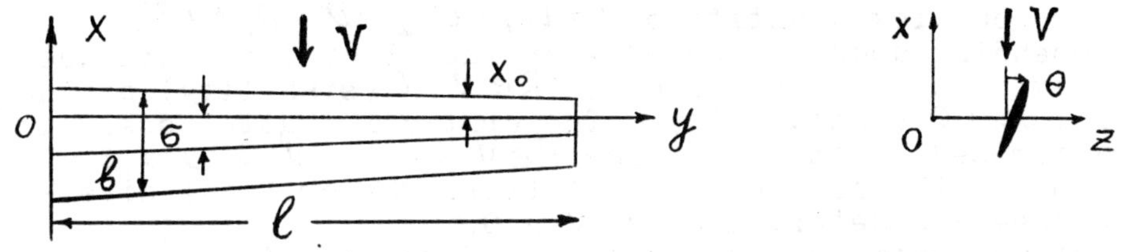

Fig. 9, Elastic wing in incompressible flow

Here, EI and GJ denote the bending and torsional rigidities, m and I_m are the mass and the moment of inertia per unit length, σ is the distance between the axis of inertia and the elastic axis, and L_a and M_a are the aerodynamic lift force and the pitching moment about the elastic axis per unit length. To describe the aerodynamic load we use the so-called strip theory and stationary hypothesis [30, 31], according to which

$$L_a = c_y^\alpha \rho b V^2 \left[\theta + \frac{b}{V}\left(\frac{3}{4} - \frac{x_0}{b}\right)\frac{\partial\theta}{\partial\tau} - \frac{1}{V}\frac{\partial w}{\partial\tau}\right] \tag{7.2}$$

$$M_a = c_m^\alpha \rho b^2 V^2 \left[\theta + \frac{b}{V}\left(\frac{3}{4} - \frac{x_0}{b} - \frac{\pi}{16 c_m^\alpha}\right)\frac{\partial\theta}{\partial\tau} - \frac{1}{V}\frac{\partial w}{\partial\tau}\right]$$

where b is the chord of the wing and x_0 is the distance between the leading edge of the wing and its elastic axis, and where ρ and V are the density and the speed of the flow, respectively. The boundary conditions for a cantilever beam are given by the relations

$$w = \frac{\partial w}{\partial y} = \theta = 0 \qquad \text{at} \quad y = 0 \tag{7.3}$$

$$EI\frac{\partial^2 w}{\partial y^2} = \frac{\partial}{\partial y}\left(EI\frac{\partial^2 w}{\partial y^2}\right) = GJ\frac{\partial\theta}{\partial y} = 0 \qquad \text{at} \quad y = l$$

The system of equations (7.1)-(7.3) represents a linear and homogeneous boundary value problem. The solution to this problem is found in the general form

$$w(y,\tau) = u(y)e^{\nu\tau}, \quad \theta(y,\tau) = v(y)e^{\nu\tau} \tag{7.4}$$

where ν is an eigenvalue and where $u(y)$ and $v(y)$ are the corresponding eigenfunctions. Due to the fact that the aerodynamic force is nonconservative, the eigenvalues ν are complex quantities $\nu = q + i\omega$, the functions u and v

are also complex quantities $u = u_1 + iu_2$, $v = v_1 + iv_2$.

Depending upon the speed V of the flow, the amplitude of the vibrations can decrease $Re\,\nu < 0$ (stability) or increase $Re\,\nu > 0$ (instability). The critical flutter speed is characterized by the relations $Re\,\nu = 0$, $Im\,\nu = \omega \neq 0$, where ω is the frequency of flutter. The critical divergence speed is determined by $\nu = 0$.

Let us consider the equations of motion of the wing at the flutter point. For this purpose we substitute (7.4) into the system (7.1)-(7.3) with $\nu = i\omega$. Taking $V = V_f$ we get a system of equations for the eigenfunction $u(y)$ and $v(y)$ in the form [1, 2]

$$Lf = \begin{pmatrix} L_{11} & L_{12} \\ L_{21} & L_{22} \end{pmatrix} \begin{pmatrix} u \\ v \end{pmatrix} = 0 \tag{7.5}$$

where L_{ij} are linear differential operators

$$\begin{aligned} L_{11} &= \frac{d^2}{dy^2}\left(EI\frac{d^2}{dy^2}\right) - m\omega^2 + i\omega c_y^{\alpha}\rho b V_f, \\ L_{12} &= m\sigma\omega^2 - c_y^{\alpha}\rho b V_f^2 - i\omega c_y^{\alpha}\rho b^2 V_f\left(\frac{3}{4} - \frac{x_0}{b}\right), \\ L_{21} &= m\sigma\omega^2 + i\omega c_m^{\alpha}\rho b^2 V_f, \\ L_{22} &= -\frac{d}{dy}\left(GJ\frac{d}{dy}\right) - I_m\omega^2 - c_m^{\alpha}\rho b^2 V_f^2 - i\omega c_m^{\alpha}\rho b^3 V_f\left(\frac{3}{4} - \frac{x_0}{b} - \frac{\pi}{16 c_m^{\alpha}}\right) \end{aligned} \tag{7.6}$$

The boundary conditions for u and v follow from (7.3)

$$u = \frac{du}{dy} = v = 0 \quad \text{at} \quad y = 0 \tag{7.7}$$

$$EI\frac{d^2u}{dy^2} = \frac{d}{dy}\left(EI\frac{d^2u}{dy^2}\right) = GJ\frac{dv}{dy} = 0 \quad \text{at} \quad y = l$$

We will now consider the divergence problem. For this purpose we put $\nu = 0$ in (7.4)-(7.6), whereby we get the self-adjoint, positive - definite eigenvalue problem

$$\frac{d}{dy}\left(GJ\,\frac{dv_d}{dy}\right)+c_m^{\alpha}\rho b V_d^2 v_d=0 \tag{7.8}$$

$$v_d(0)=0, \qquad GJ\,\frac{dv_d}{dy}=0 \quad \text{at} \quad y=\ell$$

Here, $v_d(y)$ denotes the eigenfunction at divergence, and the critical divergence speed V_d is determined by the lowest eigenvalue of the problem (7.8).

Let us now introduce the control function $h(y)$. We assume that the cross-section of the wing is a thin-walled closed profile of arbitrary geometry. If all the thicknesses of the profile are changed in h times, then all mass and stiffness characteristics of the profile will be also changed in h times, while σ and x_o remain unchanged. We shall therefore write

$$EI(y)=EI_o(y)h(y), \quad GJ(y)=GJ_o(y)h(y)$$
$$I_m(y)=I_{m_o}(y)h(y), \quad m(y)=m_o(y)h(y) \tag{7.9}$$

where EI_o, GJ_o, I_{m_o} and m_o are some fixed stiffness and mass functions. The function $h(y)$ serves as a non-dimensional control function. For physical reasons we must assume $h(y)\geqslant 0$. A variation of this function leads to the new distributions of masses and stiffnesses and therefore influences the critical flutter and divergence speeds. Our first aim is to determine the influence of a small variation of the control function on the critical flutter and divergence speeds. Our second aim is to raise the critical speed by means of suitable variations of the control function $h(y)$ keeping the total mass of the wing constant.

7.2 Sensitivity analysis

Let us find the increment of the critical speeds due to a variation of the control function $h(y)$.

For this purpose we introduce the adjoint to (7.5), (7.7) flutter problem [1, 2]

$$L^T p=\begin{pmatrix} L_{11} & L_{21} \\ L_{12} & L_{22}\end{pmatrix}\begin{pmatrix}\phi \\ \psi\end{pmatrix}=0 \tag{7.10}$$

The operators L_{ij} are defined by the expressions (7.6). The functions $\Phi(y)$ and $\Psi(y)$ are complex quantities, $\Phi = \Phi_1 + i\Phi_2$, $\Psi = \Psi_1 + i\Psi_2$. The boundary conditions have the form

$$\Phi = \frac{d\Phi}{dy} = \Psi = 0 \qquad \text{at} \quad y = 0$$

$$EI\frac{d^2\Phi}{dy^2} = \frac{d}{dy}\left(EI\frac{d^2\Phi}{dy^2}\right) = GJ\frac{d\Psi}{dy} = 0 \qquad \text{at} \quad y = \ell \tag{7.11}$$

It can be shown that the critical speed and frequency of flutter of the problem (7.10), (7.11) coincide with those of the problem (7.5)-(7.7) because these problems are adjoint to each other. The considered problems are linear and homogeneous with respect to the vector-functions f and ρ. Hence, any solution is determined up to an arbitrary complex multiplier.

We proceed now to calculate the variations. Consider first the main flutter problem (7.5), (7.7) taking relations (7.9) into account. An increment $\delta h(y)$ leads to increments of V_f, ω and the complex vector function $f(y)$. The increments $\delta h(y)$, δV_f and $\delta\omega$ are real quantities, and the variation of the complex function f has the form

$$\delta f(y) = \begin{pmatrix} \delta u \\ \delta v \end{pmatrix} = \begin{pmatrix} \delta u_1 + i\delta u_2 \\ \delta v_1 + i\delta v_2 \end{pmatrix}$$

Now for the problems (7.5) and (7.7) we obtain the equation in variations [1, 2]

$$K(\delta h)f + L_{V_f} f\,\delta V_f + L_\omega f\,\delta\omega + L\,\delta f = 0 \tag{7.12}$$

$$\delta u = \frac{d\delta u}{dy} = \delta v = 0 \qquad \text{at} \quad y = 0$$

$$EI_0\,\delta h\frac{d^2 u}{dy^2} + EI\frac{d^2\delta u}{dy^2} = \frac{d}{dy}\left(EI_0\,\delta h\frac{d^2 u}{dy^2} + \right. \tag{7.13}$$

$$+ EI \frac{d^2 \delta u}{dy^2}\Big) = GJ_0 \delta h \frac{dv}{dy} + GJ \frac{d\delta v}{dy} = 0 \quad \text{at} \quad y = \ell$$

where the matrices L_{V_f} and L_ω are produced by the matrix L, see (7.5) and (7.6), by formal differentiation with respect to the variables V_f and ω. The matrix operator $K(\delta h)$ is given by the expression

$$K(\delta h) = \begin{pmatrix} \frac{d^2}{dy^2}\left(EI_0 \delta h \frac{d^2}{dy^2}\right) - m_0 \omega^2 \delta h, & m_0 \sigma \omega^2 \delta h \\ m_0 \sigma \omega^2 \delta h, & -\frac{d}{dy}\left(GJ_0 \delta h \frac{d}{dy}\right) - I_{m_0} \omega^2 \delta h \end{pmatrix}$$

Further, we multiply (7.12) by the vector function $p^T(y) = (\phi(y), \psi(y))$, where p is the solution to the adjoint flutter problem (7.10), (7.11) and integrate between 0 and ℓ

$$\int_0^\ell [p^T K(\delta h) f + (p^T L_{V_f} f)\delta V_f + (p^T L_\omega f)\delta\omega + p^T L \delta f]\, dy = 0 \qquad (7.14)$$

Integrating by parts with the use of the boundary conditions (7.11), (7.13), we find that the last term of the integrand in (7.14) vanishes

$$\int_0^\ell p^T L \delta f \, dy = \int_0^\ell \delta f^T L^T p \, dy = 0$$

Here, the last equality follows from (7.10). The first term in (7.14) can be rewritten in the form [1, 2]

$$\int_0^\ell p^T K(\delta h) f \, dy = \int_0^\ell H \delta h \, dy \,, \qquad (7.15)$$

$$H = EI_0 \frac{d^2 u}{dy^2} \frac{d^2 \phi}{dy^2} + GJ_0 \frac{dv}{dy} \frac{d\psi}{dy} + \omega^2 p^T \begin{pmatrix} -m_0 & m_0 \sigma \\ m_0 \sigma & -I_{m_0} \end{pmatrix} f$$

Using the notation

$$A = \int_0^\ell (p^T L_{V_f} f)\,dy\,, \quad B = \int_0^\ell (p^T L_\omega f)\,dy \tag{7.16}$$

the equation (7.14) gets the form

$$\int_0^\ell H\delta h\,dy + A\delta V_f + B\delta\omega = 0 \tag{7.17}$$

Note that the function H is a complex function of the real variable y and that the constants A and B are complex quantities. Let us multiply (7.17) by $\bar{B}$, complex conjugate to B, and separate the imaginary part. Because $Im\,(B\bar{B})=0$, and δV_f and $\delta\omega$ are both real quantities, we get the following expression for the variation [1, 2]

$$\delta V_f = \int_0^\ell g\delta h\,dy\,, \quad g = -\frac{Im\,(H\bar{B})}{Im\,(A\bar{B})} \tag{7.18}$$

It follows that the function g is the gradient of the functional for the critical flutter speed with respect to the control function h.

The variation of the flutter frequency can be obtained from (7.17) in a similar manner [1, 2]

$$\delta\omega = \int_0^\ell t\delta h\,dy\,, \quad t = -\frac{Im\,(H\bar{A})}{Im\,(B\bar{A})} \tag{7.19}$$

Thus, in order to calculate the gradients g and t, it is necessary to solve the main and the adjoint flutter problems (7.5), (7.7) and (7.10), (7.11), and to determine the complex vector functions $f(y)$ and $p(y)$ and the real quantities V_f and ω. From (7.15), (7.16) we can then obtain the complex constants A and B and the complex function H, and hence determine the gradients g and t. Note that the eigenfunctions f and p are determined up to arbitrary complex multipliers, because the main and the adjoint flutter problems are both linear and homogeneous, but the gradients g and t remain unchanged and independent of norms of the eigenfunctions.

Now we derive the gradient function of the critical divergence speed with respect to the control function $h(y)$. Since the divergence problem (7.8) is selfadjoint and positive-definite, the Rayleigh's minimum principle is valid

$$V_d^2 = \min_{v} \frac{\int_0^l GJ\left(\frac{dv}{dy}\right)^2 dy}{\int_0^l c_m^\alpha \rho b^2 v^2 dy} \tag{7.20}$$

Here, the function v must satisfy the kinematic boundary condition $v(0) = 0$ and be continuously differentiable. Variation of (7.20) readily gives [1, 2]

$$\delta V_d = \int_0^l e\delta h\, dy, \quad e = \frac{GJ_0\left(\frac{dv_d}{dy}\right)^2}{2V_d \int_0^l c_m^\alpha \rho b^2 v_d^2 dy} \tag{7.21}$$

Note that because the divergence problem is self-adjoint, it is not necessary to introduce an adjoint problem for determination of the gradient $e(y)$.

The method described above for determining the gradients for the critical speeds may also be used to obtain the gradients with respect to some other independent functions or parameters of the problem. For example, the derivative of the critical flutter speed with respect to the density ρ of the flow is given by the formula

$$\frac{\partial V_f}{\partial \rho} = -\frac{Im\,(C\bar{B})}{Im\,(A\bar{B})}$$

Here the quantities A and B are defined in (7.16) and C is given by

$$C = \int_0^l (p^T L_\rho f)\, dy$$

where the matrix L_ρ is obtained by differentiating L with respect to ρ.

Knowing the gradients of the critical flutter and divergence speeds with respect to distributed and discrete parameters we can improve the characteristics of aeroelastic stability of a structure in a rational manner.

8. BENDING-TORSIONAL FLUTTER. OPTIMIZATION OF AEROELASTIC STABILITY

8.1 Optimization problem

Consider the problem of maximizing the critical speed at which stability is lost, assuming the total mass of structure material to be given. Mathematically, this problem is formulated by the expressions [1, 2]

$$\max_{h \in \Omega} \min \left[V_f(h), V_d(h) \right] = \min \left[V_f(h^*), V_d(h^*) \right]$$
$$\Omega = \left\{ h(y) : M(h) = \int_0^\ell h(y) m_o(y) dy = M_o \right\} \tag{8.1}$$

Thus, this is the problem of determining the mass distribution $h^*(y)$, which, within the constraint of given total mass M_o, maximizes the smaller of the critical flutter and the critical divergence speeds.

The variation of the functional of the total mass is

$$\delta M = \int_0^\ell m_o(y) \delta h(y) dy$$

so the gradient of the mass functional is simply given by $m_o(y)$. Taking into account the gradients g and e of the critical flutter and divergence speeds, obtained in the previous chapter, and using the results of maximin approach, we obtain the necessary optimality conditions for the optimal mass distribution [1, 2]

$$\mu_1 e(y) + (1-\mu_1) g(y) + \mu_2 m_o(y) = 0$$
$$\mu_1 = 0, \quad if \quad V_f(h^*) < V_d(h^*) \tag{8.2}$$
$$\mu_1 = 1, \quad if \quad V_d(h^*) < V_f(h^*)$$

$$0 \leq \mu_1 \leq 1 \ , \ if \ \ V_d(h^*) = V_f(h^*)$$

Here, the Lagrange's multipliers μ_1 and μ_2 are defined by the isoperimetric conditions of the problem.

In the numerical solution of the optimization problem (8.1) it is necessary to vary the initial distribution $h(y)$ in order to increase the smaller critical speed for loss of stability. Assuming that the initial distribution $h(y)$ satisfies the condition $M(h) = M_0$ we may take an improved design variation in the form

$$\delta h(y) = \alpha(y)[\mu_1 e(y) + (1-\mu_1)g(y) + \mu_2 m_0(y)] \tag{8.3}$$

where $\alpha(y)$ is a so-called "gradient step", that is, an arbitrary positive function chosen by the researcher. The two as yet unknown multipliers μ_1 and μ_2 are defined by the isoperimetric conditions.

Consider first the case $V_f(h) < V_d(h)$. Taking $\mu_1 = 0$ and defining μ_2 via substituting (8.3) into the condition $\delta M = 0$ we obtain

$$\mu_1 = 0 \ , \qquad \mu_2 = -\int_0^\ell \alpha g m_0 \, dy \Big/ \int_0^\ell \alpha m_0^2 \, dy \tag{8.4}$$

Analogously, for the case $V_d(h) < V_f(h)$ we get

$$\mu_1 = 1 \ , \quad \mu_2 = -\int_0^\ell \alpha e m_0 \, dy \Big/ \int_0^\ell \alpha m_0^2 \, dy \tag{8.5}$$

Finally, if we have $V_f(h) = V_d(h)$, we can substitute (8.5) into the conditions $\delta V_f = \delta V_d$ and $\delta M = 0$, and obtain the following system of linear equations

$$\begin{aligned} \mu_1 \int_0^\ell \alpha(e-g)^2 dy + \mu_2 \int_0^\ell \alpha(e-g) m_0 \, dy &= -\int_0^\ell \alpha(e-g) g \, dy \\ \mu_1 \int_0^\ell \alpha(e-g) m_0 \, dy + \mu_2 \int_0^\ell \alpha m_0^2 \, dy &= -\int_0^\ell \alpha g m_0 \, dy \end{aligned} \tag{8.6}$$

It is easily seen that the determinant of this system is a

Gram determinant. It is equal to zero only when $e-g$ and m_0 are linearly dependent functions.

In the case of $V_f = V_d$ it can be shown that the variations δV_f and δV_d can be expressed by

$$\delta V_f = \delta V_d = \int_0^{\ell} \alpha [\mu_1 e + (1-\mu_1) g + \mu_2 m_0]^2 dy \geq 0$$

which implies that our algorithm meets the condition $M(h) = M_0$ and increases the critical speed of instability at each step of variation. The same holds good for the cases of $V_f < V_d$ and $V_d < V_f$. If gradient procedure (8.3)-(8.6) is converged then the necessary optimality conditions (8.2) are satisfied.

At each step of the gradient procedure for computing the improved mass and stiffness distribution we have to solve the main and the adjoint flutter problems. Solution to the main flutter problem (8.5), (8.7) is performed by the method of successive iterations described in [30].

The adjoint flutter problem is solved by the same method with only insignificant differences in the computer program. A comparison of the values obtained for V_f and ω in these two problems yields an effective check of the accuracy of the computations.

Also the divergence problem (7.8), see chapter 7, is to be solved at each step of the gradient procedure. The solution of this problem is performed by the method of successive iterations and gives us the critical divergence speed V_d, the eigenfunction $v_d(y)$ and the gradient function $e(y)$.

The numerical procedure (with n designating the iteration number) consists of the following iteration steps [1, 2].

1) For the first iteration ($n = 1$), assume a distribution $h^{(n)}(y)$ that satisfies the constraint $M(h^{(n)}) = M_0$. For $n > 1$, apply the distribution obtained by the end of the previous iteration.

2) Solve the main and the adjoint flutter problems and obtain the eigenfunctions and their derivatives together with the values $V_f^{(n)}$ and $\omega^{(n)}$.

3) Determine the gradient function $g^{(n)}(y)$ according to (7.15)-(7.18).

4) Solve the divergence problem and obtain the critical

quantity V_d along with the eigenfunction v_d and its first derivative. According to the expression (7.21) determine the gradient function $e^{(n)}(y)$

5) Compare the critical values $V_f^{(n)}$ and $V_d^{(n)}$ and determine the constants $\mu_1^{(n)}$ and $\mu_2^{(n)}$.

6) Determine the variation $\delta h^{(n)}$ by(8.3) and compute the distribution function for the subsequent iteration as $h^{(n+1)} = h^{(n)} + \delta h^{(n)}$.

8.2 Numerical results

As a numerical example, let us consider a rectangular wing with uniform initial distributions EI_o, GJ_o, I_{m_o}, m_o, b, ℓ, σ and x_o. These parameters, together with the quantities c_y^{α} and c_m^{α} are taken to be equal to those in [30], wing N 3. Due to the constant m_o, the isoperimetric condition $M(h) = M_o$ takes the form $\int_0^{\ell} h(y)dy = 1$. In the numerical calculations the interval [0,1] was divided into N = 20,40 equal subintervals, and numerical integration was performed in applying the trapezoidal rule.

We consider first the uniform initial distribution $h^{(o)}(y) = 1$, which corresponds to uniform mass and stiffness distributions along the wing span. For this distribution, the critical divergence speed is much greater than the flutter speed, $V_f = 29.4 < V_d = 59.4$. Hence, the optimization algorithm described in the foregoing is reduced to maximization of flutter speed at fixed total mass.

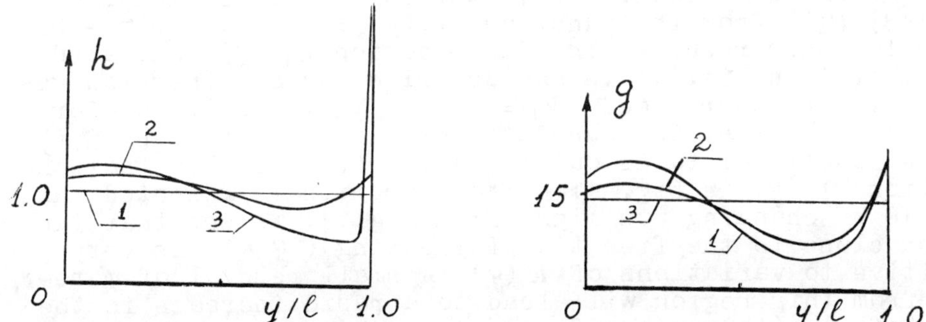

Fig.10, Distribution $h(y)$ and gradient function $g(y)$ during iterations

Fig. 10 shows how the distribution $h(y)$ and the gradient $g(y)$ of the flutter speed develop during iterations. It is clearly indicated that the distribution which corresponds to a local maximum of the critical flutter speed tends to infinity when $y \to 1$. Physically, this means that a concentrated mass should be located at the free tip of the wing in order to increase the critical speed. For a correct statement of the problem, an upper constraint $h(y) \leq h_{max}$ should therefore be specified for the distribution function. In the present case we take $h_{max} = 4.5$, and the function $h_o(y)$ and the appropriate sensitivity function $g_o(y)$ that correspond to the local maximum of the critical flutter speed are indicated by the number 3 in Fig. 10.

During the iteration process the critical flutter speed increases from the value $V_{f(o)} = 29.4$, which corresponds to the initial distribution $h^{(o)}(y) = 1$, and up to the value $V_f = 30.9$ for the distribution $h_o(y)$. At the same time, the flutter frequency changes from the initial value 107.1 and up to 112.4. Thus, the increase of the flutter speed is rather small: only about 5%. The critical divergence speeds for the distributions $h(y)$ shown in Fig. 10 are much greater than the flutter speeds.

The above distribution $h_o(y)$ is obtained not only if we start the iterations from $h^{(o)}(y) = 1$, but also if we use other initial functions such as, e.g. $h^{(o)}(y) = 1.95-1.9y$.

However, it turns out that the distribution $h_o(y)$ only corresponds to a local maximum of the critical flutter speed at given total mass. Thus, starting the iteration process (8.3), (8.4) from the function $h_1(y) = 2.7(1-y)^2 + 0.1$ leads to other results. The distribution $h_1(y)$ indicated by number 1 in Fig. 11 is characterized by a rather low value of the flutter speed $V_f = 29.1$ ($V_d = 57.5$), but for this distribution, the absolute value of the gradient $g_1(y)$ of the functional turns out to be large. Note also that the function $g_1(y)$ in Fig. 11 differs from that presented in Fig. 10 by changing its sign. From Fig. 11 we see that the region close to the free tip of the wing ($y \to 1$) is very sensitive to variations of $h(y)$. A small removal of material from this region will lead to a rapid increase in the value of V_f (thin tip effect).

Since the gradient $g_1(y)$ attains negative values in the region [0,1], we may conclude that V_f can be increased if we reduce the total mass of the wing. Therefore, some distributions $h(y)$ exist for which a reduction of mass

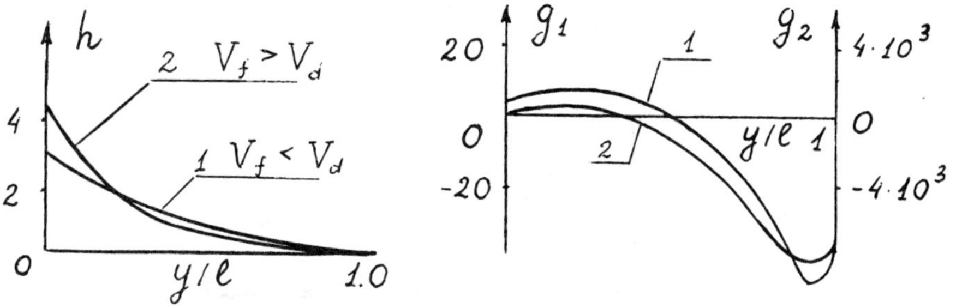

Fig.11, Distributions and gradients

does not contradict an increase of the critical flutter speed. Note that according to (7.21) the gradient of the divergence speed is always positive, so an increase in the amount of material will always lead to an increase of the divergence speed.

A few iterations (8.3), (8.4) started from the distribution $h_1(y)$ lead to an essential increase in the critical flutter speed. The distribution $h_2(y)$ indicated by the number 2 in Fig. 2 is associated with the value V_f = 52.1, but note that the divergence speed V_d for this distribution turns out to be smaller, $V_d = 47.8 < V_f$. Hence, the wing with this distribution of material loses its stability by divergence.

Thus, we see that the iteration process (8.3), (8.4) of maximizing the critical flutter speed does not lead to a maximization of the smallest of the critical speeds. In order to obtain the true optimal solution it is therefore necessary to use further the iterative formulas (8.3), (8.6) after attaining the equality $V_f = V_d$. This process converges to the optimal distribution $h^*(y)$ shown in Fig.12, where it should be noted that $h^*(1) = 0$. The critical speed of aeroelastic instability attains the value $V_f = V_d = 48.8$, with the frequency of flutter equal to $\omega^* = 265.1$ [1, 2]. Thus, the critical speed is increased by 66 pct. relative to the critical value for the uniform distribution.

We may conclude that the optimal distribution $h^*(y)$ for our problem is characterized by equal critical values

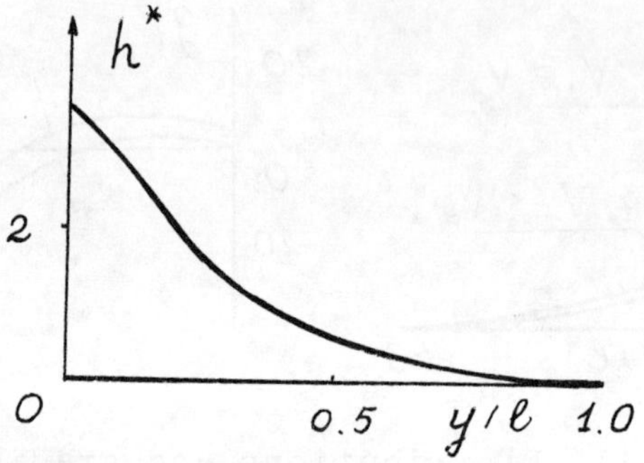

Fig.12, Optimal distribution $V_f = V_d = 48.8$

of flutter and divergence speeds. Physically this means that the optimal structure will experience two different types of loss of stability (dynamic and static) at the critical point. Recall that similar bimodal effect was also observed for the panel flutter problem, considered in the chapter 6. In conservative (selfadjoint) case this effect corresponds to multiplicity of eigenvalues.

Besides the optimization problem considered the other optimization problems can be stated. Let us consider, for example, the problem of distributing nonstructural mass along the wing span. Fuel, electronic equipment and some loads may be considered as nonstructural masses. It is assumed that the nonstructural mass doesn't change the stiffness properties of the structure, it affects only inertia properties of the wing. So, we may introduce the control function

$$m(y) = m_o(y)(1+h(y)),\ I_m(y) = I_{m_o}(1+h(y))$$

By means of the mathematical theory of optimal control it is easy to show that the Hamiltonian of this problem is linear in the control function. This implies that the correct formulation of the optimization problem must include prescribed upper and lower limits h_{min} and h_{max}, respectively, and the optimal distribution h^* is of the bang-bang type [1, 2].

Fig. 13 shows the optimal distributions with one and two switching points and the corresponding gradient functions.

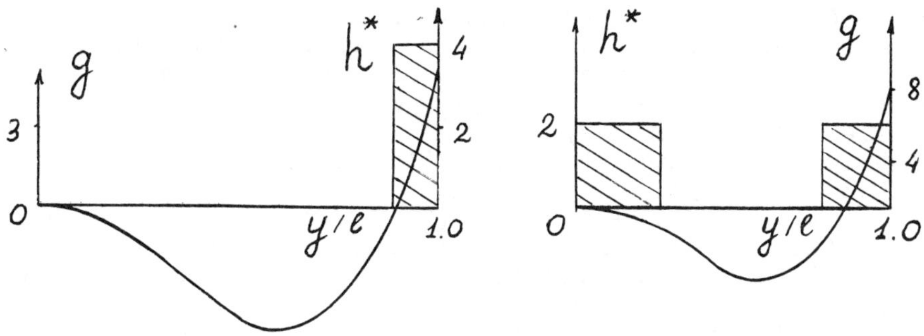

Fig.13, Optimal bang-bang distributions and gradient functions

Optimization problems under flutter conditions for wings in discrete formulation were studied in [7].

REFERENCES

1. Seyranian, A.P.: Sensitivity analysis and optimization of aeroelastic stability characteristics, Institute of Problems in Mechanics, USSR Academy of Sciences, N 162, Moscow 1980 (in Russian).
2. Seyranian, A.P.: Sensitivity analysis and optimization of aeroelastic stability, Int. J. Solids Struct., 18, (1982), 791-807.
3. Seyranian, A.P.: Optimization of structures subjected to aeroelastic instability phenomena, Arch. Mech., 34 (1982), 133-146.
4. Seyranian, A.P.: Optimization of stability of a plate in supersonic gas flow, Mechanics of Solids, 15 (1980), 5, 141-147.
5. Seyranian, A.P.: Homogeneous functionals and structural optimization problems. Int. J. Solids Struct., 15 (1979), 749-759.
6. Seyranian, A.P. and Sharanyuk, A.V.: Sensitivity and optimization of critical parameters in dynamic stability problems, Mechanics of Solids, 18 (1983), 5, 174-183.
7. Seyranian, A.P. and Sharanyuk, A.V.: Optimization of flutter characteristics, Izv. AN Armenian SSR, Mekhanica,

37 (1984), 5, 38-51 (in Russian).
8. Seyranian, A.P. and Sharanyuk, A.V.: Sensitivity analysis of vibrational frequencies of mechanical systems, Mechanics of Solids, 22 (1987), 2, 34-38.
9. Pedersen, P. and Seyranian, A.P.: Sensitivity analysis for problems of dynamic stability. Int. J. Solids Struct., 19 (1983) 315-335.
10. Bolotin, V.V.: Nonconservative Problems of the Theory of Elastic Stability, Pergamon Press, Oxford 1963.
11. Ziegler, H.: Principles of Structural Stability, Blaisdell, Waltham, Mass. 1968.
12. Leipholz, H.: Stability of Elastic Systems, Sijthoff and Noordhoff, Amsterdam 1980.
13. Hanaoka, H. and Washizu, K.: Optimum design of Beck's column, Comput. Struct., 11 (1980), 473-480.
14. Claudon, J.-L. and Sunakawa, M.: Optimizing distributed structures for maximum flutter load, AIAA J., 19 (1981), 957-959.
15. Pierson, B.L. and Hajela, O.: Optimal aeroelastic design of an unsymmetrically supported panel, J. Struct. Mech., 8 (1980), 331-346.
16. Sheu, C.Y. and Prager, W.: Recent developments in optimal structural design, Appl. Mech. Rev., 21 (1968), 985-992.
17. Poston, T. and Stewart, I.: Catastrophe Theory and its Applications, Pitman 1978.
18. Thompson, J.M.T.: Instabilities and Catastrophes in Science and Engineering, John Wiley and Sons, New York 1982
19. Braţus, A.S. and Seyranian, A.P.: Bimodal solutions in optimization problems of eigenvalues, PMM, 47 (1983) 451-457.
20. Bratus, A.S., Seyranian, A.P.: Sufficient optimality conditions in optimization problems of eigenvalues, PMM, 48 (1984), 466-474.
21. Seyranian, A.P.: Multiple eigenvalues in optimization problems, PMM, 51 (1987), 349-352.
22. Olhoff, N.: Optimal Design with Respect to Structural Eigenvalues, in: Proc. 15th Inter. Congress of Theor. Appl. Mech., Toronto, 1980, North-Holland, Toronto 1981, 133-149.
23. Haug, E.J. and Cea, I., eds.: Optimization of Distributed Parameter Structural Systems, Sijthoff and Noordhoff, Alphen aan den Rijn, Netherlands, 1981, 2 vols.

24. Eschenauer, H., Olhoff, N., eds.: Optimization Methods in Structural Design, B.-I. Wissenschaftsverlag, Wien 1983.
25. Gajewski, A. and Życzkowski, M.: Optimal Structural Design under Stability Constraints, Nijhoff, Dordrecht 1987.
26. Medvedev, N.G.: Some spectral singularities in optimal problem of shell of nonuniform thickness, Dokl. Ukr. Academy of Sciences, A, 9 (1980), 59-63 (in Russian).
27. Seyranian, A.P.: A solution of a problem of Lagrange, Sov. Phys. Dokl., 28 (1983), 7, 550-551.
28. Seyranian, A.P.: On Lagrange's problem, Mechanics of Solids, 19 (1984), 2, 101-111.
29. Masur, E.F.: Optimal structural design under multiple eigenvalue constraints, Inter. J. Solids Struct., 20 (1984), 211-231.
30. Grossman, E.P. Flutter, Transactions of CAHI, 284, Moscow (1937) (in Russian).
31. Fung, Y.C. An Introduction to the Theory of Aeroelasticity, Wiley, New York, 1955.